Welding Skills Workbook

Fifth Edition

atp AMERICAN TECHNICAL PUBLISHERS
Orland Park, Illinois 60467-5756

Jonathan F. Gosse

5 6 7 8 9 – 15 – 9 8 7 6 5 4 3 2

Printed in the United States of America

ISBN 978-0-8269-3085-9

This book is printed on recycled paper.

Contents

section | one
Introduction to Welding

section | two
Oxyacetylene Welding (OAW)

section | three
Shielded Metal Arc Welding (SMAW)

section | four
Gas Tungsten Arc Welding (GTAW)

section | five
Gas Metal Arc Welding (GMAW)

section | six
Other Welding and Joining Processes

section | seven
Weld Evaluation and Testing

section | eight
Welding Technology

Section-Based Activities

Section-Based Exams

Appendix

Know Your Welding Symbols

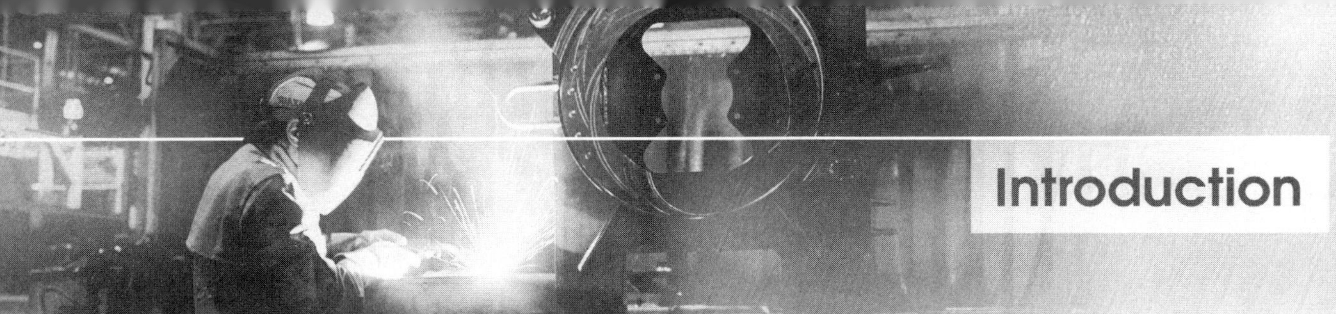

Introduction

Welding Skills Workbook covers material presented in the corresponding chapter of *Welding Skills*. The workbook includes review questions, a know your welding symbols feature, section-based activities, section-based exams, and an appendix.

Review Questions

Read the assigned chapter in the textbook. Without using the textbook, answer the review questions in the corresponding workbook chapter. After answering the review questions, check your answers using the textbook. Identify and correct any wrong answers. Your instructor will have specific directions for the next assigned activity.

Fillet Weld Arrow Side

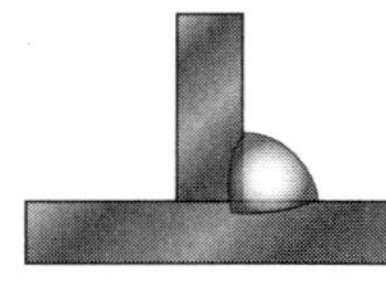

DESIRED WELD

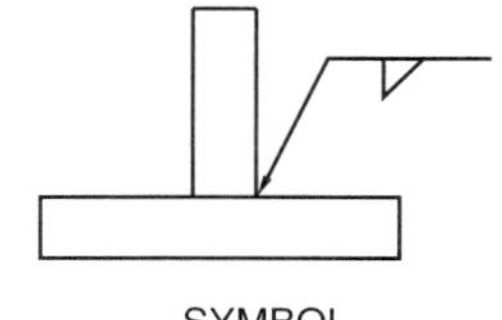

SYMBOL

Know Your Welding Symbols

At the bottom of the first page of the review questions for each chapter, a welding symbol and its meaning are introduced. This allows progressive learning of common welding symbols used, which are described in Chapter 45 of *Welding Skills*. Questions related to the welding symbols introduced in each chapter are included in the review questions.

Section-Based Activities

Section-based activities follow the review questions and feature short answer questions, matching questions, and hands-on activities corresponding to topics covered within the chapters of a section. Welding exercises contained in the section-based activities detail specific welding tasks.

Section-Based Exams

Section exams are at the back of the workbook. Scores on these tests indicate understanding of the material. Before starting a section exam, study the corresponding review questions and chapters in the textbook. The section exams should be completed without using the textbook.

Appendix

The appendix provides welding symbols commonly used in the welding industry.

Safety

Before beginning any exercise, secure permission from the instructor. Observe all safety precautions as stated in the textbook and as specified by the instructor. Under no circumstances should any machine or equipment be used without proper authorization.

How to Answer Questions

For True-False questions, circle T if the statement is true and F if the statement is false.

(T) F **1.** Eye protection should be worn when welding.

(T) F **2.** An accident must always be reported no matter how slight it may be.

T (F) **3.** Oxygen should be used to ventilate a closed container before it is welded.

For Multiple Choice questions, place the letter of the correct answer in the blank next to the question.

___D___ **1.** When welding, ___ should be worn to prevent injury to the welder.
A. safety glasses
B. fire resistant clothing
C. welding gloves
D. all of the above

___C___ **2.** ___ properties refer to the behavior of metals under applied loads.
A. Physical
B. Chemical
C. Mechanical
D. all of the above

___B___ **3.** When welding ___, a respirator is required to protect the welder from toxic fumes.
A. aluminum
B. stainless steel
C. steel
D. all of the above

For Matching questions, select the correct answer for each number. Place the letter of the answer in the blank next to the question.

___B___ **1.** ___ can be harmful to eyes and skin.

___A___ **2.** A welding machine must have the proper ___ to prevent electrical shock.

___C___ **3.** Operating with currents above the rated cable capacity causes ___.

A. ground
B. ultraviolet rays
C. overheating

Identify the following welding positions.

___C___ **1.** Vertical

___D___ **2.** Horizontal

___A___ **3.** Overhead

___B___ **4.** Flat

Ⓐ Ⓑ Ⓒ Ⓓ

Chapter 1 Review

An Essential Skill

Name ______________________________ **Date** ____________________

True-False

T F **1.** GMAW soon became the process of choice for high-production welding because it used continuous solid-wire electrodes.

T F **2.** An autogenous weld is a fusion weld made with filler metal.

T F **3.** Welding can be used to repair farm, mining, and construction equipment.

T F **4.** Oxyfuel gas welding processes use heat from the combustion of a mixture of oxygen and a fuel gas for welding.

T F **5.** The gas metal arc welding process does not require a shielding gas.

T F **6.** Welding as an occupation does not require specific training.

T F **7.** A welder generally has to pass a certification test given by an employer, government agency, or inspection authority.

T F **8.** Welding requires good eyesight and eye-hand coordination.

T F **9.** Gas tungsten arc welding was developed to weld magnesium and aluminum on World War II fighter planes.

T F **10.** The powdered core of a flux-cored electrode contains elements that break down in the heat of the arc to form shielding gas.

T F **11.** The gas tungsten arc welding process was used before the oxyacetylene welding process was developed.

T F **12.** Submerged arc welding is automated and most often used to join thick metals requiring deep penetration, such as in heavy steel plate fabrication.

T F **13.** Shielded metal arc welding is an arc welding process that produces an arc between a consumable, coated electrode and the workpiece, creating a weld pool.

Know Your Welding Symbols

Arrow Side/Other Side of Welding Symbol

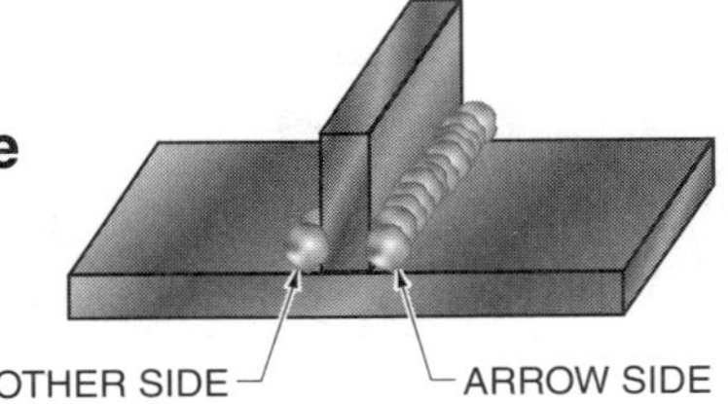

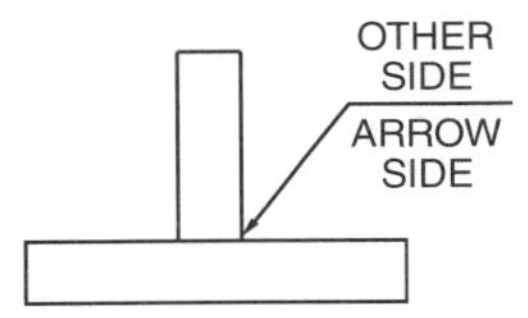

Multiple Choice

_______________ **1.** ___ welding is the most commonly used oxyfuel gas welding process.
A. Oxyacetylene
B. Propane
C. Hydrogen
D. Methylacetylene-propadiene

_______________ **2.** Transferred arc ___ is used for welding high-strength, thin metal.
A. SMAW
B. GTAW
C. FCAW
D. PAW

_______________ **3.** Identify the arrow side of the reference line shown.
A. A
B. B
C. either A or B
D. none of the above

A
B

_______________ **4.** ___ arc welding is commonly used to weld carbon, low-alloy and stainless steels, and cast iron.
A. Plasma
B. Gas metal
C. Submerged
D. Flux cored

_______________ **5.** The ___ welding process is used primarily for high-production welding.
A. oxyacetylene
B. shielded metal arc
C. gas tungsten arc
D. gas metal arc

_______________ **6.** Forge welding was first used in ___.
A. AD 1863
B. AD 1475
C. 2000 BC
D. 1915 BC

_______________ **7.** The gas metal arc welding process ___.
A. uses oxygen as a shielding gas
B. deposits weld metal at a slower rate than shielded metal arc welding
C. uses a continuous wire electrode
D. can be used with or without filler metal

_______________ **8.** The ___ welding process does not use shielding gas to protect the weld area.
A. gas metal arc
B. plasma arc
C. shielded metal arc
D. submerged arc

_______________ **9.** ___ refers to the way the molten filler metal crosses the arc to become part of the weld.
- A. Shielding
- B. Flux
- C. Fabrication
- D. Metal transfer

_______________ **10.** Identify the other side of the reference line shown.
- A. A
- B. B
- C. either A or B
- D. none of the above

A

B

_______________ **11.** ___ arc welding does not require external shielding gas, which makes it more portable than other welding processes.
- A. Plasma
- B. Gas tungsten
- C. Flux cored
- D. Shielded metal

_______________ **12.** Flux-cored arc welding was introduced in the ___.
- A. 1930s
- B. 1940s
- C. 1950s
- D. 1960s

_______________ **13.** A welding ___ is a person with a college degree and professional certification who is qualified to specify necessary weld requirements.
- A. instructor
- B. engineer
- C. supervisor
- D. inspector

_______________ **14.** ___ welding is the most common method of welding metals.
- A. Resistance
- B. Arc
- C. Forge
- D. Oxyfuel gas

_______________ **15.** A(n) ___ welding machine fuses metals together by heat and pressure.
- A. resistance
- B. oxyfuel gas
- C. arc
- D. forge

Matching

______________	**1.** The oxyfuel gas welding process ___.	**A.** is limited to flat or low-curvature base metals
______________	**2.** The shielded metal arc welding process ___.	**B.** uses heat from the combustion of a mixture of oxygen and fuel gas
______________	**3.** The submerged arc welding process ___.	**C.** does not require external shielding gas
______________	**4.** The gas tungsten arc welding process ___.	**D.** uses a nonconsumable electrode

Identify the welding process shown.

______________ **5.** Resistance

______________ **6.** Arc

______________ **7.** Oxyfuel gas

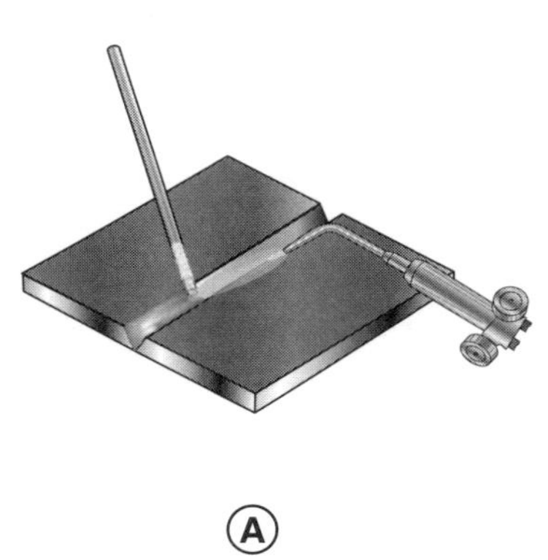

Ⓐ

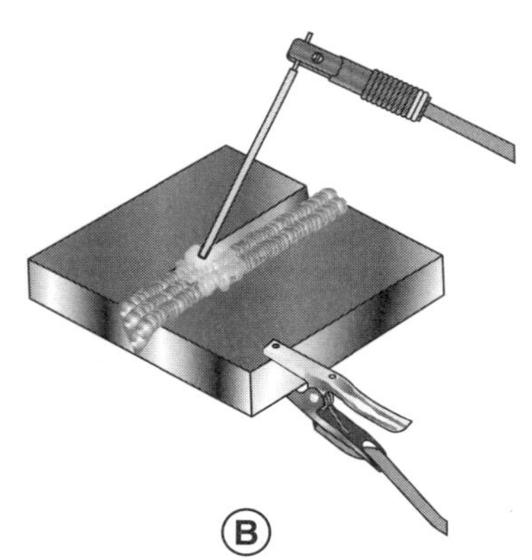

Ⓑ

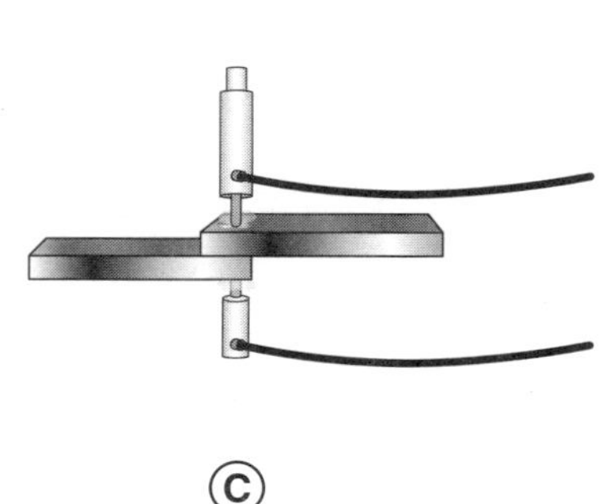

Ⓒ

Chapter 2 Review

Welding Safety

Name ______________________________ **Date** ________________

True-False

T F **1.** Employers are responsible for safety training at the job site and for ensuring that their employees are familiar with, and follow, OSHA or CCOHS regulations.

T F **2.** When arc welding, make sure there are no cable splices within 10′ of the work.

T F **3.** All accidents should be reported and documented, regardless of how minor they may be.

T F **4.** Any focus away from the job may result in an accident, causing injury to the welder and/or damage to property and equipment.

T F **5.** Attempting to operate a piece of equipment without instruction may not only damage the equipment, but it could also result in serious injury.

T F **6.** Malfunctioning welding equipment should be repaired by the welder as quickly as possible.

T F **7.** Some welding processes do not require a well-ventilated area.

T F **8.** When welding is completed, the welding machine should be turned OFF.

T F **9.** Faulty insulation, improper grounding, and incorrect operation and maintenance of electrical equipment are typical sources of danger from electric shock.

T F **10.** First degree burns are burns that penetrate all three layers of skin and cause permanent tissue damage.

T F **11.** A fire extinguisher should be accessible in locations where welding is performed.

T F **12.** Never weld or cut drums, barrels, or tanks until they have been properly cleaned and vented.

Fillet Weld Arrow Side

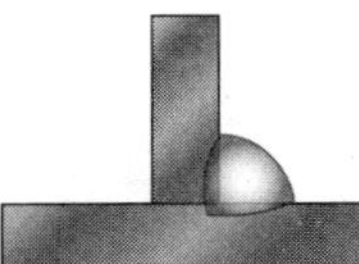

DESIRED WELD

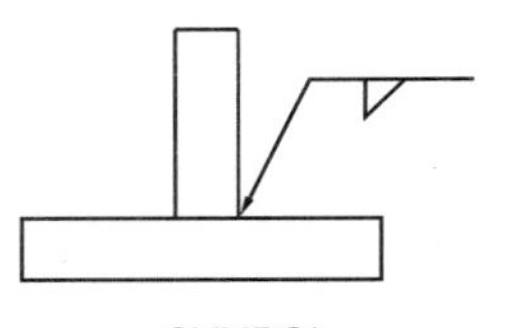

SYMBOL

T F **13.** Even when ventilation is provided, a welding helmet equipped with a respirator should be worn when welding metals that produce toxic fumes.

T F **14.** A permit is required to enter confined spaces containing atmospheric hazards.

T F **15.** An MSDS provides information about every hazardous component comprising 1% or more of a material's content and is used by a manufacturer, importer, or distributor to relay chemical hazard information to the employee.

T F **16.** If a fire erupts, the nozzle of the fire extinguisher should be aimed at the base of the fire.

T F **17.** Areas under 1000 cu ft per welder, or that have ceilings lower than 16′, require forced ventilation.

Multiple Choice

______________ **1.** The maximum safe operating pressure for acetylene is ___ psi.
A. 2
B. 5
C. 10
D. 15

______________ **2.** A respirator is used when welding metals that give off ___ fumes.
A. corrosive
B. flammable
C. toxic
D. combustible

______________ **3.** During ___ cleaning, the container should be grounded to minimize the possibility of static charge buildup and spark charges.
A. steam
B. hot chemical
C. chemical
D. mechanical

______________ **4.** When using oxygen and acetylene cylinders, open the valves ___.
A. quickly
B. slowly
C. with a pair of pliers
D. at least three times

______________ **5.** A(n) ___ permit should be obtained for jobs that pose a high risk of fire.
A. confined space entry
B. MSDS
C. hot work
D. hazardous location

______________ **6.** ___ produces flame temperatures in the range of 5800°F to 6300°F when mixed with oxygen.
A. Methylacetylene-propadiene
B. Hydrogen
C. Propane
D. Acetylene

_______________ **7.** All wind or air movement (ventilation) should be ___ the body.
A. above
B. in front of
C. behind
D. across

_______________ **8.** ___ should be worn to protect the face from metal particles when grinding.
A. Safety glasses
B. Arc welding masks
C. Goggles
D. Face shields

_______________ **9.** Stainless steels require special attention to ventilation because they produce ___ fumes when welded.
A. zinc
B. hexavalent chromium
C. manganese
D. beryllium

_______________ **10.** What is the meaning of the welding symbol shown?
A. fillet weld arrow side
B. fillet weld other side
C. slot weld other side
D. slot weld arrow side

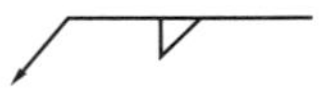

_______________ **11.** A Class ___ fire may be caused by most combustible materials, such as wood, paper, rubber, plastic, and cloth.
A. A
B. B
C. C
D. K

Matching

_______________ **1.** Do not weld on containers unless they have been properly ___.

_______________ **2.** Repairs to welding equipment should be made with the ___ off.

_______________ **3.** Pieces of metal that have been ___ should be handled with pliers and appropriate welding gloves.

_______________ **4.** A welding machine must have a proper ___ to prevent electrical shock.

_______________ **5.** ___ can be harmful to eyes and skin.

A. welded
B. ground
C. vented and cleaned
D. power
E. radiation

________	**6.** Welding should only be performed in ___ areas to prevent injury from welding fumes.	**A.** hot
________	**7.** Examples of ___ spaces include tanks, silos, storage bins, hoppers, vaults, pits, and trenches.	**B.** ventilated
________	**8.** Only ___ personnel should designate the container cleaning method.	**C.** confined
________	**9.** The welding machine must be properly ___ to prevent injury from electrical shock.	**D.** grounded
________	**10.** Gloves should be worn at all times to prevent the hands from being burned by ___ metal.	**E.** qualified

The following statements refer to the proper cleaning methods used prior to welding or cutting a container.

________	**11.** Using ___ on cylinders, regulators, and connections can deteriorate seals, gaskets, and hoses.	**A.** oil
________	**12.** If acetylene cylinders are stored on their sides, ___ can leak into valves, regulators, and hoses.	**B.** amperage
________	**13.** Filter plate shade numbers for arc welding are selected based on ___.	**C.** wet areas
________	**14.** Welding while standing in ___ increases the potential for electric shock.	**D.** acetone

________	**15.** A Class ___ fire extinguisher is identified by the color red inside a square.	**A.** A
________	**16.** A Class ___ fire extinguisher is identified by the color green inside a triangle.	**B.** B
________	**17.** A Class ___ fire extinguisher is identified by the color blue inside a circle.	**C.** C
________	**18.** A Class ___ fire extinguisher is identified by the color yellow inside a star.	**D.** D

Name ______________________________ Date ________________

True-False

T F 1. The four basic welding positions are flat, horizontal, vertical, and overhead.

T F 2. Weld joint design is influenced by the cost of preparing the joint.

T F 3. Joint penetration includes the weld reinforcement measurement.

T F 4. A fillet is one of the five basic weld joints.

T F 5. A square groove butt joint requires little or no edge preparation.

T F 6. A lap joint is a weld joint in which two members are set approximately level to each other and are positioned edge-to-edge.

T F 7. A single-V groove butt joint is used on material from ⅜″ to ¾″ thick.

T F 8. For joint efficiency in a lap joint, an overlap greater than three times the thickness of the thinnest member is recommended.

T F 9. Welding in overhead position is easier to learn than welding in flat position.

T F 10. Proper fit-up ensures that joint members are in correct alignment, have the correct edge preparation, and have the required root opening for proper penetration and sufficient weld reinforcement.

T F 11. The base metal is the metal or alloy that is to be welded.

T F 12. A pipe weld requires a 60° to 75° groove angle.

Fillet Weld Other Side

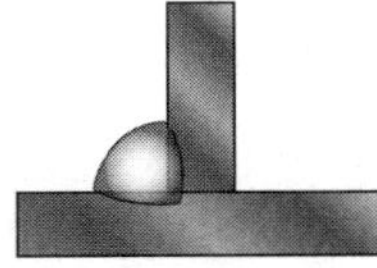

DESIRED WELD

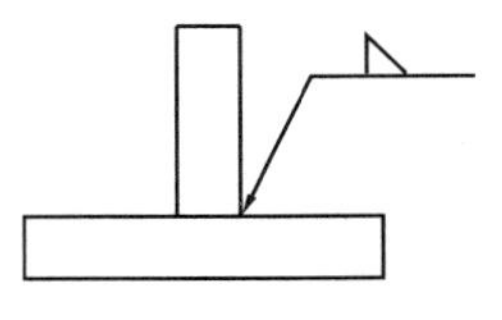

SYMBOL

Multiple Choice

_______________ **1.** A ___ weld is a weld made in a circular hole in one member to fuse it to the other member.
- A. spot
- B. stud
- C. back
- D. plug

_______________ **2.** The ___ is the area where filler metal intersects the base metal and extends the furthest into the weld joint.
- A. weld root
- B. weld leg
- C. root pass
- D. joint root

_______________ **3.** Identify the welding symbol shown.
- A. fillet weld arrow side
- B. fillet weld other side
- C. fillet weld both sides
- D. slot weld arrow side

Matching

_______________ **1.** The ___ is the metal to be welded.

_______________ **2.** A weld that results from a weld pass is a ___.

_______________ **3.** A ___ is the shape within the deposited bead caused by the movement of the welding heat source.

_______________ **4.** A single progression of welding along a joint is a ___.

_______________ **5.** A ___ is a depression at the termination of the weld bead.

A. weld pass
B. base metal
C. ripple
D. weld bead
E. crater

_______________ **6.** The ___ is the depth of the weld metal from the weld face into the joint.

_______________ **7.** The ___ is the amount of weld metal in excess of that required to fill the joint.

_______________ **8.** The portion of the groove face within the joint root is the ___.

_______________ **9.** The distance from toe to toe across the face of the weld is the ___.

_______________ **10.** The junction of the base metal and the weld face is the ___.

A. root face
B. joint penetration
C. weld toe
D. weld width
E. weld reinforcement

______________	**11.** The exposed surface of the weld bounded by the weld toes is the ___.	**A.** actual throat
______________	**12.** The ___ is the portion of a weld joint where joint members are closest.	**B.** weld face
______________	**13.** The shortest distance from the face of a fillet weld to the weld root is the ___.	**C.** joint root
______________	**14.** The ___ is the distance from the joint root to the toe of a fillet weld.	**D.** root opening
______________	**15.** The ___ is the distance between joint members at the joint root before welding.	**E.** weld leg

Identify the weld joints shown.

______________ **16.** T-joint

______________ **17.** Lap joint

______________ **18.** Edge joint

______________ **19.** Butt joint

______________ **20.** Corner

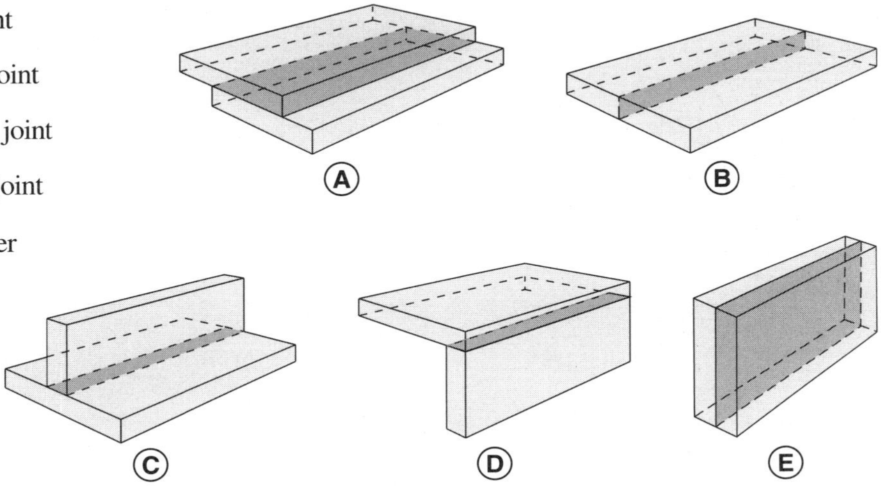

Ⓐ Ⓑ Ⓒ Ⓓ Ⓔ

Identify the weld types shown.

______________ **21.** Single-J-groove

______________ **22.** Fillet

______________ **23.** Single-V-groove

______________ **24.** Plug

______________ **25.** Slot

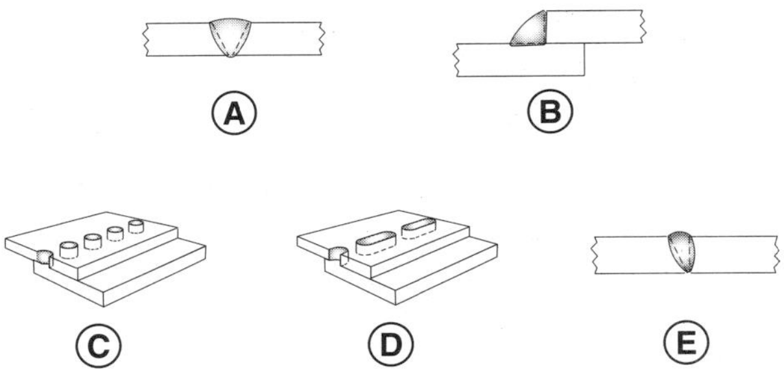

Ⓐ Ⓑ Ⓒ Ⓓ Ⓔ

Identify the types of butt joint designs shown.

_______________ **26.** Square groove

_______________ **27.** Single-bevel groove

_______________ **28.** Single-U groove

_______________ **29.** Double-V groove

_______________ **30.** Single-V groove

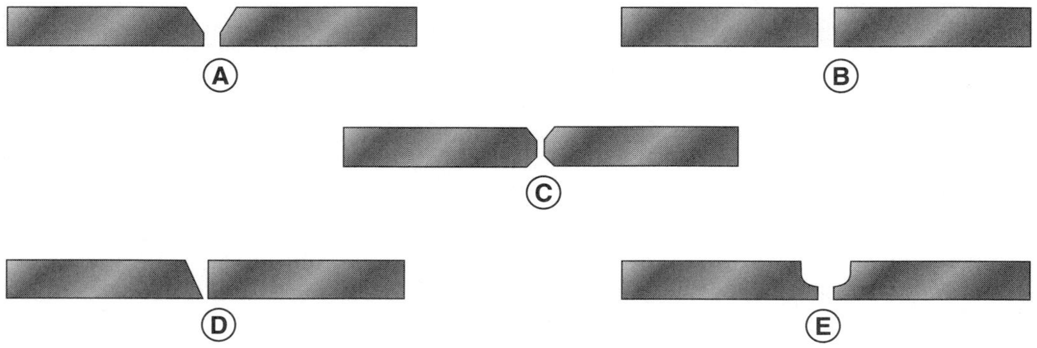

A B C D E

Identify the types of T-joint designs shown.

_______________ **31.** Square groove

_______________ **32.** Single-bevel groove

_______________ **33.** Double-bevel groove

_______________ **34.** Single-J groove

_______________ **35.** Double-J groove

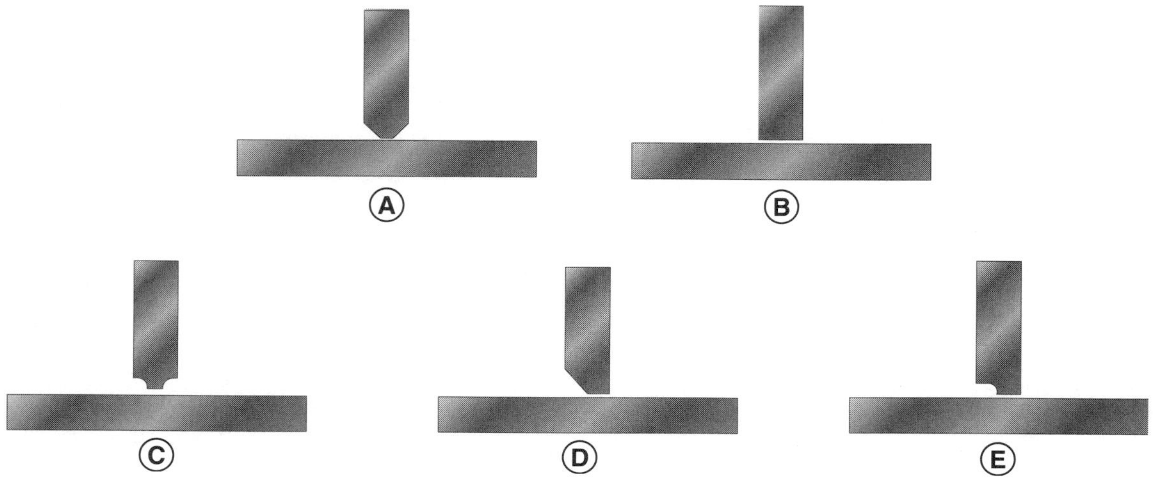

A B C D E

Identify the weld joint designs shown.

____________________ **36.** Lap joint

____________________ **37.** Flush corner joint

____________________ **38.** Half-open corner joint

____________________ **39.** Full-open corner joint

____________________ **40.** Single-bevel edge joint

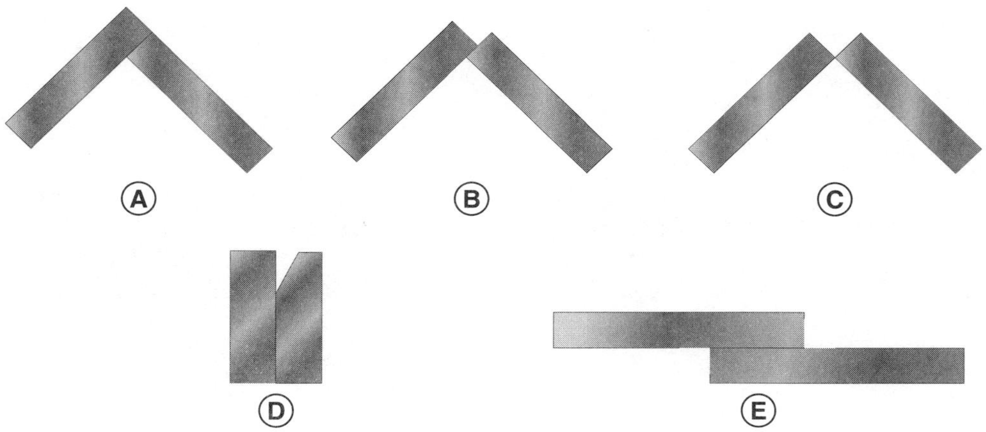

Identify the parts of the weld shown.

____________________ **41.** Base metal

____________________ **42.** Depth of fusion

____________________ **43.** Weld bead

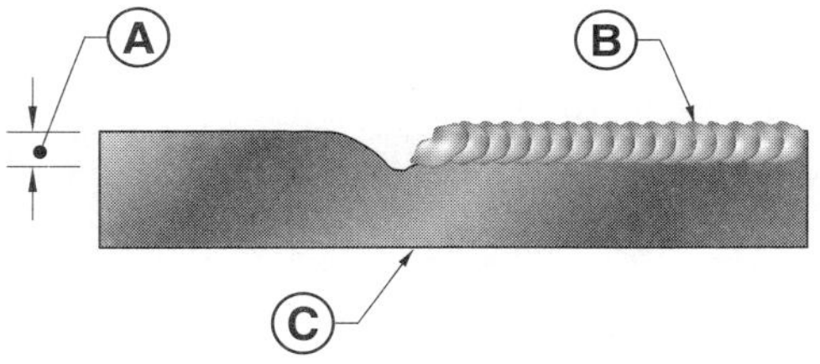

Identify the parts of the T-joint shown.

____________________ **44.** Weld leg

____________________ **45.** Joint root

____________________ **46.** Actual throat

____________________ **47.** Weld face

____________________ **48.** Weld toe

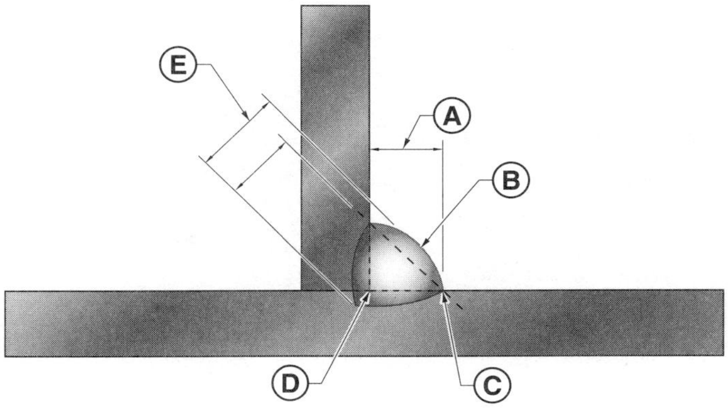

Identify the parts of the groove weld shown.

_____________________ **49.** Root face

_____________________ **50.** Weld reinforcement

_____________________ **51.** Root opening

_____________________ **52.** Weld width

_____________________ **53.** Groove face

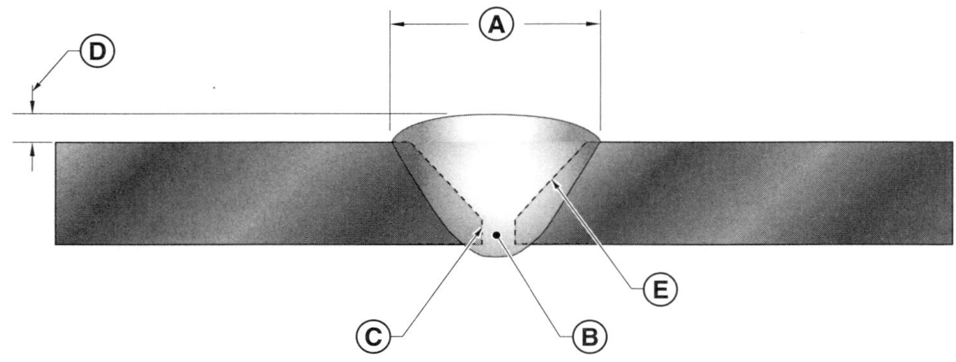

Identify the welding positions shown.

_____________________ **54.** Flat

_____________________ **55.** Horizontal

_____________________ **56.** Vertical

_____________________ **57.** Overhead

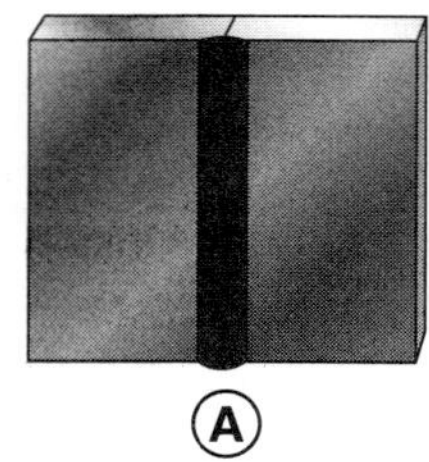

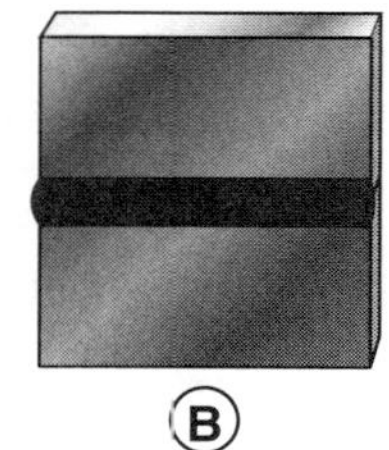

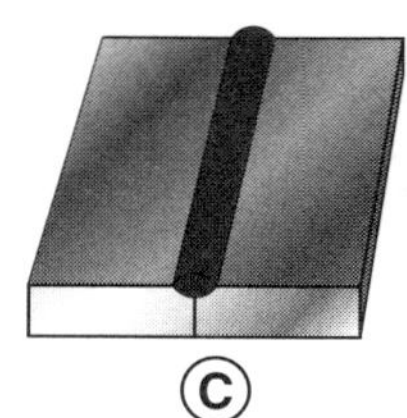

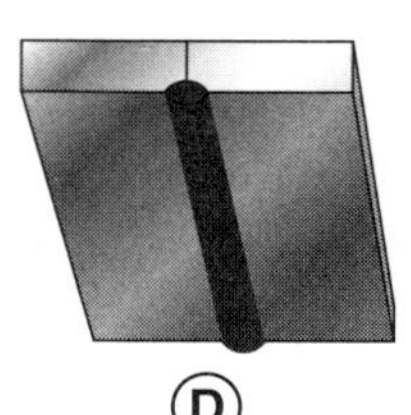

Name ______________________ Date ______________

True-False

T F **1.** The oxygen cylinder valve is opened with a wrench.

T F **2.** The operating pressure for acetylene should never exceed 15 psi.

T F **3.** Acetylene has no detectable odor.

T F **4.** A cylinder cart should be used to move gas cylinders.

T F **5.** Two types of oxyacetylene welding torches are the injector and the medium-pressure.

T F **6.** Welding tip size is governed by the diameter of its opening.

T F **7.** Left-hand threads are on all fittings for acetylene gas.

T F **8.** Oxygen cylinders are packed with a porous material to store oxygen safely.

T F **9.** Cylinders should be kept in a vertical position.

T F **10.** A pair of pliers should be used to install the tip on the welding torch.

T F **11.** A sparklighter should be used to light the oxyacetylene torch.

T F **12.** MAPP gas is more sensitive to shock than acetylene.

T F **13.** Needle valves on the torch regulate the flow of oxygen and acetylene at the torch.

T F **14.** Two types of regulators are the single-stage and the two-stage.

T F **15.** Welding hoses can be safely repaired with tape.

Fillet Weld Both Sides

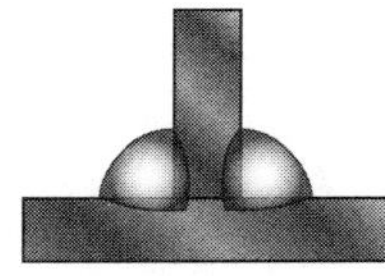

DESIRED WELD

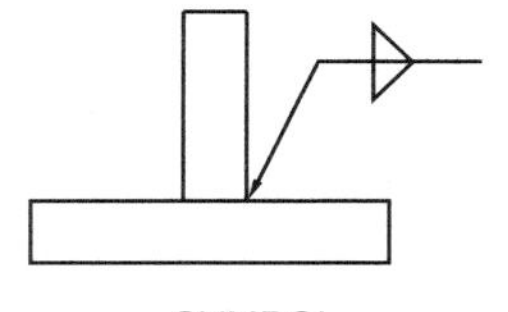

SYMBOL

T F **16.** Regulators should be handled extremely carefully when being removed from the cylinder.

T F **17.** Right-hand threads are always used on connections for acetylene hoses.

T F **18.** Oxygen and acetylene cylinders should be chained at all times during use and when stored.

T F **19.** Cylinders should not be exposed to any open fire, heat, or sparks from the torch.

T F **20.** A match should be used to ignite the oxyacetylene flame.

T F **21.** Acetylene cylinders are packed with a porous material that is saturated with acetone to allow the safe storage of acetylene.

T F **22.** An acetylene regulator has a red line above 15 psi, indicating a dangerous operating range.

T F **23.** The atmosphere is comprised of approximately 30% oxygen.

Multiple Choice

______________ **1.** What is the meaning of the welding symbol shown?
- A. fillet weld other side
- B. fillet weld both sides
- C. slot weld both sides
- D. slot weld other side

______________ **2.** Acetylene gas ___.
- A. has a very distinctive, nauseating odor
- B. must be used in a mixture with MAPP gas
- C. is yellow in color
- D. is formed by the mixture of carbon and nitrogen

______________ **3.** A ___ is caused by the flame going out suddenly on the torch.
- A. flashback
- B. firewall
- C. creeping regulator
- D. backfire

______________ **4.** The skills required for ___ are often compared with those developed by practicing oxyacetylene welding.
- A. SAW
- B. GMAW
- C. GTAW
- D. SMAW

_______________ **5.** The combination of oxygen and ___ generates a low-temperature flame used primarily for welding thin sections of metal, usually aluminum, for which low temperatures are required.
A. propane
B. methylacetylene-propadiene
C. nitrogen
D. hydrogen

_______________ **6.** A ___ allows the flow of liquid or gas in one direction only.
A. flash arrestor
B. check valve
C. needle valve
D. tip

Matching

_______________ **1.** In an acetylene cylinder, ___ stabilizes acetylene and allows the cylinder to be pressurized to 250 psi.

_______________ **2.** A(n) ___ prevents damage to the oxygen valve.

_______________ **3.** Cylinders are charged with ___ to a pressure of 2200 psi.

_______________ **4.** A(n) ___ prevents a flashback from reaching the regulator and the acetylene cylinder.

_______________ **5.** At operating pressures over 15 psi, ___ becomes unstable.

A. flash arrestor
B. oxygen
C. acetylene
D. protector cap
E. acetone

_______________ **6.** A ___ is used to squeeze the hose around the nipple.

_______________ **7.** A ___ transports oxygen.

_______________ **8.** A ___ transports acetylene gas.

_______________ **9.** A ___ is used for the connector on an oxygen hose.

_______________ **10.** A ___ is used for the connector on an acetylene hose.

A. red hose
B. green hose
C. clamp
D. right-hand thread
E. left-hand thread

____________________ **11.** A protector cap guards against damage to the ___.

____________________ **12.** The ___ is used to regulate the flow of oxygen and acetylene delivered to the torch from the regulator.

____________________ **13.** Multiple cylinders are connected in line in a(n) ___ to supply acetylene to the welding stations.

____________________ **14.** A(n) ___ controls the flow of gas from the cylinder.

____________________ **15.** A(n) ___ controls the flow of gas at the welding torch.

A. manifold system
B. regulator
C. adjusting screw
D. needle valve
E. cylinder valve

____________________ **16.** A ___ is used to ignite the torch flame.

____________________ **17.** A large oxygen cylinder contains ___ cubic feet of oxygen.

____________________ **18.** An oxygen cylinder is pressurized to ___ psi.

____________________ **19.** A ___ mixes oxygen and acetylene in a mixing chamber.

____________________ **20.** A ___ prevents a flashback from reaching the manifold system.

A. medium-pressure torch
B. sparklighter
C. 244
D. flash arrestor
E. 2200

Identify the parts of the welding torch shown.

____________________ **21.** Check valve

____________________ **22.** Mixing chamber

____________________ **23.** Oxygen inlet

____________________ **24.** Acetylene inlet

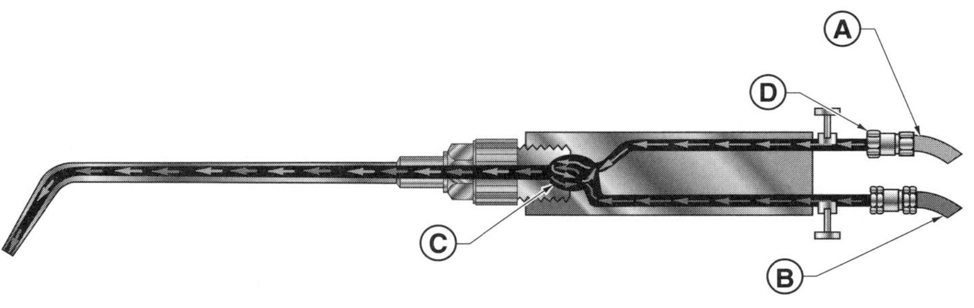

Chapter 5 Review

OAW – Setup and Operation

Name ______________________________ Date ____________________

True-False

T F 1. Applying soapy water to an oxyacetylene outfit with a brush can be used to locate leaks.

T F 2. The cylinder valve is cracked to remove foreign matter lodged in the outlet nozzle.

T F 3. Open the cylinder valves to purge the welding hoses.

T F 4. The size of a welding tip depends on the amount of pressure used.

T F 5. The welder should stand to one side of the regulator when opening the cylinder valves.

T F 6. The acetylene needle valve should be opened fully to ensure quick lighting of the torch.

T F 7. A torch should be lit with the tip pointing downward.

T F 8. No pressure should be left in the working pressure gauges when the welding unit is shut down.

T F 9. An oxidizing flame has a slight feather extending from the inner cone of the flame.

T F 10. A neutral flame is used for most welding operations.

T F 11. Popping is caused by insufficient gas flow to the welding torch.

T F 12. A neutral flame results from a one-to-one mixture of oxygen and acetylene.

Single-V-Groove Weld

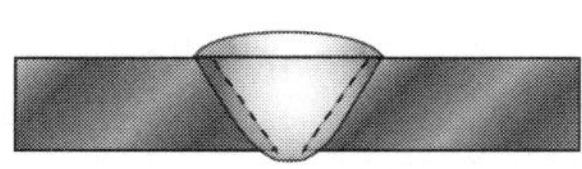

DESIRED WELD

SYMBOL

Multiple Choice

________________ **1.** An oxidizing flame resembles the neutral flame slightly, but has an inner cone that is shorter and more pointed with an almost ___ color rather than brilliant white.

A. yellow
B. orange
C. green
D. purple

________________ **2.** A(n) ___ flame is sometimes used for brazing.

A. reducing
B. neutral
C. carburizing
D. oxidizing

________________ **3.** A ___ flame has a tendency to depress the molten surface and cause the metal to spatter around the edges of the weld pool.

A. quiet
B. backfiring
C. flashback
D. harsh

________________ **4.** ___ is the last step to take when shutting off the entire welding unit after all welding is complete.

A. Shutting off both the acetylene and oxygen cylinder valves
B. Removing pressure on the working gauges by opening the needle valves until the lines are drained, then promptly closing the needle valves
C. Releasing the adjusting screws on the pressure regulators
D. Cracking the cylinder valves by opening and closing them quickly to remove foreign matter

________________ **5.** What is the meaning of the welding symbol shown?

A. single-V-groove weld other side
B. single-V-groove weld arrow side
C. fillet weld arrow side
D. butt weld other side

Matching

_______________ **1.** The welder must ___ the welding hoses to remove all residual gases.

_______________ **2.** The ___ is determined by the thickness of the metal being welded.

_______________ **3.** The ___ is submerged in water to check for leaks.

_______________ **4.** A(n) ___ hose connection always has left-hand threads as indicated by the notched nut.

_______________ **5.** To remove foreign matter or dirt, the cylinder valves should be ___ before mounting the regulators.

A. welding hose
B. welding torch tip
C. purge
D. acetylene
E. cracked

_______________ **6.** A(n) ___ flame results from a one-to-one mixture of gases.

_______________ **7.** A(n) ___ flame has a slight excess of fuel gas.

_______________ **8.** A welding torch marked ___ indicates that it is connected to the acetylene check valve.

_______________ **9.** A(n) ___ flame has a slight excess of oxygen.

_______________ **10.** A welding torch marked ___ indicates that it is connected to the oxygen.

A. AC
B. OX
C. neutral
D. oxidizing
E. carburizing

_______________ **11.** Testing for leaks requires the use of ___.

_______________ **12.** When lighting the torch, the ___ should point downward.

_______________ **13.** Stand to one side of the regulator when opening a(n) ___.

_______________ **14.** Close the ___ needle valve first when shutting off the welding torch.

_______________ **15.** The first gas ignited when preparing to weld is ___.

A. cylinder valve
B. welding tip
C. soapy water
D. oxygen
E. acetylene

A. backfire
B. flashback
C. popping
D. carburizing
E. oxygen

__________ **16.** An acetylene feather is present beyond the inner cone of a(n) ___ flame.

__________ **17.** A(n) ___ occurs when the flame goes out with a loud pop.

__________ **18.** A flame burning inside the torch with a shrill hissing noise is a(n) ___.

__________ **19.** A(n) ___ noise made by the torch indicates that there is an insufficient amount of gas flowing to the tip.

__________ **20.** The first gas shut off when welding is complete is ___.

Identify the flames shown.

__________ **21.** Oxidizing flame

__________ **22.** Carburizing flame

__________ **23.** Neutral flame

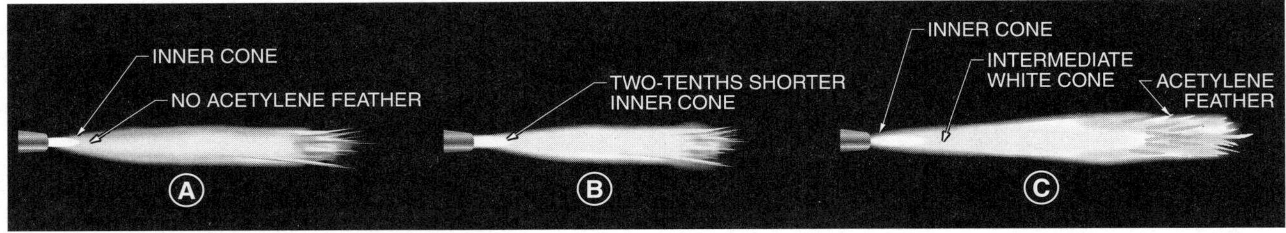

Identify the flames shown.

__________ **24.** Excess acetylene

__________ **25.** Neutral flame

__________ **26.** Acetylene burning in atmosphere

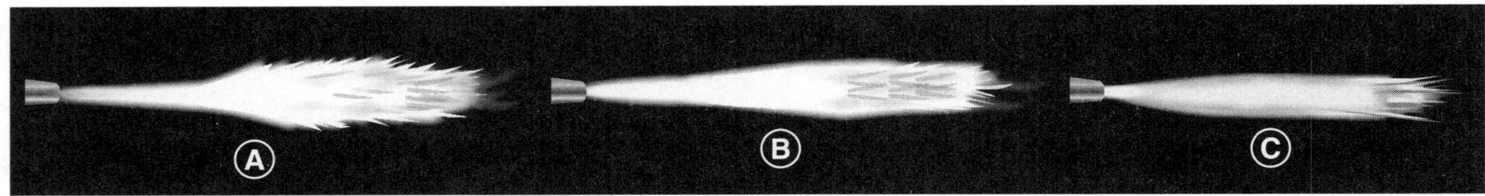

Chapter **6** Review

OAW – Flat Position

Name ______________________________ **Date** ____________________

True-False

T F **1.** The travel speed of a torch must be reduced if the weld pool becomes too large.

T F **2.** A welding torch should be held at a 45° angle to the weld joint.

T F **3.** When carrying a weld pool without filler metal, the inner cone of the flame should be held approximately 1/8″ from the workpiece.

T F **4.** Filler metal with a diameter approximately equal to the thickness of the base metal being welded should be used.

T F **5.** When welding in flat position, the filler metal is held at a 90° angle.

T F **6.** The properties of the metal being welded determine the type of filler metal to be used.

T F **7.** A hole in the weld joint can result from holding the flame in one location for too long.

T F **8.** Filler metal will stick to base metal that is too hot.

T F **9.** Workpieces must be tacked at regular intervals before welding to maintain the root opening.

T F **10.** The heat of the torch should be concentrated on the base metal when depositing weld beads with filler metal.

T F **11.** The filler metal should be positioned in the weld pool along the entire length of the joint when welding.

T F **12.** Filler metals range in diameter from 1/16″ to 3/8″.

Square-Groove Weld

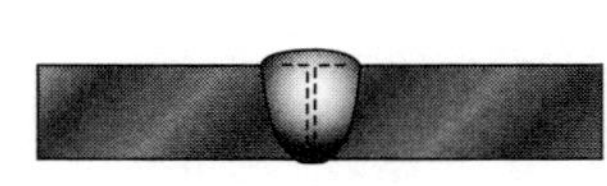

DESIRED WELD

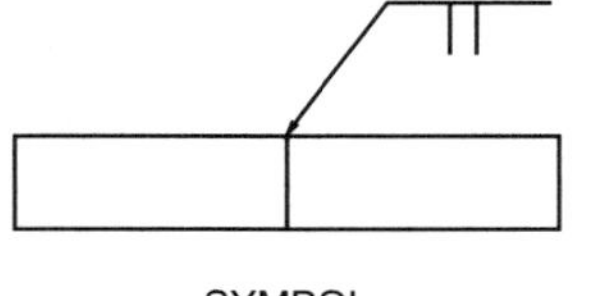

SYMBOL

T F **13.** A semicircular or circular torch movement should be used when depositing weld beads.

T F **14.** Preheat and postheating are necessary when welding gray cast iron.

T F **15.** The thermal conductivity of aluminum is approximately four times that of steel.

T F **16.** All aluminum to be welded is preheated to prevent distortion from expansion and to minimize cracks.

T F **17.** In backhand welding, the welding torch is directed away from the direction of travel.

T F **18.** When penetration is correct, the underside of the seam should show that fusion has taken place completely through the joint.

T F **19.** To allow for the high thermal conductivity of aluminum, use a welding tip that is slightly smaller than the one used for steel of the same thickness.

T F **20.** Oxyacetylene welding is most commonly used for welding heavy steel.

Multiple Choice

______________ **1.** When welding a T-joint in flat position, the torch should be held at a ___° angle to the workpiece.
A. 30
B. 45
C. 60
D. 75

______________ **2.** A ___ requires a greater amount of filler metal than other joints.
A. lap joint
B. butt joint
C. T-joint
D. corner joint

______________ **3.** The correct diameter filler metal is approximately ___ the thickness of the base metal.
A. one-quarter
B. one-half
C. equal to
D. twice

______________ **4.** Carrying a weld pool without filler metal requires that the inner cone of the flame be held approximately ___″ from the workpiece.
A. 1⁄16
B. 1⁄8
C. 3⁄16
D. 1⁄4

________ **5.** ___ is very flimsy and weak when hot.
A. Steel
B. Carbon
C. Cast iron
D. Aluminum

________ **6.** When backhand welding, the torch angle is approximately ___° above the plate.
A. 30
B. 45
C. 60
D. 75

________ **7.** The ___ of the base metal determines the diameter of the filler metal used.
A. composition
B. thickness
C. surface
D. temperature

________ **8.** Aluminum that is greater than ⅜″thick should be prepared as a ___ joint with a notched root face.
A. flanged
B. lap
C. single-V butt
D. double-V butt

________ **9.** Torch technique is determined by metal ___ or by depth of bevel.
A. surface
B. temperature
C. composition
D. thickness

________ **10.** When welding heavy steel more than ½″ thick, ___.
A. prepare the edges as a single-V bevel
B. prepare the edges as a double-V bevel
C. use a single pass to obtain maximum penetration
D. use a forehand welding technique

________ **11.** ___ is a material that hinders or prevents the formation of undesirable substances in molten metal.
A. Filler metal
B. Iron oxide
C. Flux
D. Oxyacetylene

Multiple Choice

__________________ **1.** For horizontal position groove welds, angle the filler rod about ___° above the joint with no side angle.

A. 15
B. 20
C. 30
D. 45

__________________ **2.** Filler metal is added ___ to prevent undercutting.

A. closer to the bottom edge of the joint
B. with a circular motion
C. closer to the top edge of the joint
D. with a side-to-side motion

__________________ **3.** Heat directed to the weld joint should be ___ to prevent the weld pool from falling out of the joint when welding in overhead position.

A. increased
B. reduced
C. kept the same
D. removed

__________________ **4.** What is the meaning of the welding symbol shown?

A. single-V-groove weld
B. double-bevel-groove weld
C. double-V-groove weld
D. fillet weld both sides

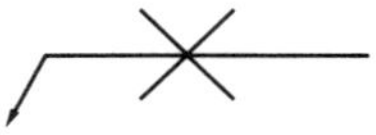

Name ______________________________ Date ______________________

True-False

T F 1. Hot start provides an extra boost of current to help establish an arc when using electrodes that are hard to start.

T F 2. A constant-current welding machine is used for shielded metal arc welding.

T F 3. The heat generated when using shielded metal arc welding ranges from 6000°F to 10,000°F.

T F 4. Current is the amount of electron flow through an electrical circuit.

T F 5. The actual voltage used to provide the welding current ranges from 220 V to 440 V.

T F 6. SMAW cannot weld carbon steel, low- or high-alloy steel, stainless steel, cast iron, or ductile iron.

T F 7. In North America, alternating current is rated at 50 Hz.

T F 8. Resistance is the force that causes current to move in a circuit.

T F 9. Direct current is electrical current that flows in one direction only.

T F 10. A transformer changes AC current from one level to another.

T F 11. Arc voltage is generally between 10 V to 35 V.

T F 12. An engine-driven welding machine's initial purchase, operating, and maintenance costs are lower than those of a static welding machine of the same current output per hour of operation.

T F 13. A welding machine with a 60% duty cycle can operate at rated output for 6 min out of every 10 min without overheating.

Single-Bevel-Groove Weld

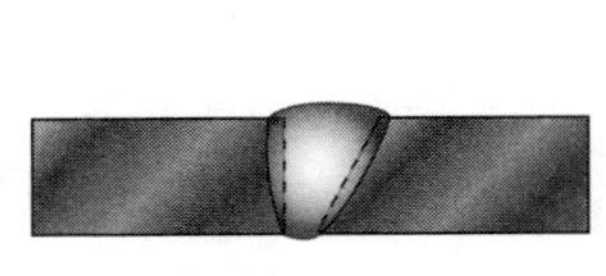

DESIRED WELD

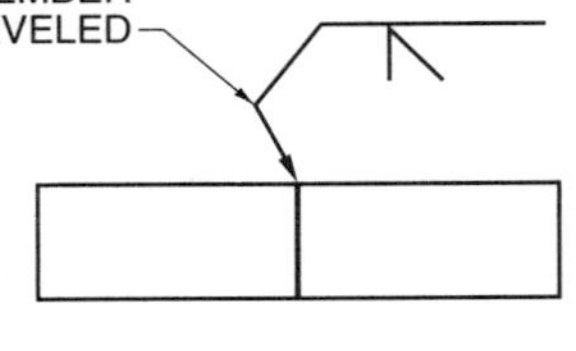

SYMBOL

T F **14.** The jaws of an electrode holder are insulated to prevent accidental flashing.

T F **15.** The workpiece lead is connected to the workpiece or to the workbench.

T F **16.** Safety glasses should be worn under the welding helmet at all times when working in the shop.

T F **17.** Pant cuffs are useful to protect the welder's shoes from molten metal.

T F **18.** SMAW is the most efficient process in terms of the amount of time it takes to lay a weld.

T F **19.** Duty cycle is the percentage of time during a 24-hour period that a welding machine can be operated at its rated load without exceeding the temperature limits of the insulation on the component parts.

T F **20.** DC current used for welding can only be direct current electrode positive (DCEP).

Multiple Choice

____________________ **1.** A(n) ___ is the basic unit of measurement of resistance.
A. ampere
B. farad
C. ohm
D. volt

____________________ **2.** After the arc is struck, the voltage drops to ___.
A. dynamic electricity
B. reverse polarity
C. variable polarity
D. arc voltage

____________________ **3.** When welding with DCEP, electrons flow from the ___.
A. electrode to the workpiece
B. workpiece to the electrode
C. workpiece to the ground
D. ground to the electrode

____________________ **4.** A(n) ___ is used to remove slag after the weld is deposited.
A. cutting torch
B. chipping hammer
C. electrode holder
D. inverter

____________________ **5.** When performing SMAW, the general recommendation for adequate ventilation is a minimum of ___ cu ft of air flow per minute per welding machine.
A. 250
B. 750
C. 1500
D. 2000

_______________ **6.** ___ determines whether current flows into the workpiece or toward the electrode in a DC circuit.
A. Current
B. Frequency
C. Resistance
D. Polarity

_______________ **7.** What is the meaning of the welding symbol shown?
A. single-bevel-groove weld arrow side
B. single-bevel-groove weld other side
C. single-V-groove weld arrow side
D. single-V-groove weld other side

_______________ **8.** NEMA has set a standard for duty cycles based on a ___ period.
A. 6 min
B. 10 min
C. 1 hr
D. 24 hr

Matching

_______________ **1.** A(n) ___ is material through which electricity flows easily.

_______________ **2.** When the workpiece is positively charged, ___ current is being used.

_______________ **3.** When the workpiece is negatively charged, ___ current is being used.

_______________ **4.** Electrical current with positive and negative values is ___.

_______________ **5.** A(n) ___ is a unit of measure for electrical current.

A. alternating current
B. DCEP
C. conductor
D. DCEN
E. ampere

_______________ **6.** Voltage produced when a welding machine is ON but no welding is being performed is ___.

_______________ **7.** A(n) ___ problem is usually associated with using welding cables that are either too small or too long or that have been damaged.

_______________ **8.** A(n) ___ is the amount of electrical pressure in a circuit.

_______________ **9.** The opposition of the material in a conductor to the passage of an electric current is ___.

_______________ **10.** The voltage present after an arc is struck is the ___.

A. voltage
B. resistance
C. arc voltage
D. voltage drop
E. open-circuit voltage

_______________ **11.** A(n) ___ is used to remove slag.

_______________ **12.** A(n) ___ conducts current to and from the work.

_______________ **13.** The ___ is connected to the workpiece or the workbench.

_______________ **14.** The ___ should be protected by insulation.

_______________ **15.** Smoke and fumes should be properly ___.

A. welding lead
B. vented
C. chipping hammer
D. electrode holder
E. workpiece lead

_______________ **16.** A(n) ___ welding machine typically operates using a 1ϕ primary power source.

_______________ **17.** A(n) ___ can change alternating current into direct current.

_______________ **18.** Welding machines with a 100% ___ can be run continuously without overheating.

_______________ **19.** A(n) ___ can change direct current into alternating current.

_______________ **20.** The positive or negative state of an object is its ___.

A. rectifier
B. constant-current
C. polarity
D. duty cycle
E. inverter

Name ______________________ **Date** ______________

True-False

T F **1.** The E7024 electrode is used for overhead butt joints.

T F **2.** Moisture can damage the coating on an electrode.

T F **3.** An electrode should always be used until it is burned down to a 4″ stub.

T F **4.** Some electrodes are designed to be used only with direct current.

T F **5.** Fast-fill electrodes from the F1 group are used for flat position welds.

T F **6.** Fast-freeze electrodes from the F3 group are best suited for vertical and overhead welding.

T F **7.** The electrode used must produce a weld with approximately the same mechanical properties as the base metal.

T F **8.** The AWS organizes electrodes for low-carbon and low-alloy steel into four F groups based on their performance characteristics.

T F **9.** Coatings on some electrodes release oxygen and nitrogen to protect the weld metal.

T F **10.** An E6010 electrode produces a flat bead and can be used in all positions.

T F **11.** When welding in vertical and overhead position, a 3⁄16″ diameter electrode is the largest electrode that should be used.

T F **12.** Damaged coatings on electrodes do not affect the performance of the electrode.

T F **13.** Elements in the coating on an electrode act as cleaning and deoxidizing agents in the molten weld pool.

Double-Bevel-Groove Weld

BREAK INDICATES MEMBER TO BE BEVELED

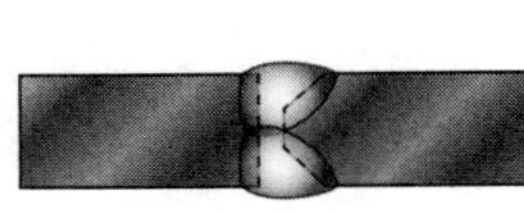

DESIRED WELD

SYMBOL

T F **14.** The first two digits of the AWS electrode classification designate special manufacturer's characteristics.

T F **15.** An E6010 electrode is designed only for use with DCEP current.

T F **16.** An E6013 electrode may be used with AC, DCEN, and DCEP current.

T F **17.** When selecting an electrode, joint design and fit-up must be considered.

T F **18.** The second to last digit of the AWS electrode classification indicates the tensile strength of an electrode.

T F **19.** AC is preferred for vertical and overhead welding.

T F **20.** The prefix E in the AWS electrode classification stands for electrode.

Multiple Choice

______________ **1.** A ___ electrode is an electrode with a high iron powder coating that has a soft arc and high deposit rates.
A. fast-freeze
B. fill-freeze
C. low-carbon steel
D. fast-fill

______________ **2.** ___ causes embrittlement and loss of toughness.
A. Silicon
B. Carbon
C. Copper
D. Phosphorus

______________ **3.** The number ___ in the second to last position of an AWS electrode classification indicates that the electrode can be used in all positions.
A. 1
B. 2
C. 3
D. 4

______________ **4.** ___ is not a consideration when selecting an electrode for SMAW.
A. Electrode diameter
B. Welding current and polarity
C. Welding position
D. Welding in the field

______________ **5.** An ___ is a fast-freeze electrode from the F3 group.
A. E6010
B. E6012
C. E6014
D. E6024

_______________ **6.** An ___ is a fast-fill electrode from the F1 group.
A. E6011
B. E7014
C. E7018
D. E7024

_______________ **7.** An ___ is a fill-freeze electrode from the F2 group.
A. E6010
B. E6011
C. E6013
D. E7028

_______________ **8.** The first two digits in the AWS electrode classification designate the minimum ___ of the weld metal in thousand pounds per square inch.
A. composition
B. bending strength
C. tensile strength
D. compressive strength

_______________ **9.** ___ electrodes are designed for welding high-sulfur and medium- or high-carbon steel as well as thick sections.
A. Low-hydrogen
B. Aluminum powder
C. Low-carbon steel
D. Iron powder

_______________ **10.** What is the meaning of the welding symbol shown?
A. flare-bevel-groove weld other side
B. double-bevel-groove weld
C. single-V-groove weld
D. double-bevel-groove weld other side

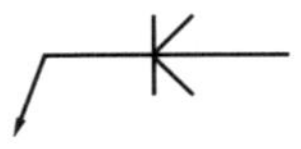

Matching

Match the welding current with the electrode.

_______________ **1.** AC, DCEP

_______________ **2.** DCEP

_______________ **3.** AC, DCEN

_______________ **4.** AC, DCEP, DCEN

A. E6010
B. E6011
C. E6012
D. E6013

Match the characteristics with the electrode.

A. Fast-freeze (F3)
B. Fill-freeze (F2)
C. Fast-fill (F1)
D. Fill-freeze (F4)

______________ **5.** E7018

______________ **6.** E6012

______________ **7.** E6010

______________ **8.** E7024

Select the answer that best completes the sentence.

A. fast-freeze (F3)
B. fill-freeze (F2)
C. fast-fill (F1)

______________ **9.** The ___ electrode produces a snappy arc with little slag.

______________ **10.** The ___ electrode includes iron powder electrodes.

______________ **11.** The ___ electrode is a general purpose electrode.

Identify the parts of the AWS electrode classification.

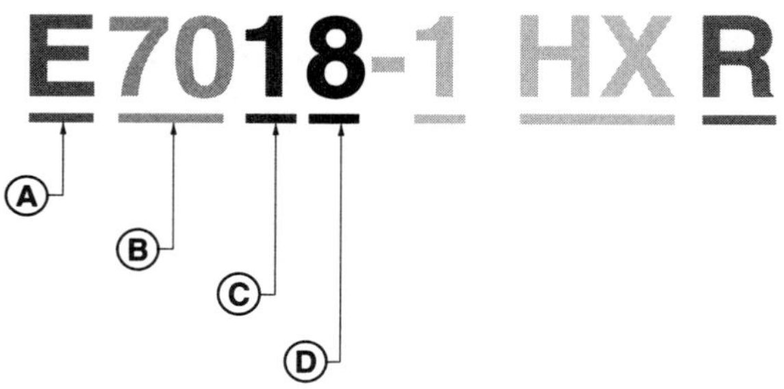

______________ **12.** Welding position

______________ **13.** Type of coating and current

______________ **14.** Tensile strength

______________ **15.** Electrode

E7018-1 HX R

A

B

C

D

Identify the parts of the SMAW process.

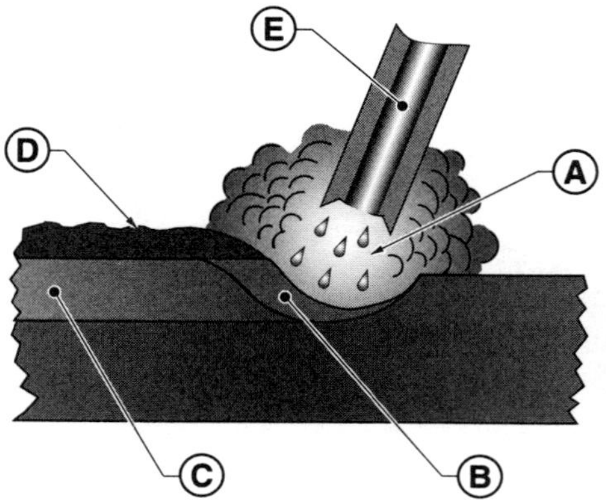

______________ **16.** Weld bead

______________ **17.** Molten weld pool

______________ **18.** Arc

______________ **19.** Electrode metal

______________ **20.** Slag

E

D

A

C

B

Name ______________________________ Date ____________________

True-False

T F 1. The jaws of an electrode holder should always be kept clean to ensure good electrical contact.

T F 2. Factors such as size and type of electrode, current, and welding position all affect the travel speed.

T F 3. The recommended current range for a specific electrode is approximate and must be adjusted as necessary to achieve a satisfactory weld.

T F 4. The bare end of the electrode is gripped in the electrode holder to ensure good electrical contact.

T F 5. Welding with an arc length that is too short will produce excessive penetration and melt-through in the base metal.

T F 6. The base metal must be free of dirt, rust, and grease.

T F 7. Welding with too much current causes the electrode to stick.

T F 8. Poor connections can either overheat and melt or produce dangerous arcs and sparks.

T F 9. When not in use, remove the electrode from the electrode holder.

T F 10. The electrode angle depends on electrode diameter and welding position.

T F 11. Arc length should be maintained at approximately the diameter of the electrode core wire.

J-Groove Weld

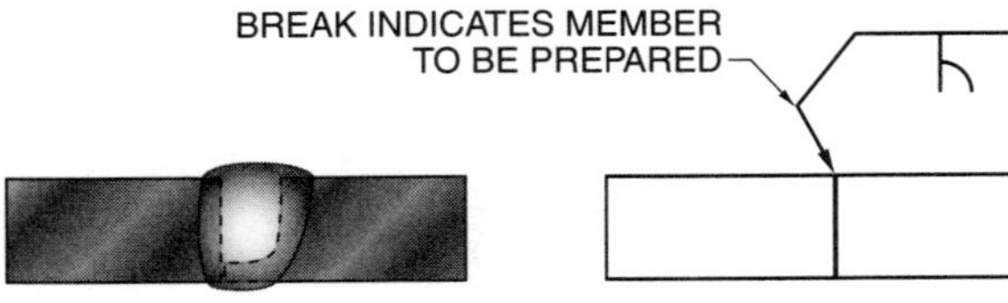

T F **12.** By increasing the current, an electrode can be made to burn off more slowly.

T F **13.** Wrap the work leads around the welding machine when not in use.

T F **14.** As the current decreases, the arc gets hotter, the electrode melts faster, and the weld bead exhibits better fusion into the base metal.

Multiple Choice

______________ **1.** ___ is the distance between the electrode core wire and the weld pool.
A. Electrode gap
B. Current length
C. Travel angle
D. Arc length

______________ **2.** The rate at which the electrode is moved along the weld joint is the ___.
A. electrode angle
B. travel speed
C. scratching speed
D. arc length

______________ **3.** Maintaining the proper ___ is essential to sustaining a strong, stable arc.
A. electrode angle
B. arc length
C. machine setting
D. travel speed

______________ **4.** The final adjustment of ___ is made after beginning the welding operation.
A. voltage
B. current
C. arc blow
D. resistance

______________ **5.** Make sure there are no ___ within 10′ of the work.
A. work leads
B. other people
C. welding machines
D. cable splices

______________ **6.** What is the meaning of the welding symbol shown?
A. Single-J-groove weld other side
B. Single-J-groove weld arrow side
C. Single-J-groove weld both sides
D. Single-V-groove weld arrow side

______________ **7.** In general, ___ should be equal to the diameter of the core wire of the electrode.
A. arc length
B. electrode gap
C. current length
D. electrode angle

Matching

______	**1.** The ___ is the angle at which the electrode is held.	**A.** machine setting
______	**2.** The ___ is the distance between the electrode core wire and the weld pool.	**B.** current
______	**3.** The ___ is the rate at which an electrode moves along the weld joint.	**C.** arc length
______	**4.** The ___ is the adjustment of a welding machine to the required current for a specific welding operation.	**D.** electrode angle
______	**5.** The ___ is the actual flow of electricity adjusted by a control on the welding machine.	**E.** travel speed

______	**6.** When using the ___, an arc is started by bringing the electrode straight downward.	**A.** scratching method
______	**7.** A ___ indicates correct current setting and arc length.	**B.** tapping method
______	**8.** A ___ is established when the electrode is angled toward the weld and away from the direction of travel.	**C.** drag travel angle
______	**9.** When using the ___, the arc is started by bringing the electrode in contact with the workpiece, similar in motion to striking a match.	**D.** crackling noise

Identify the components of the SMAW equipment shown.

_______________ **10.** Electrode holder

_______________ **11.** Workpiece connection

_______________ **12.** Welding machine

_______________ **13.** Workpiece lead

_______________ **14.** Electrode

_______________ **15.** Electrode lead

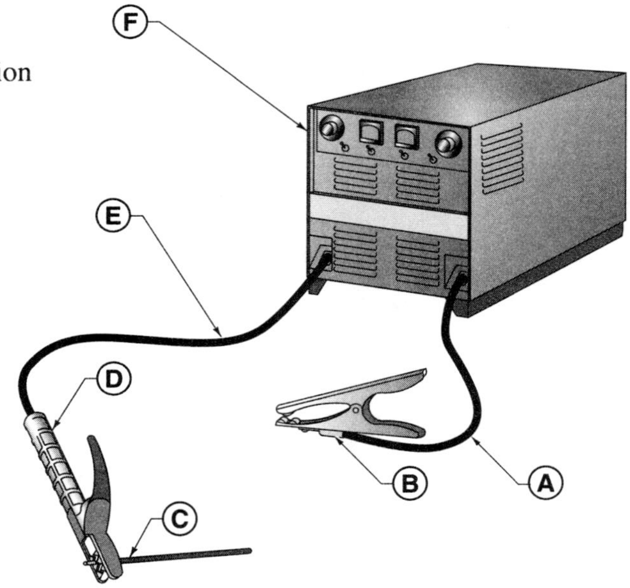

Name ______________________________ **Date** ________________

True-False

T F **1.** The arc length required is determined by the size of the electrode and the welding task.

T F **2.** A narrow weld bead with pointed ripples results if the travel speed is too slow.

T F **3.** Whipping the electrode allows better heat control in the weld pool when using an E6010 or E6011 electrode.

T F **4.** Welding with too high a current results in overlapping.

T F **5.** Undercutting is the result of welding with excessive current, traveling too fast, or using an improper work angle.

T F **6.** The layer of slag that covers a deposited bead must be removed after welding.

T F **7.** A crater is formed when the arc comes in contact with the base metal.

T F **8.** The work angle of an electrode is approximately 90° when making a groove weld in the flat position.

T F **9.** To remelt a crater, the arc should be struck approximately ½″ in front of the previously deposited bead and moved back to retrace the crater to form a new weld pool.

T F **10.** A whipping motion is used when welding with an E6013 electrode.

T F **11.** Arc blow typically occurs when welding with direct current.

T F **12.** When the welding current is too low, the result is poor fusion.

U-Groove Weld

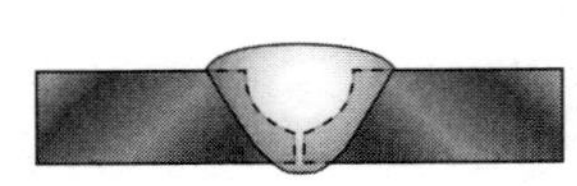

DESIRED WELD

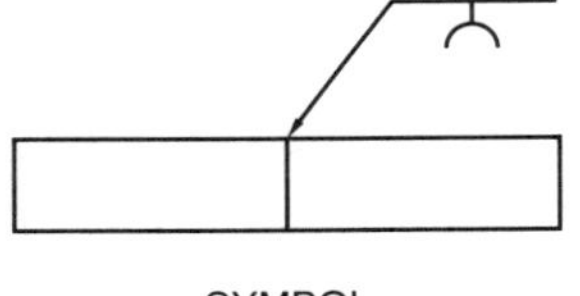

SYMBOL

Multiple Choice

______________ **1.** Arc length should be approximately ___.

A. 5⁄16″
B. 1⁄8″
C. one-third to one-half the total thickness of the workpiece
D. the diameter of the electrode core wire

______________ **2.** A ___ travel angle is a travel angle where the electrode points away from the direction of travel.

A. push
B. shallow
C. whip
D. drag

______________ **3.** The penetration of a deposited bead should equal ___ the total thickness of the weld bead.

A. two times
B. two to three times
C. one-eighth to one-third
D. one-third to one-half

______________ **4.** The ___ angle is in the line of the welding and may vary from 5° to 30°.

A. work
B. travel
C. current
D. arc

______________ **5.** ___ may occur when there is insufficient deposition of metal on a vertical plate.

A. Overlapping
B. Arc blast
C. Cold lapping
D. Undercutting

______________ **6.** Undercutting can be corrected by ___.

A. increasing the amperage
B. changing the electrode angle
C. increasing the arc length
D. decreasing travel speed

______________ **7.** Slag is removed from the weld using a(n) ___.

A. electrode holder
B. pair of pliers
C. chipping hammer
D. magnet

_______________ **8.** Welding current that is too high results in excessive ___.
A. bead height
B. travel speed
C. spatter
D. overlap

_______________ **9.** The best way to prevent arc blow is to ___.
A. reduce the welding current
B. use the back-step technique
C. use the shortest possible arc
D. use AC

_______________ **10.** What is the meaning of the welding symbol shown?
A. single-U-groove weld arrow side
B. single-bevel-groove weld other side
C. single-V-groove weld arrow side
D. single-U-groove weld other side

Matching

_______________ **1.** The weld bead will show narrow pointed ripples if the incorrect ___ is used.

_______________ **2.** The properties of the base metal determine the ___ to use.

_______________ **3.** The ___ may vary from 5° to 30° from the vertical in the line of the welding.

_______________ **4.** If the ___ is too high, the electrode will melt too quickly.

_______________ **5.** The ___ should be approximately ⅛″ when the diameter of the electrode core wire is ⅛″.

A. electrode
B. arc length
C. current
D. travel speed
E. travel angle

_______________ **6.** A(n) ___ motion is used to control the temperature of the molten weld pool.

_______________ **7.** When the current is too low, ___ occurs.

_______________ **8.** A(n) ___ is a pool in the base metal made by the arc.

_______________ **9.** A welding current that is too high results in ___.

_______________ **10.** The magnetic field produced by direct current can cause ___.

A. crater
B. whipping
C. undercutting
D. overlapping
E. arc blow

__________________ **11.** The depth of ___ should be approximately one-third to one-half the total thickness of the bead.

__________________ **12.** The ___ is measured less than 90° perpendicular to the workpiece.

__________________ **13.** The amount of ___ determines the amount of heat available to melt the base metal.

A. work angle
B. penetration
C. current

Identify the following.

__________________ **14.** Crater (weld pool)

__________________ **15.** Bead thickness

__________________ **16.** Electrode

__________________ **17.** Penetration

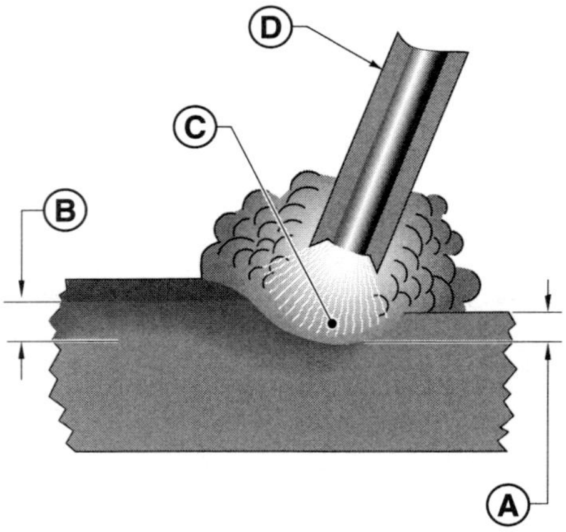

A
B
C
D

Chapter 12 Review — SMAW–Flat Position

Name ______________________________ **Date** ______________

True-False

T F **1.** Root reinforcement should not exceed fabrication code criteria or 1⁄16″ beyond the bottom surface of the joint.

T F **2.** Welding in flat position is more efficient than welding in other positions.

T F **3.** One or more intermediate weld passes may be necessary to properly complete a weld.

T F **4.** For maximum strength, a T-joint is welded on both sides.

T F **5.** Tack welds are used to cover intermediate weld passes deposited on a butt joint.

T F **6.** The cover pass is used to provide additional reinforcement and a finished appearance.

T F **7.** A weaving motion is often used when depositing a root bead.

T F **8.** A closed butt joint is suitable for welding steel that generally does not exceed 3⁄16″ thick.

T F **9.** The welding operation is simplified if the joint is in the horizontal position.

T F **10.** An open butt joint should have a root opening of about 3⁄32″ to 1⁄8″ to allow for expansion and penetration.

T F **11.** When welding workpieces of different thicknesses, more heat should be directed to the thinner workpiece for proper heat control.

T F **12.** Problems that occur during SMAW are often the result of improper settings on the welding machine.

Flare-V-Groove Weld

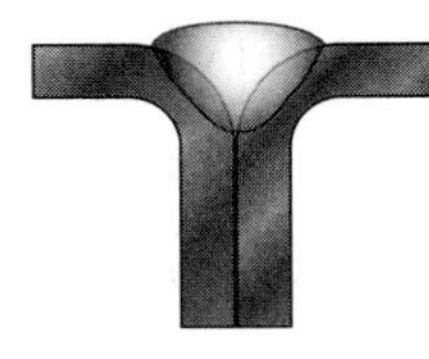

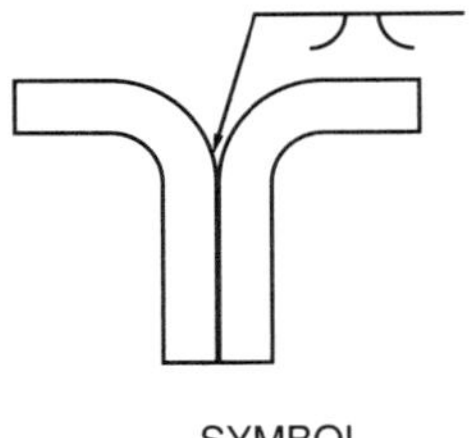

SYMBOL

Matching

_______________ **1.** No beveling or machining is necessary when using a(n) ___.

_______________ **2.** A(n) ___ is used when joined structural pieces must have a flat surface.

_______________ **3.** A(n) ___ is used to construct rectangular objects such as tanks.

_______________ **4.** A(n) ___ is used to increase bead width.

_______________ **5.** A(n) ___ is used to build up worn surfaces or shafts.

A. lap joint
B. surfacing weld
C. outside corner
D. butt joint
E. weaving technique

Identify the parts of the V-butt joint shown.

_______________ **6.** Intermediate weld pass

_______________ **7.** Root pass

_______________ **8.** Cover pass

_______________ **9.** Tack weld

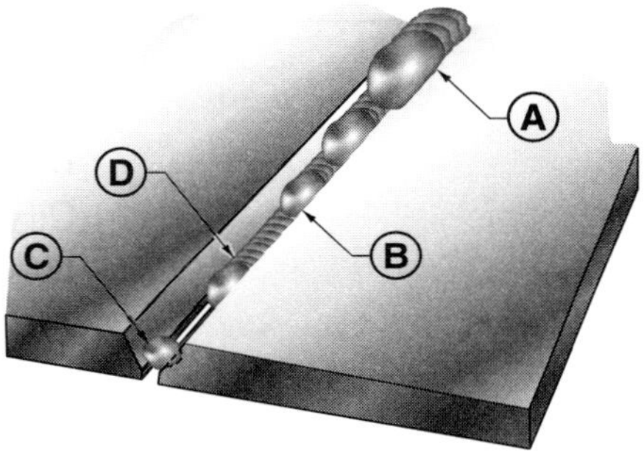

A
D
B
C

Identify the following weaving motions.

_______________ **10.** Figure eight

_______________ **11.** Rotary

_______________ **12.** Crescent

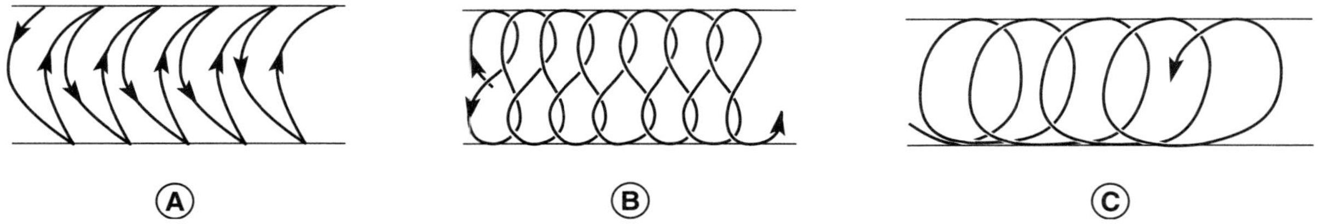

A
B
C

Chapter 13 Review
SMAW–Horizontal Position

Name ______________________________ **Date** ____________________

True-False

T F **1.** A fill-freeze or fast-freeze electrode is best suited for horizontal welding.

T F **2.** A weld is considered to be in the horizontal position when the workpiece is in a horizontal position and the weld joint is approximately vertical.

T F **3.** Overlap occurs when the weld pool runs down to the lower side of the bead and solidifies on the surface without actually penetrating the base metal.

T F **4.** More current is required when welding in horizontal position.

T F **5.** Weaving the electrode when welding in horizontal position provides worse heat control.

T F **6.** A 35° work angle is used when depositing a single-pass lap joint.

T F **7.** Overlaps and undercuts strengthen a weld.

T F **8.** A fill-freeze electrode from the F2 group or a fast-freeze electrode from the F3 group should be used for horizontal welding.

T F **9.** The recommended travel angle of an electrode when depositing a weld on a butt joint in horizontal position is 20°.

T F **10.** The recommended work angle of an electrode when depositing a weld on a butt joint in horizontal position is 20°.

Flare-Bevel-Groove Weld

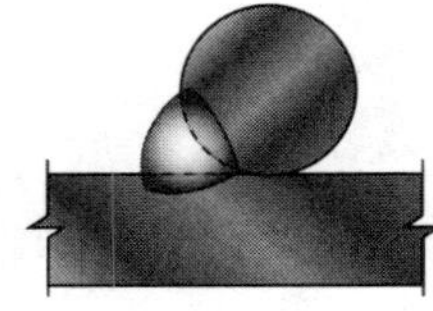
DESIRED WELD

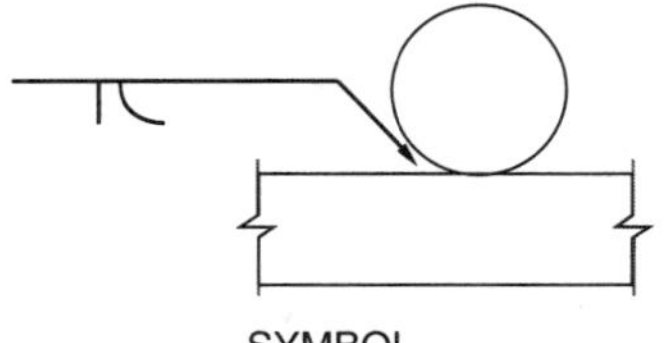
SYMBOL

Multiple Choice

_______________ **1.** When welding a butt joint in horizontal position, the electrode should be held at a work angle of ___ below perpendicular.
- A. 5° to 10°
- B. 15° to 20°
- C. 25° to 30°
- D. 35° to 40°

_______________ **2.** A sagging weld pool usually leaves a(n) ___ on the top side of the weld seam and an improperly shaped bead.
- A. overlap
- B. inclusion
- C. burn-through
- D. undercut

_______________ **3.** What is the meaning of the welding symbol shown?
- A. square groove weld arrow side
- B. single-bevel-groove weld arrow side
- C. flare-V-groove weld arrow side
- D. flare-bevel-groove weld arrow side

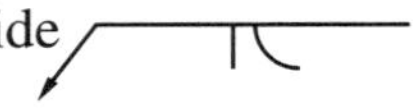

Matching

_______________ **1.** A(n) ___ is a welding defect caused by molten metal running down from a bead and solidifying without penetrating.

_______________ **2.** A(n) ___ occurs when a weld pool sags out of a weld seam.

_______________ **3.** The ___ type of electrode is best suited for horizontal welding.

A. fast-freeze
B. overlap
C. undercut

Name ______________________________ **Date** ____________________

True-False

T	F	**1.** Fast-freeze electrodes are recommended for uphill welding.
T	F	**2.** Downhill welding obtains greater penetration than uphill welding.
T	F	**3.** A light dragging motion with a short arc is used when downhill welding with an E7018 electrode.
T	F	**4.** A 20° to 35° travel angle is recommended for downhill welding.
T	F	**5.** SMAW vertical position uphill is recommended for metal more than ¼″ thick.
T	F	**6.** A whipping motion is used with an E7018 electrode.
T	F	**7.** The width of the weld bead can be increased by using a weaving motion.
T	F	**8.** When welding in vertical position using an E7018 electrode, increase the current to levels higher than used in flat position.
T	F	**9.** When welding uphill, start at the top of the workpiece.
T	F	**10.** When using a whipping motion, the arc should be broken so the metal may cool.
T	F	**11.** A whipping motion is recommended when using an E7018 LS electrode for vertical welding.
T	F	**12.** To ensure a strong weld, slag must be completely removed after each pass of a multiple-pass joint.

Slot Weld

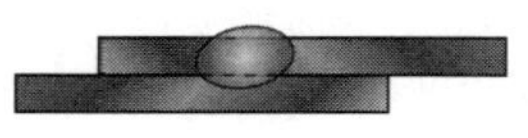

DESIRED WELD

SYMBOL

Multiple Choice

__________ **1.** A ___ travel angle is used when downhill welding.
A. 5°
B. 10°
C. 10° to 15°
D. 15° to 30°

__________ **2.** Downhill welding is used for welding light-gauge metal because the ___ is shallow.
A. arc length
B. whipping motion
C. penetration
D. travel angle

__________ **3.** A(n) ___ electrode is recommended for welding in vertical position.
A. fast-freeze
B. iron powder
C. fast-fill
D. low-carbon steel

__________ **4.** When using a ___ motion, point the electrode ahead of the weld pool briefly so it can solidify without breaking the arc.
A. whipping
B. weaving
C. dragging
D. rotary

__________ **5.** What is the meaning of the welding symbol shown?
A. square groove weld
B. slot weld other side
C. bevel-groove weld other side
D. slot weld arrow side

Matching

__________ **1.** A(n) ___ progression is used for welding light-gauge metals.

__________ **2.** A fast-freeze electrode and a(n) ___ motion are commonly used for uphill welding.

__________ **3.** In a(n) ___ weld, the weld axis is approximately vertical.

__________ **4.** The width of the weld bead is controlled using a(n) ___ motion.

__________ **5.** A(n) ___ progression provides deeper penetration.

A. downhill
B. uphill
C. vertical
D whipping
E. weaving

Name ______________________________ Date ____________________

True-False

T F 1. Shirtsleeves should be rolled up when overhead welding.

T F 2. An electrode holder is held with knuckles facing upward and the palm facing downward.

T F 3. A 75° work angle is used when welding in overhead position.

T F 4. A headcap is recommended when welding in overhead position.

T F 5. Stand to the side rather than directly under the arc to avoid injury from hot metal spatter when welding in overhead position.

T F 6. A long arc length is used to prevent molten metal from falling out of the weld pool when welding in overhead position.

T F 7. To start welding in overhead position, hold the electrode at a right angle to the joint.

T F 8. A positioner is used to secure workpieces in the correct position for practicing welds in overhead position.

T F 9. In the overhead position, the weld pool has a tendency to drop out of the joint, making it harder to produce a consistent bead with complete penetration.

T F 10. When welding in overhead position, the electrode holder can be held in one hand.

Spot Weld

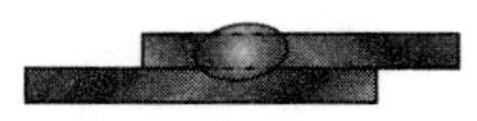

DESIRED WELD

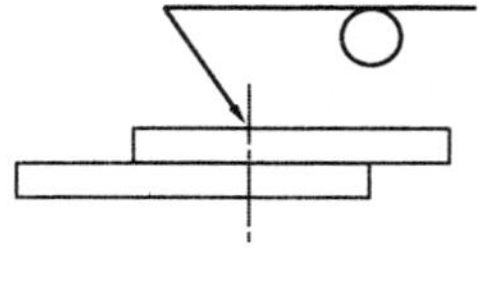

SYMBOL

Multiple Choice

______ **1.** A ___ drag angle is used when welding in overhead position.
A. 5° to 10°
B. 10° to 15°
C. 20° to 30°
D. 40° to 50°

______ **2.** ___ electrodes are recommended for welding in overhead position.
A. Fast-freeze
B. Fast-fill
C. Iron-powder
D. Low carbon-steel

______ **3.** A welder must hold an electrode at a ___° work angle when overhead welding.
A. 45
B. 75
C. 90
D. 115

______ **4.** When welding in overhead position, a welder should not ___.
A. grip the electrode holder so the knuckles are up and the palm is down
B. stand to the side rather than directly underneath the weld
C. increase the arc length to obtain maximum penetration
D. use both hands to hold the electrode holder if necessary

______ **5.** To reduce the cable weight when welding in overhead position, ___.
A. drape it over a shoulder
B. support it with a free hand
C. drape it over the welding machine
D. clamp it to the workpiece

______ **6.** What is the meaning of the welding symbol shown?
A. plug weld arrow side
B. spot weld arrow side
C. butt weld other side
D. bevel-groove weld arrow side

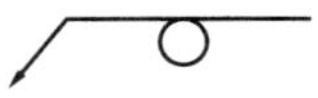

Chapter 16 Review

GTAW–Equipment

Name ______________________________ Date ________________

True-False

T F 1. Filler metals for GTAW use the same classification system as filler metals used for GMAW.

T F 2. Filler metal is necessary on all welds made with the GTAW process.

T F 3. DCEP allows deeper penetration than DCEN.

T F 4. A remote current control allows the welder to adjust current while welding.

T F 5. Postflow is the flow of shielding gas after the arc is extinguished.

T F 6. The chemical symbol for tungsten is W.

T F 7. When welding with DCEN, the electrode is positive and the work is negative.

T F 8. Alloyed tungsten electrodes last longer and have a higher current capacity than pure tungsten electrodes.

T F 9. For thin materials, the diameter of the filler metal should equal the thickness of the metal to be welded or should be slightly smaller.

T F 10. The addition of helium to argon shielding gas improves penetration.

T F 11. Filler metals for oxyacetylene can be used for GTAW as long as they are copper-coated.

T F 12. The GTAW process uses a nonconsumable tungsten electrode to supply welding current to the workpiece.

Seam Weld

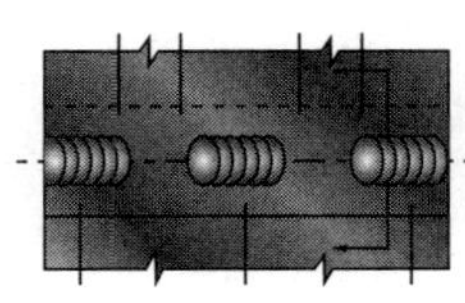

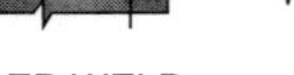

DESIRED WELD

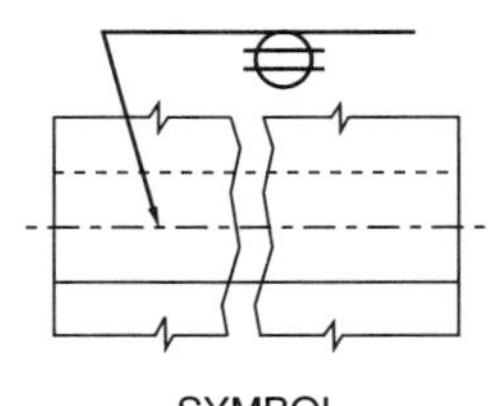

SYMBOL

T F **13.** GTAW requires a constant-current welding power source capable of producing direct current, alternating current, and high frequency.

T F **14.** Argon is the most commonly used as a shielding gas in the GTAW process.

T F **15.** The amount of electrode stickout is determined by the type of weld joint.

T F **16.** The filler metal used should have the same composition as the base metal.

T F **17.** Shielding gas is not required after the arc is extinguished.

T F **18.** A flowmeter regulates the amount of shielding gas flowing to the torch.

T F **19.** A gas nozzle directs the shielding gas from the torch to the weld area.

T F **20.** Helium has a higher ionization potential than argon.

T F **21.** The heated end of the filler metal should be kept in the flow of shielding gas.

Multiple Choice

______________ **1.** Inverters equipped with ___ can use tapered electrodes for welding with AC.
A. high frequency
B. slope controls
C. frequency controls
D. amplitude controls

______________ **2.** DCEN ___ than DCEP when using GTAW.
A. requires slower travel speed
B. uses larger diameter electrodes
C. provides deeper penetration
D. produces wider welds

______________ **3.** When welding using alternating current, EP provides good ___.
A. cleaning action
B. penetration
C. arc starting
D. welds in ferrous metals

______________ **4.** ___ time is the period of time the shielding gas continues to flow after the arc is stopped.
A. Flow
B. Postflow
C. Dwell
D. Purge

______________ **5.** A ___ end on the electrode is typically used when welding with alternating current.
A. tapered
B. hemispherical
C. square
D. concave

__________ **6.** Of the shielding gases used for GTAW, ___ allows the deepest penetration.
- A. carbon dioxide
- B. argon
- C. helium
- D. nitrogen

__________ **7.** What is the meaning of the welding symbol shown?
- A. plug weld
- B. spot weld
- C. slot weld
- D. seam weld

Matching

__________ **1.** In ___, electrons flow from the electrode to the work.

__________ **2.** ___ stands for alternating current.

__________ **3.** ___ is used to start an arc without touching an electrode to the work.

__________ **4.** In ___, electrons flow from the work to the electrode.

A. High frequency
B. AC
C. DCEP
D. DCEN

__________ **5.** ___ is rarely used in GTAW.

__________ **6.** ___ is a combination of DCEP and DCEN.

__________ **7.** When using a conventional square wave transformer rectifier, ___ is used to maintain the welding arc.

__________ **8.** ___ provides deep penetration.

A. AC
B. DCEP
C. DCEN
D. High frequency

__________ **9.** The most commonly used shielding gas is ___.

__________ **10.** A(n) ___ is used to prevent turbulence of the gas stream.

__________ **11.** A(n) ___ directs the shielding gas to the weld.

__________ **12.** The electrode is prepared with a tapered end when welding with ___.

__________ **13.** The electrode is typically prepared with a hemispherical end when welding with ___.

A. gas nozzle
B. gas lens
C. argon
D. AC
E. DCEN

____________	**14.** The ___ must have the same composition as the base metal.	**A.** postflow
____________	**15.** The ___ controls the amount of gas flow in cubic feet per hour (cfh).	**B.** flowmeter
____________	**16.** The ___ is determined by the current to be used and the thickness of the base metal.	**C.** filler metal
____________	**17.** The ___ control includes a timer that maintains gas flow after the arc is extinguished.	**D.** shielding arc

____________	**18.** The ___ is designed for welding on light-gauge metals at less than 200 A current.	**A.** stickout
____________	**19.** The ___ is adjusted as necessary for different joints.	**B.** water-cooled torch
____________	**20.** The ___ is designed for welding at current levels above 200 A.	**C.** air-cooled torch
____________	**21.** Air-cooled torches use ___ to cool the electrode and the internal components of the torch.	**D.** shielding gas

Match each description to the appropriate letter.

____________ **22.** Gas nozzle

____________ **23.** Electrode extension

____________ **24.** Collet

____________ **25.** Stickout

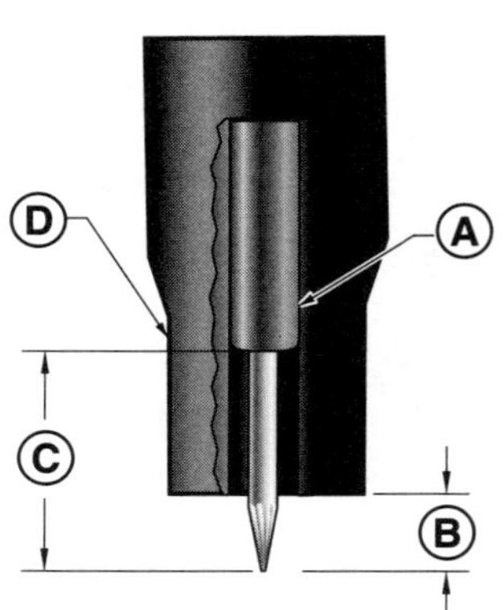

Chapter 17 Review

GTAW– Procedures

Name ______________________________ **Date** ____________________

True-False

T F **1.** One method of adding filler metal to a weld is to use the dip technique with an in-and-out motion.

T F **2.** When using a scratch start, the electrode is withdrawn quickly from the base metal after the arc is struck.

T F **3.** When welding a lap joint, filler should be added when the weld pool forms a V-shaped notch.

T F **4.** Downhill welding is preferred for welding with GTAW in the vertical position.

T F **5.** On thin-gauge metals sensitive to atmospheric contamination, a backing bar is used to protect the root side of the weld.

T F **6.** When using HF start, the electrode does not touch the workpiece when starting the arc.

T F **7.** In the semiautomatic process, the torch is controlled automatically while the operator feeds the filler metal into the weld pool.

T F **8.** T-joints on thin metals can be welded with or without filler metal.

T F **9.** Heat sink is an indentation on the side of the joint opposite the weld.

T F **10.** In hot wire welding, filler metal is automatically fed from a wire feeder.

T F **11.** Pulsed GTAW uses two levels of welding current.

T F **12.** Capacitor discharge start is an arc starting method that requires the electrode to touch the work.

Back Weld

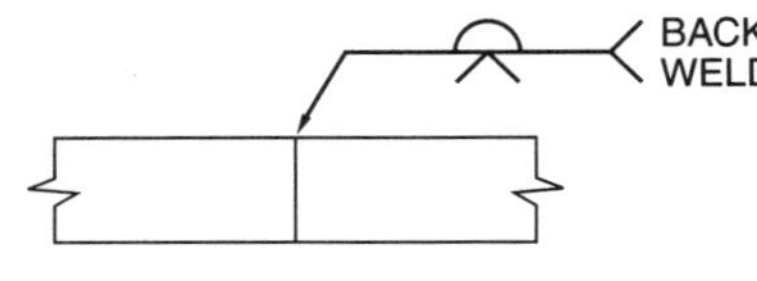

SYMBOL

T F 13. When welding in overhead position, the current should be increased 5% to 10% from that used in flat position.

T F 14. When welding a T-joint, filler metal is required regardless of the thickness of the metal.

T F 15. Copper bars are suitable backing bars for most applications.

T F 16. The filler metal should be dragged in the weld pool to increase the weld metal deposited.

Multiple Choice

______________ 1. Electrode stickout is ⅛″ to 3⁄16″ for welding ___ joints.
- A. corner
- B. edge
- C. lap
- D. butt

______________ 2. In the ___ process of application, the weld operation is done by hand.
- A. automatic
- B. semiautomatic
- C. mechanized
- D. manual

______________ 3. A ___° work angle is required to deposit a fillet weld in a T-joint.
- A. 30
- B. 45
- C. 80 to 85
- D. 90

______________ 4. Filler metal should be held at a ___ angle above the surface of the work when welding a butt joint.
- A. 5° to 10°
- B. 10° to 15°
- C. 15° to 20°
- D. 20° to 30°

______________ 5. What is the meaning of the welding symbol shown?
- A. single-V-groove weld with backing
- B. fillet weld with back weld
- C. single-V-groove weld with back weld
- D. single-bevel-groove weld with backing

BACK WELD

______ **6.** For butt joints on thin-gauge metals sensitive to atmospheric contamination, ___ are used to protect the root side of the weld.
   A. backing bars
   B. heat sinks
   C. HAZs
   D. oils

______ **7.** In the ___ technique, filler metal is held in the joint with a slight downward pressure as the torch is moved with a steady push.
   A. dip
   B. downward
   C. lay-wire
   D. push

Matching

______ **1.** Applying a weld using the ___ process allows the operator to control the direction and speed of travel, while the filler metal is fed automatically.

______ **2.** Weld size, weld length, rate of travel, and starting and stopping of the weld operation are controlled by equipment under the observation of the operator in the ___ process.

______ **3.** Welding is done by hand in the ___ process.

______ **4.** Constant observation and adjustment of controls by an operator are not required in the ___ process.

A. manual
B. semiautomatic
C. mechanized
D. automatic

______ **5.** Filler metal must be used to weld a(n) ___ regardless of the thickness of the metal.

______ **6.** A(n) ___ is not recommended on metals thicker than ¼″.

______ **7.** The number of passes required for a(n) ___ depends on the size of the groove angle and the thickness of the metal.

______ **8.** A(n) ___ is suitable only on thin-gauge metal.

______ **9.** For thin metals up to ⅛″, the square ___ is the easiest to prepare.

A. corner joint
B. T-joint
C. edge joint
D. lap joint
E. butt joint

____________________	**10.** In GTAW-P, welding is done during ___.	**A.** current
____________________	**11.** Background ___ maintains the arc and allows the weld pool to cool slightly.	**B.** pulse per second
____________________	**12.** The number of times that current peaks in one second is ___.	**C.** percent ON time
____________________	**13.** The length of time that peak current is maintained before it drops to background current is ___.	**D.** heat input
____________________	**14.** Pulsed current produces deep penetration with less overall ___, making it easier to control the weld pool.	**E.** peak current

Chapter 18 Review
GTAW–Applications

Name ______________________________ Date ______________

True-False

T F **1.** Filler metals for welding stainless steels are alloyed to prevent cracking problems.

T F **2.** The electrode positive part of the AC cycle produces a cleaning action that removes the oxide coating on aluminum.

T F **3.** When welding, the fumes from copper and copper alloys can be toxic.

T F **4.** High-carbon steels are weldable but require preheat and postheating.

T F **5.** Filler metal for GTAW is the same as the filler metal used for SMAW.

T F **6.** Stainless steels, especially those in the 300 austenitic series, are commonly welded with GTAW using DCEP.

T F **7.** The welding characteristics of magnesium are comparable to those of carbon steel.

T F **8.** A forehand welding technique usually produces the best results when welding copper and copper alloys.

T F **9.** The procedure for welding all types of stainless steels with GTAW is the same.

T F **10.** Acetone is commonly used for cleaning contaminants on aluminum.

Melt-Through

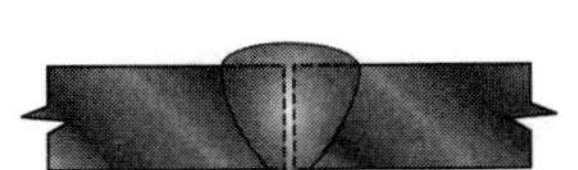

DESIRED WELD

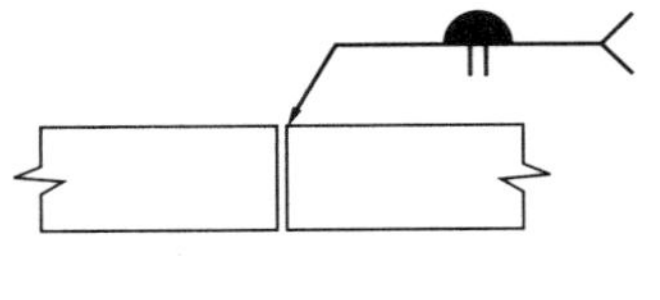

SYMBOL

Multiple Choice

_______________ **1.** When welding aluminum with GTAW, the best results are obtained by using ___ shielding gas.
A. 75% argon / 25% CO_2
B. 98% argon / 2% CO_2
C. 100% helium
D. 100% argon

_______________ **2.** Wrought aluminum alloys in the ___ series are readily weldable.
A. 2000
B. 3000
C. 6000
D. 7000

_______________ **3.** When welding thin-gauge aluminum, ___ should be used to minimize distortion.
A. filler metal
B. stainless steel backing bars
C. copper backing bars
D. postheating

_______________ **4.** A nonferrous metal is a metal that does not contain ___.
A. copper
B. aluminum
C. magnesium
D. iron

_______________ **5.** What is the meaning of the welding symbol shown?
A. square groove weld with melt-through
B. flare-groove weld with melt-through
C. double-groove weld
D. single-bevel-groove weld

Chapter 19 Review
GMAW–Equipment

Name ______________________________ **Date** ____________________

True-False

T F **1.** A pull type wire feeder is used for welding with soft and small-diameter electrodes.

T F **2.** In a constant-voltage welding machine, arc length remains constant with changes in electrode extension.

T F **3.** A constant-voltage welding machine is commonly used for GMAW.

T F **4.** Alternating current is used for welding in overhead position on mild steel.

T F **5.** The constant-voltage welding machine has a slightly sloping volt-ampere curve.

T F **6.** Inadequate gas shielding is one cause of porosity.

T F **7.** Poor arc starting is a symptom of a worn contact tip.

T F **8.** If electrode extension is increased, welding current is decreased.

T F **9.** The letter S in the electrode designation ER70S-3 stands for short circuiting transfer.

T F **10.** Voltage is increased by increasing the wire feed speed.

T F **11.** In semiautomatic GMAW, the operator sets the welding parameters, and the welding gun is controlled by a mechanical device.

T F **12.** DCEP produces deep penetration and excellent cleaning action.

T F **13.** Argon is commonly used as a shielding gas when welding aluminum with GMAW.

Surfacing Weld

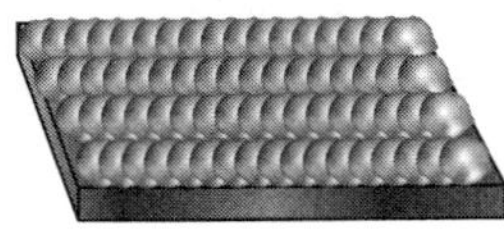

DESIRED WELD

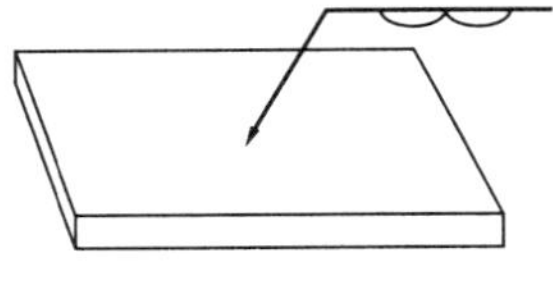

SYMBOL

Matching

________	**1.** ___ results in inadequate shielding.	**A.** A shielding gas
________	**2.** ___ automatically advances the welding wire from the wire spool, through the welding gun cable liner and welding gun, to the arc.	**B.** A wire feeder
________	**3.** ___ is not recommended for the GMAW process.	**C.** A contact tip
________	**4.** ___ prevents oxygen, nitrogen, and hydrogen from contaminating the weld.	**D.** Gas drift
________	**5.** ___ energizes the electrode wire when the welding gun trigger is activated.	**E.** DCEN

Identify the parts of the GMAW gun assembly.

________ **6.** Trigger

________ **7.** Contact tip

________ **8.** Gas nozzle

________ **9.** Conductor tube

________ **10.** Insulator

________ **11.** Gas diffuser

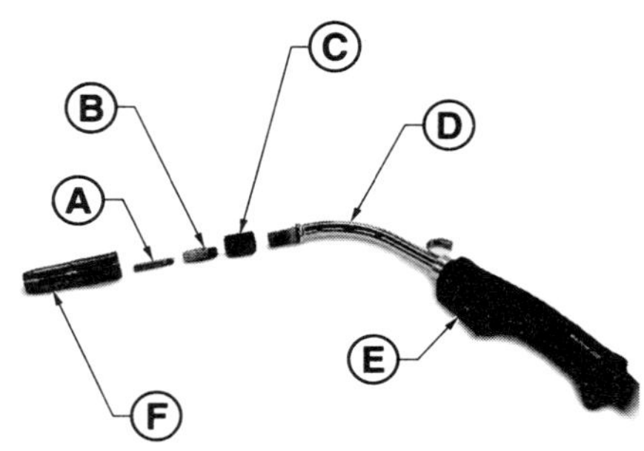

The Lincoln Electric Company

Chapter 20 Review — GMAW–Procedures

Name ______________________ **Date** ______________

True-False

T F **1.** Globular transfer is often used for welding metal more than ½″ thick.

T F **2.** Crater cracks can be caused by improperly filled craters.

T F **3.** Short circuiting transfer is used most frequently at current levels less than 200 A.

T F **4.** GMAW is not well suited for use in windy areas.

T F **5.** The edges of metals thicker than ¼″ should be beveled when using GMAW.

T F **6.** A drag angle is a travel angle where the electrode wire points away from the direction of travel.

T F **7.** Surface porosity in a weld is caused by insufficient current.

T F **8.** Starting an arc becomes increasingly difficult as electrode extension increases.

T F **9.** GMAW produces a heavy slag coating that is easily removed.

T F **10.** DCEN is used for most GMAW applications.

T F **11.** Pulsed spray transfer is used to weld a wide variety of metal thicknesses.

T F **12.** With pulsed spray transfer, current changes from DCEP to AC each cycle.

T F **13.** Globular transfer produces a stable arc, excellent weld appearance, and little weld spatter.

T F **14.** Spray transfer produces high deposition rates with high welding-wire efficiencies.

T F **15.** Globular transfer can be used in all welding positions.

Edge Weld

DESIRED WELD

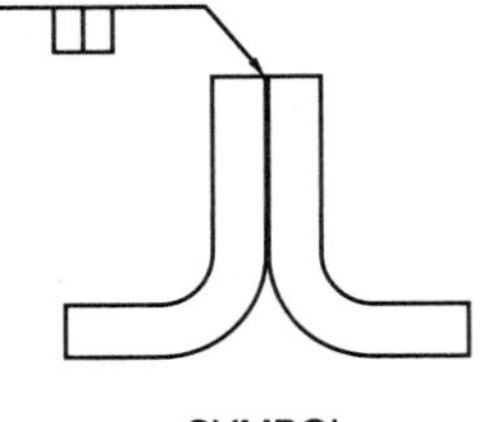

SYMBOL

T F **16.** The spray transfer transition current varies depending on the type and diameter of the welding wire and the composition of the shielding gas.

T F **17.** Two variables that affect droplet transfer in short circuiting transfer are slope and inductance.

T F **18.** Incomplete fusion (overlap) occurs when the weld pool flows onto unwelded base metal without fusing.

T F **19.** Whiskers in a weld joint are caused by pushing a welding wire past the leading edge of the weld pool.

T F **20.** Push angle refers to the torch pointing toward the beginning of the weld as the bead is deposited.

Multiple Choice

______________ **1.** Greater penetration is obtained using the ___ technique.
A. pushing
B. pulling
C. weaving
D. crescent

______________ **2.** ___ transfer is the transfer of molten metal in large droplets from the welding wire across an arc to the workpiece.
A. Spray
B. Short circuiting
C. Globular
D. Pulsed spray

______________ **3.** The type of metal transfer that occurs at lower current and voltage levels with smaller diameter welding wires is ___ transfer.
A. spray
B. pulsed spray
C. globular
D. short circuiting

______________ **4.** In order for spray transfer to occur, CO_2 levels cannot exceed ___%.
A. 5
B. 12
C. 18
D. 25

______________ **5.** Spray transfer for nonferrous metals requires ___ shielding gas.
A. 95% argon / 5% O_2
B. 90% argon / 10% CO_2
C. 98% argon / 2% CO_2
D. 100% argon

______________ **6.** What is the meaning of the welding symbol shown?
A. flare-groove
B. double groove
C. edge
D. plug or slot

Matching

_______________ **1.** ___ results directly from atmospheric contamination.

_______________ **2.** ___ is caused by excessive heat in the weld zone.

_______________ **3.** ___ occurs when the arc does not melt the base metal completely, causing molten metal to flow outside the weld area.

_______________ **4.** ___ results from a lack of heat input.

_______________ **5.** ___ is caused by moisture in the shielding gas.

A. Overlap
B. Surface porosity
C. Subsurface porosity
D. Insufficient penetration
E. Melt-through

_______________ **6.** ___ transfer requires high current levels and an argon rich shielding gas, and is practical for heavy-gauge metal.

_______________ **7.** ___ transfer occurs at the rate of a few droplets per second.

_______________ **8.** ___ transfer is a transfer process in which current is pulsed from a low background level to a peak level above the spray transfer transition current.

_______________ **9.** ___ transfer is most practical for welding thin-gauge base metals.

A. Spray
B. Globular
C. Short circuiting
D. Pulsed spray

_______________ **10.** The ___ is the travel angle applied when the welding gun is pointed away from the weld toward the end of the joint.

_______________ **11.** The ___ and penetration rates must be monitored to ensure that proper penetration is taking place.

_______________ **12.** The ___ is the travel angle applied when the welding gun is angled back toward the beginning of the weld.

_______________ **13.** The ___ is the force that squeezes the droplet of metal from the wire.

A. drag angle
B. push angle
C. travel speed
D. electromagnetic pinch

Identify the parts of spray transfer.

__________________ **14.** Molten droplets

__________________ **15.** Arc

__________________ **16.** Welding wire

__________________ **17.** Weld

__________________ **18.** Bright inner cone

Identify the steps of the short circuiting process.

__________________ **19.** A droplet separates and transfers to the weld pool as the current peaks.

__________________ **20.** The current builds, causing the electromagnetic pinch force to form a droplet of filler metal.

__________________ **21.** The welding wire touches the weld pool, causing a short circuit.

__________________ **22.** The voltage increases and the arc reignites when the droplet separates.

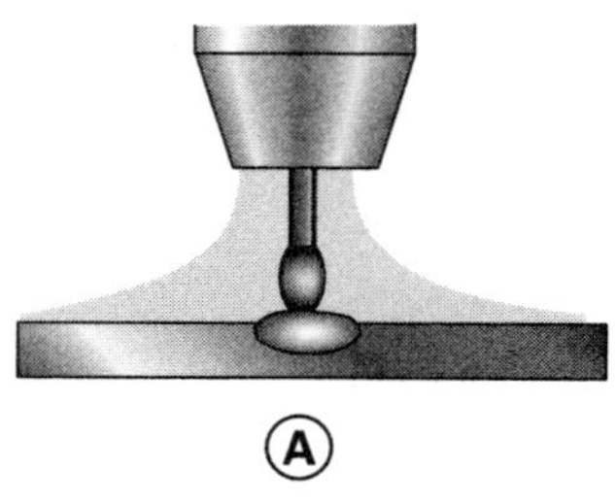

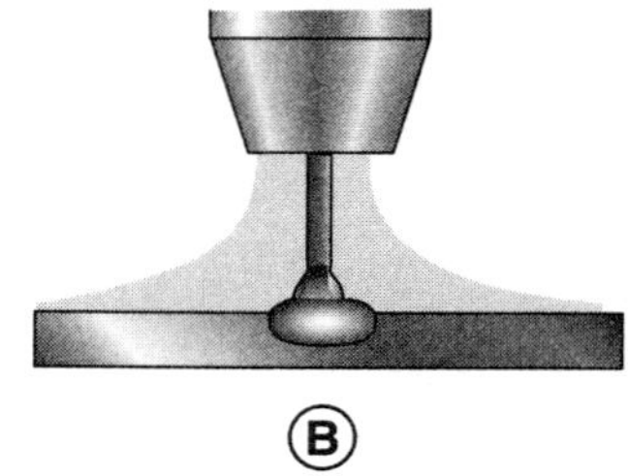

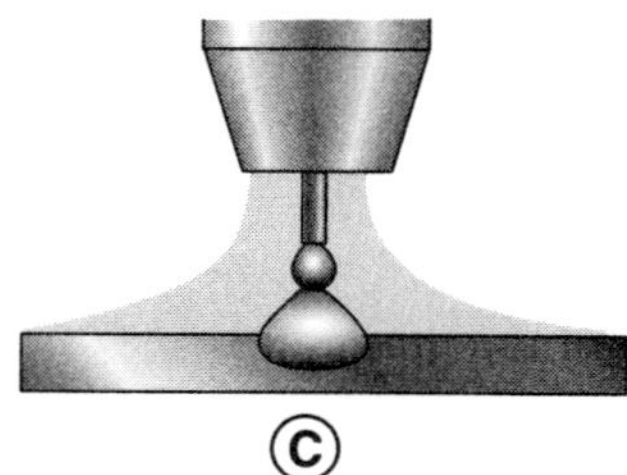

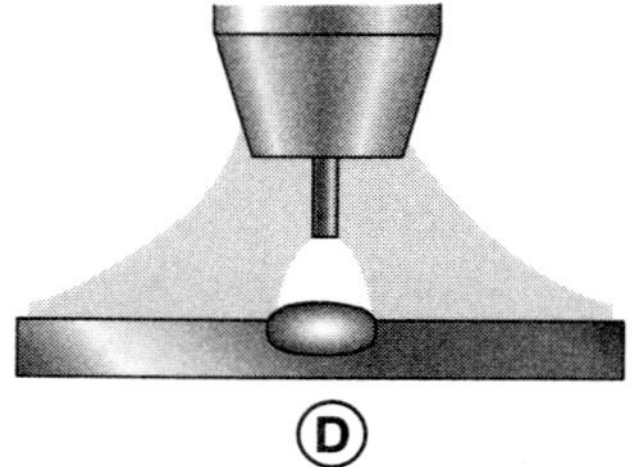

Chapter **21** Review

GMAW– Applications

Name ______________________________ **Date** ____________________

True-False

T F **1.** Steel from ¼″ to ½″ thick may be butt welded with no edge preparation.

T F **2.** The backhand technique is generally used for welding stainless steel.

T F **3.** Using short circuiting transfer on aluminum produces a colder arc than is produced with spray transfer.

T F **4.** When using GMAW on aluminum, a lower current than that used for other metals is required.

T F **5.** A 95% Ar/5% O_2 shielding gas mixture provides higher heat input when welding aluminum between 1″ and 2″ thick.

T F **6.** When welding stainless steel more than ¼″ thick using GMAW, a slight side-to-side movement should be used.

T F **7.** Spray transfer is commonly used for thin stainless steel in overhead position.

T F **8.** Using GMAW on copper is usually restricted to the deoxidized types of copper.

T F **9.** The preferred shielding gas for welding thin copper is argon.

T F **10.** An argon back-up gas may be used when welding stainless steel to prevent air from contacting the underside of the weld, causing embrittlement.

Double-Flare-Bevel-Groove Weld

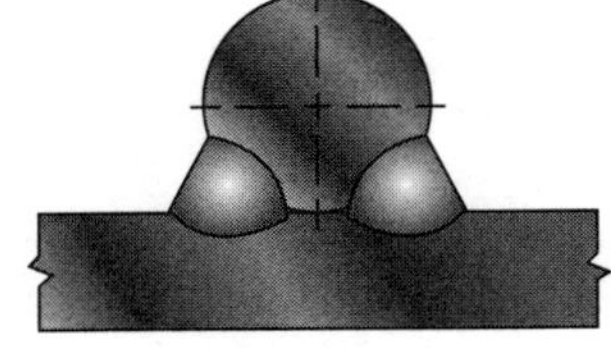

DESIRED WELD

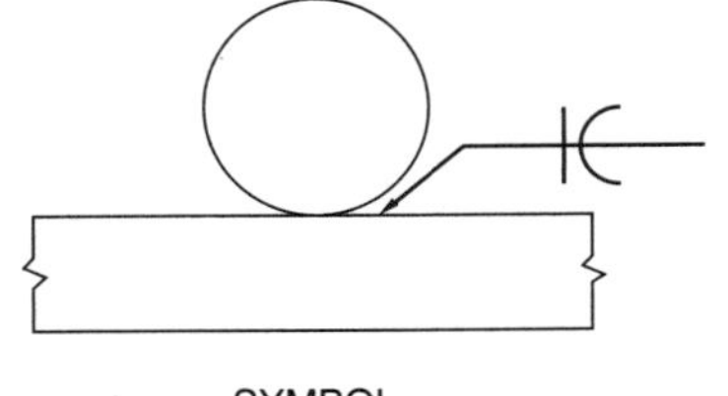

SYMBOL

Multiple Choice

_______________ **1.** For short circuiting transfer on carbon steels and low-alloy steels, the preferred shielding gas is ___.
- A. straight CO_2
- B. straight helium
- C. a mixture of 75% argon and 25% CO_2
- D. a mixture of 95% argon and 5% CO_2

_______________ **2.** A 50° to ___° groove angle is required on carbon steel up to 1″ thick.
- A. 55
- B. 60
- C. 75
- D. 90

_______________ **3.** For carbon steel welded without edge preparation, a ___″ root opening is recommended.
- A. 1⁄32
- B. 1⁄16
- C. 1⁄8
- D. 1⁄4

_______________ **4.** ___ is the shielding gas most commonly used for aluminum up to 1″ thick.
- A. Oxygen
- B. Helium
- C. Carbon dioxide
- D. Argon

_______________ **5.** Neither oxygen, hydrogen, nor CO_2 should ever be used for welding ___ with GMAW, not even in trace amounts.
- A. copper
- B. carbon steel
- C. stainless steel
- D. aluminum

_______________ **6.** What is the meaning of the welding symbol shown?
- A. fillet weld
- B. flare-V-groove weld
- C. double-flare-bevel-groove weld
- D. square-groove weld

Name ______________________________ **Date** ____________________

True-False

T F **1.** Tubular electrodes are used for FCAW.

T F **2.** A typical electrode extension for self-shielded electrodes ranges from ¼″ to 1¾″.

T F **3.** FCAW-G commonly uses carbon dioxide as a shielding gas.

T F **4.** Some flux cored electrodes can be used without the addition of a shielding gas.

T F **5.** FCAW welding most commonly uses AC current for flat position.

T F **6.** For FCAW, welding current is adjusted by increasing or decreasing the wire feed speed at the wire feeder.

T F **7.** The FCAW-S process requires a shielding gas from an external source.

T F **8.** Gas flow rates can range from 30 cfh to 45 cfh.

T F **9.** FCAW produces less distortion than SMAW as well as smooth, uniform beads.

T F **10.** The T in the electrode classification for flux cored electrodes means that the electrode is tough.

T F **11.** Too much drive roll tension can collapse tubular electrodes, so FCAW uses knurled drive rolls.

T F **12.** When using the Z-weave technique for FCAW, move up the joint in large increments, and stop momentarily at each toe to prevent undercut.

Fillet Weld Size

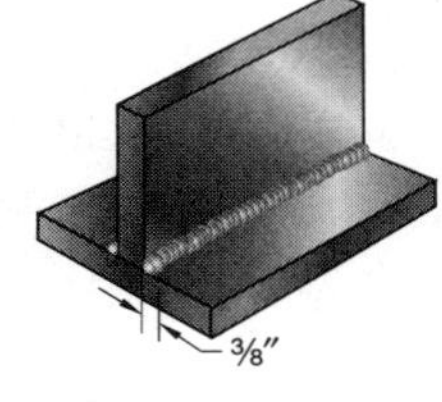

DESIRED WELD

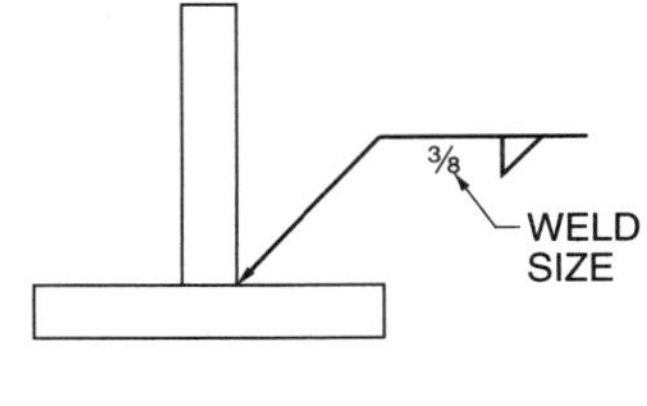

SYMBOL

Multiple Choice

______________ **1.** Unlike GMAW, FCAW uses a ___ with flux in its core.
A. wire feeder
B. power source
C. tubular electrode
D. workpiece connection

______________ **2.** Flux cored electrodes are ___ electrodes.
A. slag covered
B. tubular
C. spatter free
D. nonconsumable

______________ **3.** A gaseous shield created by the tubular electrode protects the weld area in the ___ welding process.
A. GMAW-P
B. FCAW-S
C. FCAW-G
D. SMAW

______________ **4.** A(n) ___ slag system is a slag system that promotes spray transfer with low spatter and fast-freezing slag.
A. basic
B. alloying
C. denitrifying
D. acid

______________ **5.** FCAW uses a(n) ___ welding power source.
A. direct current, CV
B. direct current, CC
C. alternating current, CV
D. alternating current, CC

______________ **6.** The higher the ___, the higher the welding current.
A. wire feed speed
B. arc length
C. shielding gas flow
D. electrode temperature

______________ **7.** What is the meaning of the welding symbol shown?
A. fillet weld, other side with ¼″
B. fillet weld, arrow side with ¼″ equal legs
C. fillet weld, arrow side with ¼″ fillet weld length
D. fillet weld, other side with ¼″ fillet weld length

¼

______________ **8.** In the ___ process, shielding gas is provided exclusively by the flux core within the electrode.
A. FCAW-S
B. FCAW-G
C. SMAW
D. GMAW-S

______________ **9.** ___ equipment should be used to keep smoke and fumes from entering a welder's helmet.
A. Gas shield
B. Post heating
C. Fume extraction
D. Oxygen

______________ **10.** A common gas mixture for FCAW-G is ___, which produces a spray transfer with small droplet size.
A. 100% argon
B. 98% argon and 2% CO_2
C. 75% argon and 25% CO_2
D. 50% argon and 50% CO_2

______________ **11.** ___ are added to the flux core to produce desired chemical and mechanical properties in the deposited weld metal, such as hardness, strength, toughness, ductility, and corrosion resistance.
A. Deoxidizers
B. Gas formers
C. Denitrifiers
D. Alloying elements

Matching

Identify the parts of the FCAW-S process.

______________ **1.** Weld bead

______________ **2.** Flux core

______________ **3.** Weld pool

______________ **4.** Electrode sheath

______________ **5.** Slag

______________ **6.** Arc

______________ **7.** Gaseous shield

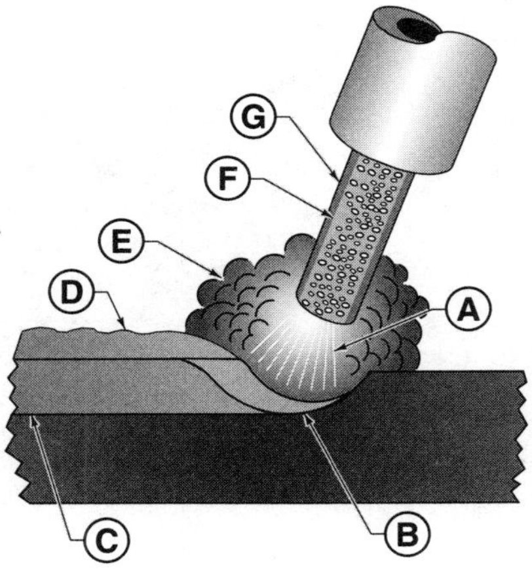

List the welding current and shielding gas required for the following flux cored electrodes.

8. EXXT-2 ______

9. EXXT-4 ______

10. EXXT-6 ______

11. EXXT-7 ______

12. EXXT-10 ______

Identify the equipment used for FCAW.

______ **13.** Flux cored electrode wire

______ **14.** CV power source

______ **15.** Heat shield

______ **16.** Shielding gas (for FCAW-G)

______ **17.** Shielding gas regulator (for FCAW-G)

______ **18.** Flowmeter (for FCAW-G)

______ **19.** Water-coolant system (if required)

______ **20.** Welding gun

______ **21.** Workpiece lead

______ **22.** Welding gun cable

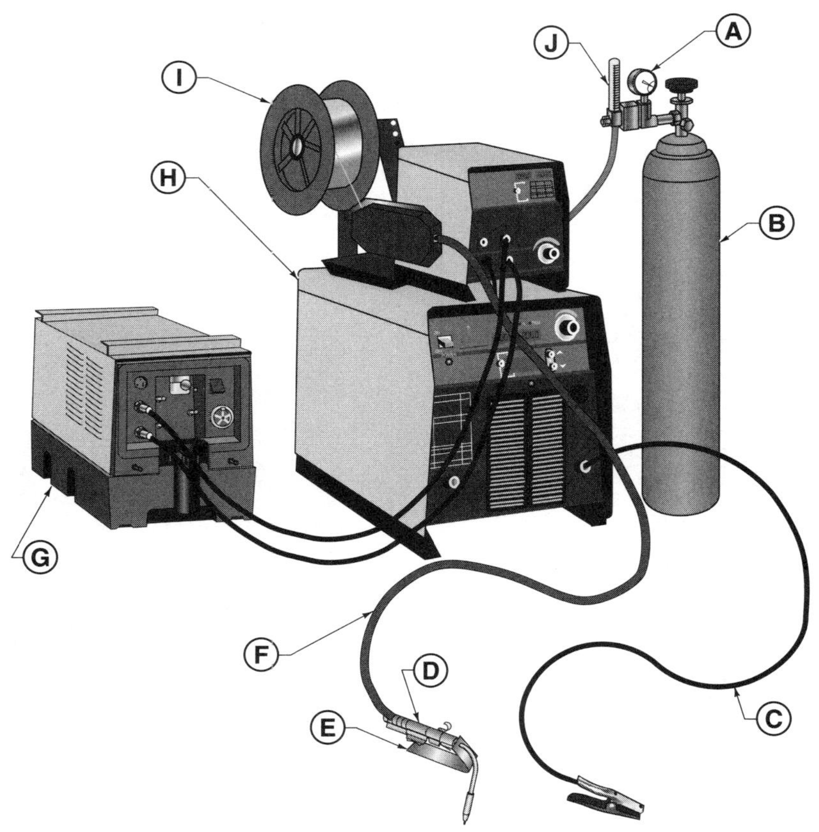

Name ______________________________ Date ________________

True-False

T F **1.** Braze welding should be performed in flat position.

T F **2.** When brazing, the lowest effective brazing temperatures possible should be used to minimize the effects of heat on the base metal.

T F **3.** Uniform capillary action is only possible when surfaces are completely free of foreign substances, such as dirt, oil, grease, and oxide.

T F **4.** Surface oxide on a metal to be brazed can be eliminated by sanding, grinding, filing, blasting, or wire brushing.

T F **5.** Filler metals suitable for brazing are those that begin to melt, or change to a liquid state, at temperatures above 370°F.

T F **6.** Some filler metals for brazing are manufactured with a flux coating.

T F **7.** The flux residue should be removed from the base metal after brazing.

T F **8.** When brazing, the base metal is melted to form a good joint.

T F **9.** Brazing can be used to join dissimilar metals.

T F **10.** Production brazing often uses a furnace as a heat source.

T F **11.** A butt joint provides the same strength as a lap joint.

T F **12.** A brazing filler metal must be molten before it flows into a joint.

T F **13.** If a base metal is not hot enough when braze welding, the filler metal will bead and roll off of the base metal as water does on a greasy surface.

Single-V-Groove Weld Size

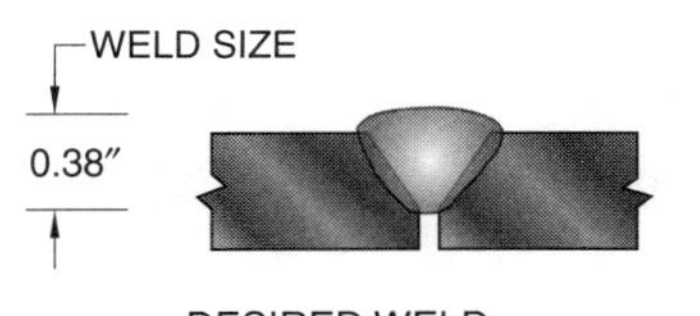

DESIRED WELD

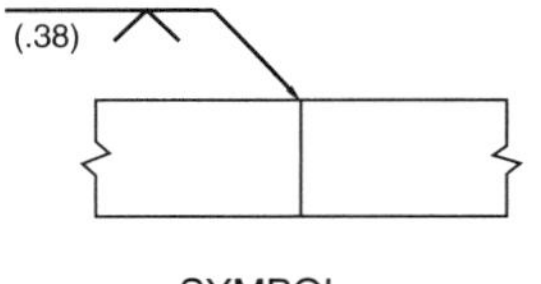

SYMBOL

T F **14.** Brazing is done at temperatures above 840°F and below the melting temperatures of the metals joined.

T F **15.** Braze welding filler metals are usually brasses, with an approximate composition of 60% copper and 40% zinc.

T F **16.** When braze welding, preheat the base metal until it turns a bright red color.

T F **17.** Braze welding uses capillary action to bond the weld pieces together.

T F **18.** The welding torch is rotated in a circular motion when braze welding.

T F **19.** Soldering requires less heat than brazing.

T F **20.** The point of a soldering copper should be coated with a thin layer of solder to form a proper joint.

Multiple Choice

______________ **1.** Braze welding should be performed in ___ position.
- A. flat
- B. horizontal
- C. vertical
- D. overhead

______________ **2.** Brazing is performed at an approximate temperature that is greater than ___°F and less than the melting point of the base metal.
- A. 480
- B. 840
- C. 1750
- D. 3200

______________ **3.** In the ___ brazing process, heat is generated in the same manner as it is in spot welding.
- A. dip
- B. manual heat
- C. induction
- D. resistance

______________ **4.** Solder containing 50% lead and 50% tin melts at ___°F.
- A. 370
- B. 471
- C. 633
- D. 840

______________ **5.** ___ soldering is a process in which two surfaces are soldered together without any solder being visible.
- A. Hard
- B. Seam
- C. Sweat
- D. Induction

______________________ **6.** A ___ is used to solder electronic components.
A. soldering copper
B. soldering pencil
C. gas torch
D. workpiece lead with a workpiece clamp

______________________ **7.** ___ is used to clean a soldering copper.
A. Acid
B. Sal ammoniac
C. Tin
D. Hot water

______________________ **8.** A ___ is a material used to outline areas that should not be brazed.
A. solder
B. paste flux
C. soldering copper
D. stopoff

______________________ **9.** A(n) ___ mixture produces a high flame temperature and is effective where higher brazing heat is required.
A. MAPP-oxygen
B. oxyacetylene
C. oxyhydrogen
D. gas-oxygen

______________________ **10.** ___ is limited to small assemblies such as wire connections or metal strips that can easily be held in fixtures.
A. Resistance brazing
B. Induction brazing
C. Furnace heating
D. Dip-brazing

______________________ **11.** The most common general-purpose solder contains ___.
A. 5% tin and 95% lead
B. 30% tin and 70% lead
C. 50% tin and 50% lead
D. 70% tin and 30% lead

______________________ **12.** An advantage of brazing over braze welding is that ___.
A. it produces stronger bonds in joints than welding does
B. the high bonding temperatures reduce the possibility of a change in base metal properties
C. it can be used to join dissimilar metals
D. it can be used to repair broken cast iron parts

______________________ **13.** What is the meaning of the dimension on the welding symbol shown?
A. bevel angle
B. weld size
C. effective throat size
D. weld size

(.38)

Identify the heating devices used for soldering shown below.

____________________ **29.** Electric soldering iron

____________________ **30.** Gas torch

____________________ **31.** Soldering copper

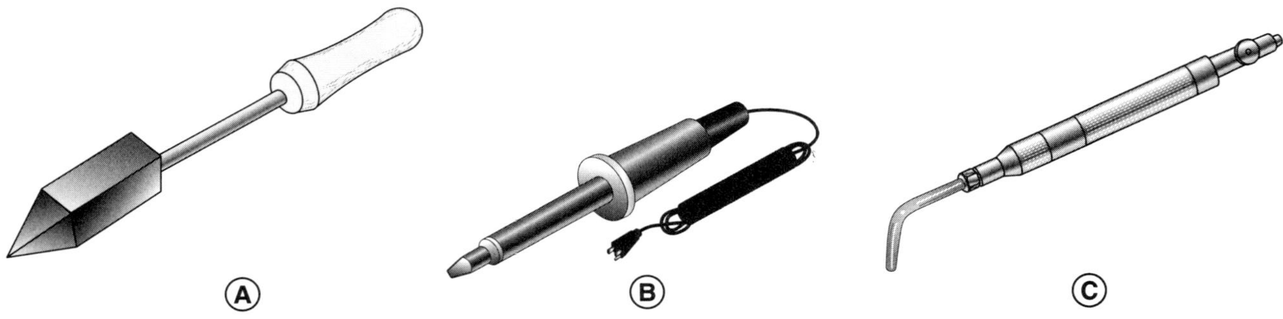

Name ______________________________ Date ________________

True-False

T F 1. Surfacing and hardfacing overlays are usually applied in the horizontal position.

T F 2. Surfacing using GMAW weld overlay allows for a faster application of the weld metal.

T F 3. The SAW weld overlay process is used when heavy deposits are required while surfacing parts.

T F 4. Plasma arc hot-wire surfacing is a variation of PAW overlay in which the surfacing material is heated to almost melting temperature and the deposited metal is melted on the surface of the workpiece.

T F 5. Wire spray coating materials are metals that can be made into flexible wire that melts in an oxyacetylene flame.

T F 6. A slightly reducing flame is recommended when surfacing with OAW weld overlay.

T F 7. Slag should be removed completely after each pass deposited on the workpiece using SMAW weld overlay.

T F 8. Surfacing should be done with the highest possible current to ensure proper penetration.

T F 9. When using SMAW weld overlay to surface a part, use a short arc length.

T F 10. A greater oscillation width results in more dilution.

T F 11. The first pass applied with thermal spraying should be applied as slowly as possible to ensure complete coverage.

Know Your Welding Symbols

Combination Welding Symbols

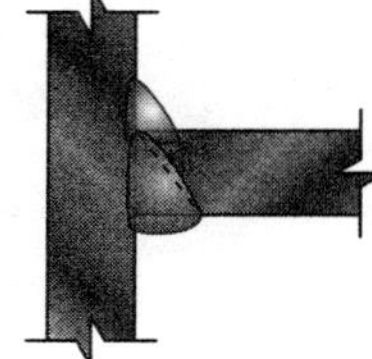

DESIRED WELD

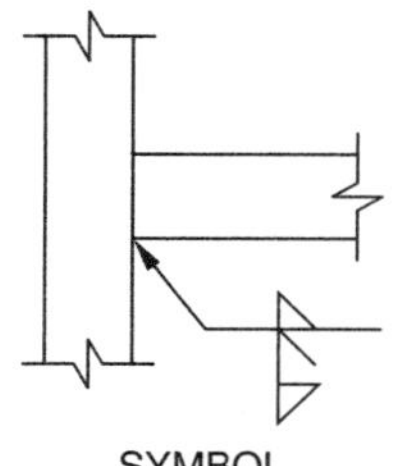

SYMBOL

T F **12.** Flame spraying can only be applied manually.

T F **13.** Thermal sprayed coatings are generally used for wear or corrosion resistance, not strength.

T F **14.** A metallurgical bond is created when two components are joined by fusion.

T F **15.** Hardfacing is only applied to the surface that is exposed to excessive wear.

Multiple Choice

______________ **1.** With SMAW weld overlay, a ___ motion is often used to prevent melt-through on thin sections.
A. crescent
B. whipping
C. drag
D. figure-eight

______________ **2.** When surfacing with PAW weld overlay, ___.
A. metal powder is carried from a hopper to the electrode holder in an argon gas stream
B. carbon dioxide is used to form the plasma
C. the surfacing powder is carried to the arc with compressed air
D. composite filler metal is typically used

______________ **3.** Surfacing with ___ is somewhat slower than with other arc welding processes, but the resulting weld overlay is of slightly higher quality.
A. PAW
B. FCAW
C. SAW
D. GTAW

______________ **4.** An OAW weld overlay can be applied to most materials, ___ alloys being one of the exceptions.
A. iron
B. magnesium
C. nickel
D. copper

______________ **5.** ___ weld overlay is used when heavy deposits are required on large areas.
A. SAW
B. SMAW
C. PAW
D. FCAW

______________ **6.** ___ is wear that cuts grooves in or deeply scratches a surface.

A. Corrosion
B. Impact damage
C. Gouging
D. Brinelling

______________ **7.** ___ is wear caused by repetitive impact collisions between two surfaces.

A. Corrosion
B. Impact damage
C. Abrasion
D. Cavitation

______________ **8.** ___ is the forming of localized cavities in metal that result from corrosion, repetitive sliding or rolling surface stresses, or poor electroplating.

A. Impact damage
B. Abrasion
C. Gouging
D. Pitting (spalling)

______________ **9.** ___ produces oxide debris and leads to pitting and fatigue failure.

A. Cavitation
B. Slurry erosion
C. Adhesive wear
D. Fretting

______________ **10.** ___ is surface damage caused by collapsing vapor bubbles in a flowing liquid.

A. Cavitation
B. Slurry erosion
C. Adhesive wear
D. Fretting

______________ **11.** Thermal spraying should not be used for ___.

A. liquid impingement
B. slurry erosion
C. adhesive wear
D. fretting

______________ **12.** Proper corrosion resistance must be ensured when using thermal spraying for ___.

A. solid particle impingement
B. slurry erosion
C. erosion
D. fretting

______________ **13.** ___ is a change in the composition of welding filler metal in a weld deposit caused by the addition of a melted base metal.
A. Atomizing
B. Penetration
C. Brinelling
D. Dilution

______________ **14.** ___ is the application of surfacing using a welding process that creates a metallurgical bond with the base metal by melting the surfacing metal.
A. Surfacing
B. Weld overlay
C. Thermal spraying
D. Spray and fuse

______________ **15.** In ___, a mixture of oxygen and a combustible gas is fed into the barrel of a spray gun with a charge of surfacing powder.
A. plasma spraying
B. flame spraying
C. high-velocity oxyfuel flame spraying
D. oxyacetylene spraying

______________ **16.** Spray and fuse (spraywelding) is a ___-step thermal spray process in which a coating is first deposited on a part and then subsequently fused by heating with a torch or by placing the part in a furnace.
A. two
B. three
C. four
D. six

______________ **17.** When preparing the surface of a porous material for thermal spraying, the porous material must be baked at temperatures from 400°F to 600°F to ___.
A. maintain the metallurgical properties before welding
B. remove contaminants from the pores
C. preheat the surface
D. improve bond strengths for ceramic coatings

______________ **18.** Each coating of thermal spraying should be between ___ thick.
A. 0.001″ and 0.003″
B. 0.003″ and 0.005″
C. 0.006″ and 0.007″
D. 0.008″ and 0.010″

______________ **19.** What is the meaning of the combination welding symbol shown?
A. single-J-groove weld with fillet welds
B. flare-groove weld with fillet welds
C. single-bevel-groove with fillet welds, both sides
D. U-groove with fillet welds, both sides

Matching

Identify the surfacing method(s) (welding, thermal spraying, rubber lining) used to prevent or repair damage caused by each listed wear type. List all that apply.

1. Erosion ______________________________

2. Gouging ______________________________

3. Cavitation ______________________________

4. Adhesive wear ______________________________

5. Brinelling ______________________________

6. Solid particle impingement ______________________________

7. Fretting ______________________________

8. Liquid impingement ______________________________

9. Impact damage ______________________________

10. Pitting (spalling) ______________________________

Identify the parts of the arc spraying equipment shown.

______________ **11.** Insulated housing

______________ **12.** Wire

______________ **13.** Base metal

______________ **14.** Spray deposit

______________ **15.** Arc

______________ **16.** Gas nozzle

______________ **17.** Reflector plate

______________ **18.** Gas

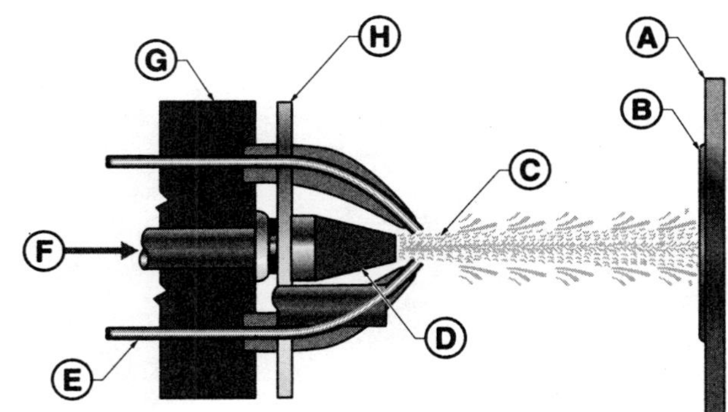

Identify the parts of the oxyacetylene metal spray torch shown.

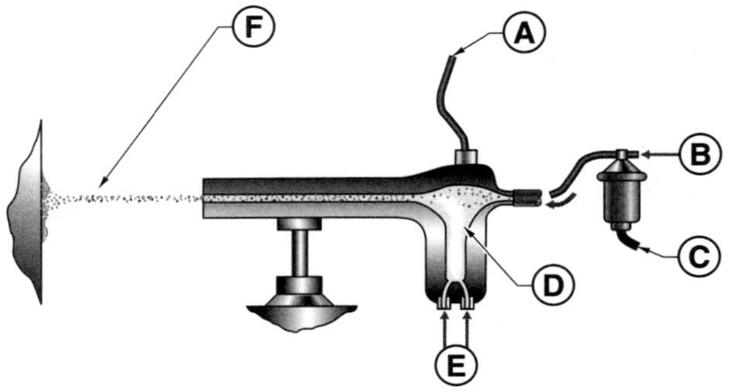

____________________ **19.** Fluidizing gas

____________________ **20.** Combustion chamber

____________________ **21.** Gas combustion jet

____________________ **22.** Powder carrier gas

____________________ **23.** Oxygen and fuel gas

____________________ **24.** Coolant water

Chapter 25 Review
Cutting Operations

Name ______________________________ **Date** ____________________

True-False

T F **1.** MAPP gas, natural gas, propane, and acetylene can be used with oxygen to cut metal.

T F **2.** A cutting torch has five needle valves to control the oxyfuel flame.

T F **3.** Preheat holes are used to blow molten metal away from a cutting section.

T F **4.** The correct oxygen and acetylene pressures of a particular torch are determined by following manufacturer recommendations.

T F **5.** To flame cut round stock, the cut should be started approximately 90° from the top edge.

T F **6.** OFC is only suitable for carbon steel applications.

T F **7.** Natural gas is commonly used for OFC because it works well for preheating and cutting materials.

T F **8.** A bar clamped parallel to the cut can serve as a guide for a straight cut when beveling.

T F **9.** When cutting cast iron, the cut can be stopped and started as necessary until complete.

T F **10.** The plasma arc cutting process can cut carbon steel up to 10 times faster than an oxyfuel mixture.

T F **11.** In a plasma arc torch, the tip of the electrode is located within the nozzle.

T F **12.** In making a bevel cut, the cutting torch is held at an angle to the metal.

Welding Process

"SMAW" indicates "Shielded Metal Arc Welding process to be used."

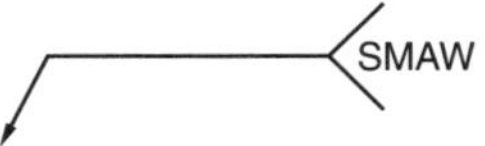

DESIRED WELD SYMBOL

T F **13.** The air jet orifices on an air carbon arc electrode holder should be positioned over the electrode for maximum air blast.

T F **14.** The air carbon arc cutting process uses compressed argon with a carbon arc to blow molten metal away from the cutting zone.

T F **15.** DCEN current is used for plasma arc cutting.

T F **16.** The width of a cut made by any cutting process is called a kerf.

Multiple Choice

______ **1.** The maximum safe working pressure of acetylene is ___ psi.
A. 5
B. 15
C. 25
D. 50

______ **2.** ___ is the most efficient orifice gas for PAC.
A. Oxygen
B. Argon
C. Shop air
D. Nitrogen

______ **3.** For beveling, hold the electrode at approximately a ___° angle, with the oxygen blast between the electrode and the metal surface.
A. 10
B. 45
C. 60
D. 90

______ **4.** The air carbon arc cutting process requires compressed air in the range of ___.
A. 10 psi to 20 psi
B. 40 psi to 80 psi
C. 100 psi to 200 psi
D. 250 psi to 350 psi

______ **5.** The electrode should extend a minimum of ___ from the electrode holder when cutting or gouging carbon steel.
A. 1
B. 2
C. 4
D. 6

______________ **6.** ___ is the result of oxygen in the air uniting with the metal, causing it to oxidize.
A. Dross
B. Rust
C. Gouging
D. Kerf

______________ **7.** ___ is a process of removing metal from large areas, such as the removal of surfacing metal, backing material, or riser pads on castings.
A. Washing
B. Beveling
C. Cutting
D. Gouging

______________ **8.** Plasma arc cutting uses ___ current.
A. DCEP
B. DCEN
C. AC
D. ACHF

______________ **9.** The air jet orifices must be positioned ___ the electrode when using the air carbon arc cutting torch.
A. above
B. under
C. over
D. around

______________ **10.** ___ cannot be cut because chromium oxide forms on the surface, which resists melting and shields the metal.
A. Cast iron
B. Copper
C. Stainless steel
D. Aluminum

______________ **11.** For cutting thick, nonferrous metals, hold the electrode in vertical position with a push angle of ___° and, with the air jet above it, move the arc up and down through the metal with a sawing motion.
A. 30
B. 45
C. 60
D. 90

______________ **12.** ___ is the gas most commonly used for oxyfuel gas cutting.
A. Acetylene
B. Natural gas
C. Propane
D. Methylacetylene-propadiene stabilized (MAPP)

______________ **13.** What does the information in the tail of the symbol indicate?
A. spot weld
B. submerged arc welding
C. shielded metal arc welding
D. spray arc transfer

SMAW

Matching

Identify the parts of the air carbon arc cutting torch shown.

__________ **1.** Air jet orifice

__________ **2.** Compressed air pushbutton

__________ **3.** Electrode

__________ **4.** Electrode release lever

Identify the parts of the plasma arc cutting circuit shown.

__________ **5.** Plasma arc

__________ **6.** Orifice gas

__________ **7.** Gas nozzle

__________ **8.** Electrode (cathode)

__________ **9.** Kerf

__________ **10.** Shielding/cooling gas

__________ **11.** Power supply

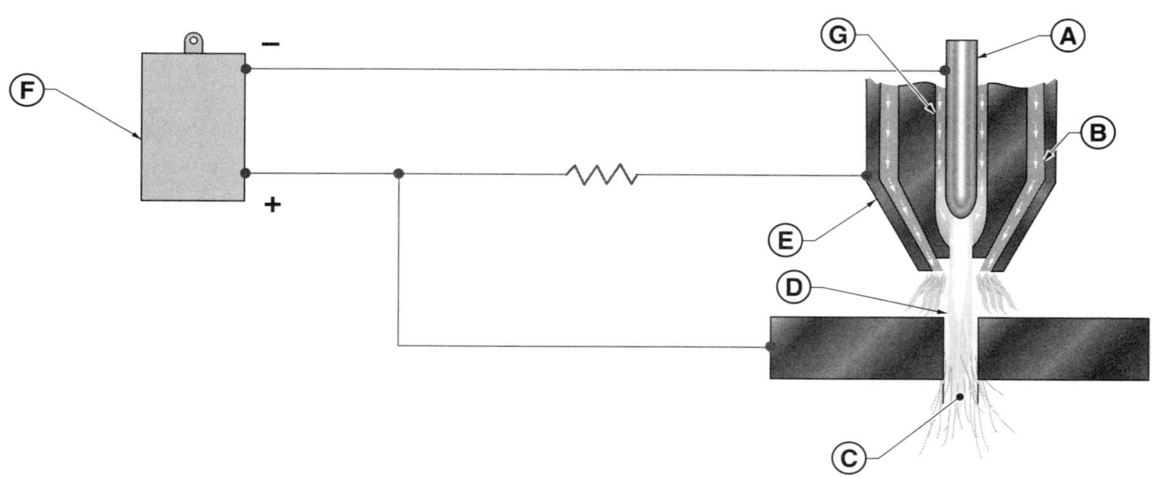

Chapter 26 Review

Repair Welding

Name ______________________________ **Date** ________________

True-False

T F **1.** Solvent cleaning includes sandblasting, sanding, and wire brushing.

T F **2.** Peening is performed prior to preheating the structure to be repaired.

T F **3.** Repair welding is used only if it is economical or if a replacement part is not available.

T F **4.** A satisfactory adhesive bond requires close contact between the surfaces to be joined.

T F **5.** Welding is a mechanical repair method.

T F **6.** Failure analysis provides an accurate explanation of the cause of a failure or loss of performance.

T F **7.** Hardness is a key indicator of mechanical properties for carbon and alloy steels.

T F **8.** Most codes rely on ASME Section IX and AWS D1.1 to describe the requirements for qualified welding procedures and welders.

T F **9.** Certain repairs may require the approval of authorized personnel, use of a qualified procedure, and/or preparation of supporting documentation.

T F **10.** Selective plating is the joining of parts with an adhesive placed between the faying (mating) surfaces.

T F **11.** Cold mechanical repair is commonly used on metals such as gray iron, ductile iron, aluminum, bronze, steel, and fabricated steel sections.

T F **12.** GTAW is used for bronze bearing surfacing weld repair.

Weld-All-Around

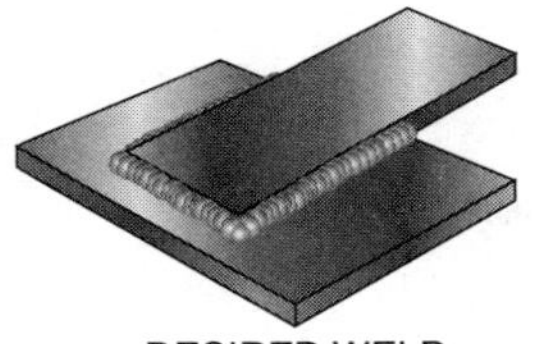

DESIRED WELD

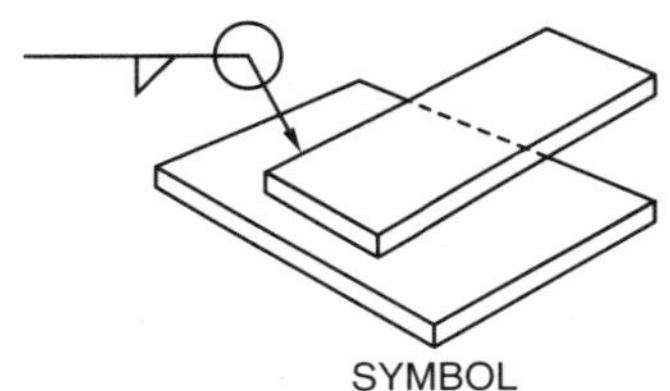

SYMBOL

T F **13.** Gas tungsten arc welding (GTAW) is typically used for surfacing weld repairs on large components.

T F **14.** Oxyacetylene welding (OAW) and shielded metal arc welding (SMAW) are often preferred for field weld repairs.

T F **15.** Larger diameter filler metal is preferred to reduce heat input for reducing distortion.

Multiple Choice

____________________ **1.** ___ analysis provides a diagnosis of the technical cause of failure using various analytical methods.

A. Fatigue
B. Procedure
C. Mechanical distortion
D. Physical failure

____________________ **2.** ___ crack when exposed to sulfur compounds, such as grease or oil.

A. Stainless steels
B. Copper alloys
C. Magnesium alloys
D. Nickel alloys

____________________ **3.** ___ repair does not involve a significant heat input.

A. Shielded metal arc welding (SMAW)
B. Mechanical
C. Welding
D. Structural weld

____________________ **4.** The process that converts an adhesive from the applied state to its final solid state is ___.

A. priming
B. curing
C. stitching
D. peening

____________________ **5.** The application of a thin, hard, chrome coating to repair minor damage is ___.

A. electroplating
B. buttering
C. laminar coating
D. blending

____________________ **6.** Cracks are most commonly located by ___ testing.

A. liquid penetrant
B. grind
C. grain alignment
D. carbon layer

________________ **7.** ___ may be required after weld repair to restore mechanical properties.
A. Tack welding
B. Preheating
C. Postheating
D. Root welding

________________ **8.** ___ requires that drillings be taken from the base metal and provides an accurate composition of the base metal.
A. Radiography
B. X-ray fluorescence
C. A spark test
D. Chemical analysis

________________ **9.** ___ is a weld repair that applies surfacing to badly worn shafts by welding snug-fitting, semi-circular forms to cover the shaft surface.
A. Sleeving
B. Wallpapering
C. Crosschecking
D. Curing

________________ **10.** A ___ joint contains insufficient adhesive to create an optimum bond.
A. pinhole
B. negative penetrant
C. stitched
D. starved

________________ **11.** ___ analysis provides a diagnosis of a technical cause of failure using experience gained from previous failures.
A. Failure modes and effects
B. Root cause
C. Physical failure
D. Selective behavior

________________ **12.** The thickness of metal required to support a load on a part is the ___.
A. corrosion allowance
B. design thickness
C. dilution dimension
D. fabrication tolerance

________________ **13.** ___ is a mechanical repair method in which a thinned, pitted, or cracked region of a part is ground away to create a gradual transition with the unaffected surface.
A. Stress bending
B. Lock pin drilling
C. Relief cutting
D. Blend grinding

_______________ **14.** A weld repair method that involves thin sheets of corrosion-resistant material welded to a corroded surface is ___.
- A. sleeving
- B. thermal lamination
- C. sheet lining
- D. surfacing

_______________ **15.** ___ is a series of parallel cracks approximately ½″ apart that occur in brittle deposits as they undergo stress relief.
- A. Crosschecking
- B. Peening
- C. Thermal segregation
- D. Linear fracture

_______________ **16.** What is the meaning of the welding symbol shown?
- A. fillet other side, weld-all-around
- B. single-bevel-groove weld arrow side, weld-all-around
- C. single-V-groove weld arrow side, weld-all-around
- D. fillet arrow side, weld-all-around

Matching

_______________ **1.** ___ is appropriate for repairing metals that are difficult to weld, such as cast iron.

_______________ **2.** ___ is useful when weld repair is too difficult or unsafe, as in a storage tank that contains flammable vapors.

_______________ **3.** ___ reduces the fatigue strength of rotating or reciprocating equipment, such as shafts.

_______________ **4.** ___ coupled with riveting increases fatigue strength.

A. Adhesive bonding
B. Cold mechanical repair
C. Electroplating
D. Blend grinding

Chapter 27 Review

Pipe Welding

Name ______________________________ **Date** ____________________

True-False

T F **1.** Groove welding is most commonly used for joining pipe.

T F **2.** Welded pipe is less expensive to produce than seamless pipe.

T F **3.** The most commonly used joint design for joining pipe is the closed butt joint.

T F **4.** Thick-wall pipe is typically welded using the downhill technique.

T F **5.** Pipe up to 3⁄16″ (0.188″) wall thickness can be welded without any joint preparation.

T F **6.** An advantage of using modified GMAW-S is that there is a high tolerance for misalignment between pipe sections.

T F **7.** Consumable inserts can be used as spacers to ensure that root openings are properly spaced.

T F **8.** Tack welds should not penetrate to the root of the groove.

T F **9.** E6010 and E6011 electrodes are commonly used to form keyholes while pipe welding.

T F **10.** An intermediate weld pass is used to give a weld joint a neat appearance.

T F **11.** Root passes are deposited in weld joints before intermediate weld passes.

T F **12.** To restart a weld on a root pass, strike the arc on the root bead about ½″ behind the feathered edge.

T F **13.** A wide groove angle is a good choice for welding thin-wall pipe.

T F **14.** Tack welds are usually placed at four locations around a pipe joint.

Know Your Welding Symbols

Scarf for Brazed Joint

DESIRED BRAZE

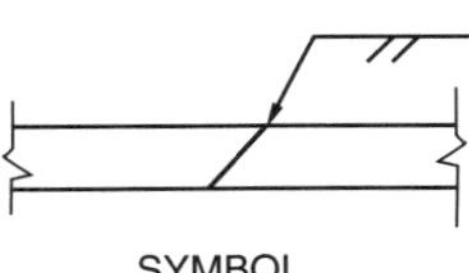

SYMBOL

T F **15.** A root pass (bead) should provide complete penetration through the wall of a pipe.

T F **16.** Line-up clamps are used to keep pipes in alignment while tack welds are made.

T F **17.** All welding positions are needed to weld sections of pipe when using position welding.

T F **18.** Intermediate weld passes are typically deposited with the largest diameter electrodes possible.

Multiple Choice

______________ **1.** Pipe with a wall thickness greater than ___″ is considered to be thick-wall pipe.
- A. 1⁄8
- B. 1⁄4
- C. 5⁄16
- D. 3⁄8

______________ **2.** A ___ is a point where a weld bead fuses into a tack weld or into another weld bead.
- A. keyhole
- B. crater
- C. tie-in
- D. tube sheet

______________ **3.** ___ is commonly used to weld the root bead in pipe joints where high quality welds are essential.
- A. SMAW
- B. GTAW
- C. FCAW
- D. GMAW-S

______________ **4.** The ___ pass is the first pass deposited after the tack welds have been made.
- A. root
- B. intermediate weld
- C. cover
- D. hot

______________ **5.** The ___ pass burns out any slag particles that may be left from the first pass.
- A. root
- B. intermediate weld
- C. cover
- D. cap

______________ **6.** Each tack weld should be about ___″ long.
- A. 1⁄16
- B. 1⁄8
- C. 3⁄8
- D. 3⁄4

______________ **7.** ___ welding requires the welder to adjust electrode angles and apply different welding techniques while welding around a pipe.
A. Roll
B. External
C. Position
D. Machine

______________ **8.** The roll welding method of pipe welding is performed by welding in ___ position.
A. 1G
B. 2FR
C. 4F
D. 6GR

______________ **9.** A ___ is a tear-shaped area of burn-through at the leading edge of the weld pool.
A. tube sheet
B. slag inclusion
C. tie-in
D. keyhole

______________ **10.** Pipe having a wall thickness of ⅛″ to ⁵⁄₁₆″ is considered to be ___ pipe.
A. rigid wall
B. thick-wall
C. thin-wall
D. double extra-heavy

______________ **11.** When welding pipe with SMAW, the root bead should be deposited using the ___ technique.
A. Z-weave
B. oval weave
C. horseshoe weave
D. whip-and-pause

______________ **12.** ___ is ideal for welding long seams in pipe because it is exceptionally fast, economical, and requires no filler metal.
A. SMAW
B. ERW
C. FCAW
D. GTAW

______________ **13.** What is the meaning of the welding symbol shown?
A. butt joint arrow side
B. butt joint other side
C. scarf for brazed joint arrow side
D. scarf for brazed joint other side

Matching

______________	**1.** The ___ pass burns out remaining particles of slag that may be in the groove.	**A.** root
______________	**2.** The ___ pass provides a neat appearance and weld reinforcement.	**B.** intermediate weld
______________	**3.** The ___ pass is typically deposited as a single weld bead.	**C.** cover

Identify the AWS Pipe Weld Joint Test Positions.

______________ **4.** 1F

______________ **5.** 2F

______________ **6.** 2FR

______________ **7.** 4F

______________ **8.** 5F

______________ **9.** 6F

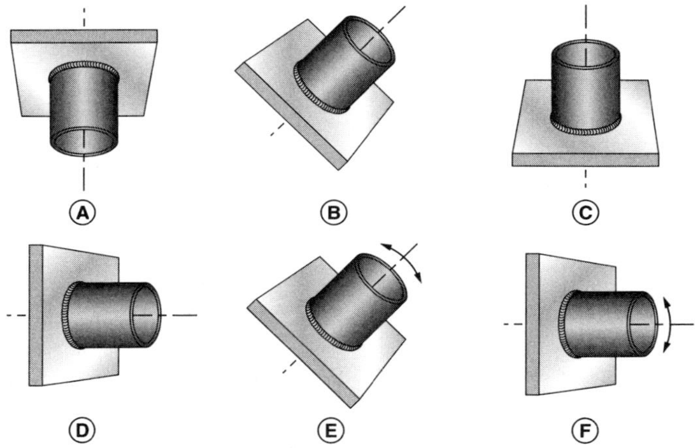

Ⓐ Ⓑ Ⓒ Ⓓ Ⓔ Ⓕ

Identify the consumable inserts.

______________ **10.** Class 2

______________ **11.** Class 3 and 5

______________ **12.** Class 4

______________ **13.** Class 1

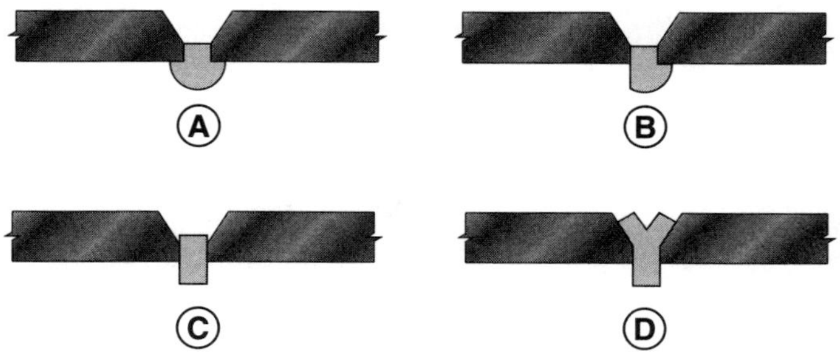

Ⓐ Ⓑ Ⓒ Ⓓ

Chapter 28 Review
Production Welding

Name ______________________________ **Date** ____________________

True-False

T F **1.** The current required for gas tungsten arc spot welding is determined by the thickness of the metal to be welded.

T F **2.** In friction welding, stored kinetic energy is used to generate the required heat for fusion.

T F **3.** Resistance welding is the most commonly used production welding process.

T F **4.** Spot welding is a type of resistance welding.

T F **5.** Spot welders require more than one electrode to perform a welding operation.

T F **6.** While multiple-impulse welding, the current flow is regulated with precise electronic control.

T F **7.** Pressure must be applied from both sides of a weld in gas tungsten arc spot welding.

T F **8.** In gas tungsten arc spot welding, good surface contact is important for sound welds.

T F **9.** Fusion in electron beam welding is achieved with a high-power-density beam focused on the area to be joined.

T F **10.** The beam-in-air electron beam welding process requires a vacuum chamber to prevent contaminants from entering the weld area.

T F **11.** Friction welding is sometimes called inertia welding.

T F **12.** The heat-affected zone in friction welding is narrow compared to that in other welding processes.

T F **13.** Laser beam welding utilizes a highly concentrated beam to generate a power intensity of more than one billion watts.

Field Weld

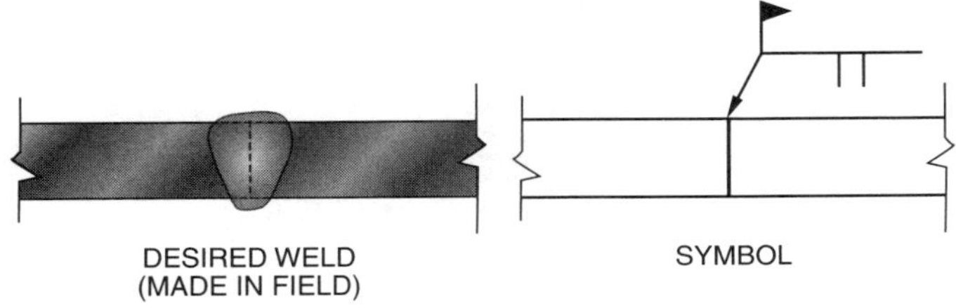

T F **14.** Argon is often used as the plasma gas in the plasma arc welding process.

T F **15.** The ultrasonic welding process uses vibratory energy to fuse weld pieces together.

T F **16.** Seam welding can be used to obtain a continuous seam.

T F **17.** Electrogas welding is used for multiple-pass welds on horizontal joints.

T F **18.** The beam-in-air electron beam welding process produces welds with characteristics similar to those in GTAW.

T F **19.** Projection welding is a form of resistance welding that produces a weld using heat obtained from the resistance of the workpiece to the welding current.

T F **20.** Ultrasonic welding uses heat to fuse metals together.

Multiple Choice

______________ **1.** In resistance welding, ___.
A. fusion takes place when pressure is applied to metal in a plastic state
B. a gas-supported flame provides fusion heat
C. electrodes do not conduct electricity to the work
D. stored kinetic energy is used to generate the required heat for fusion

______________ **2.** An advantage of ___ welding is that no melting takes place.
A. stud
B. gas tungsten arc spot
C. electrogas
D. friction stir

______________ **3.** ___ pulse welding is limited to extruded parts.
A. Aluminum
B. Galvanized iron
C. Tin plate
D. Light-gauge steel

______________ **4.** ___ welding can be used near glass or varnish-coated wires without damaging the glass or the insulating properties of the varnish.
A. Plasma arc
B. Electrogas
C. Resistance
D. Laser beam

______________ **5.** When gas tungsten arc spot welding, using ___ as a shielding gas provides deeper penetration.
A. argon
B. oxygen
C. helium
D. nitrogen

______________ **6.** ___ welding utilizes the Graham and Nelson method.
A. Beam
B. Pulsation
C. Stud
D. Submerged arc

______________ **7.** Plasma arc welding generates the heat for fusion with an electric arc that has been intensified by an injection of ___ into the arc stream.
A. ions
B. gas
C. filler metal
D. radiation

______________ **8.** Equipment necessary for ultrasonic welding includes two components: the ___.
A. power source and the electrode
B. transducer and the flowmeter
C. power source and the transducer
D. power source and the grounding mechanism

______________ **9.** The ___ welding process is commonly used to weld thick metals and where deep penetration is required.
A. plasma arc
B. ultrasonic
C. submerged
D. roll

______________ **10.** What is the meaning of the welding symbol shown?
A. single-V-groove weld arrow side, weld made in field
B. single-V-groove weld other side, weld made in field
C. square groove weld arrow side, weld made in field
D. single-bevel-groove weld arrow side

Matching

Identify the four principal elements of a standard resistance welder.

____________________ **1.** Electrode

____________________ **2.** Timing control

____________________ **3.** Frame

____________________ **4.** Electrical circuit

A. Reduces voltage and increases amperage to provide necessary heat

B. Regulates amount and duration of current

C. The main body of the machine

D. The mechanism for making and holding contact at the weld area

Identify the principal types of resistance welding.

____________________ **5.** ___ welding uses applied pressure on the material to be joined along with a charge of electricity that is sent from one electrode, through the material, to the other.

____________________ **6.** ___ welding uses roller-type electrodes to produce a continuous weld seam.

____________________ **7.** ___ welding produces a weld using heat obtained from the resistance of the workpiece to the welding current.

____________________ **8.** ___ welding uses two metal pieces moved together until an arc is established.

____________________ **9.** ___ welding uses constant pressure applied during a heating process.

A. Projection
B. Upset
C. Flash
D. Spot
E. Seam

Identify the parts of the plasma arc process shown.

____________________ **10.** Inner (hot) sheath

____________________ **11.** Outer (cool) sheath

____________________ **12.** Tungsten electrode

____________________ **13.** Plasma core

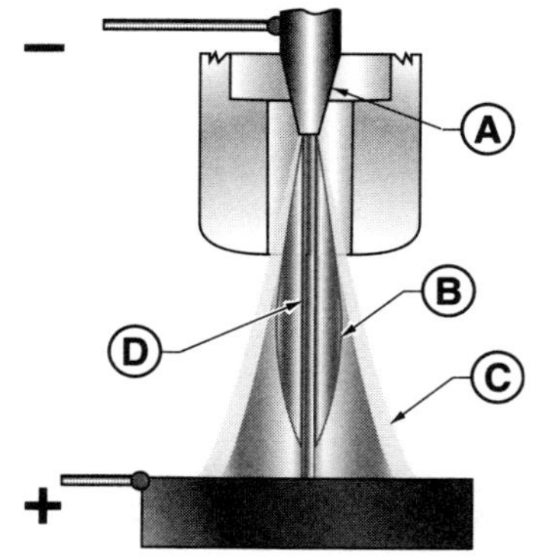

Identify the parts of the submerged arc welding process shown.

____________	**14.** Runoff tab	____________	**20.** Ground clamp
____________	**15.** Flux	____________	**21.** Flux shelf
____________	**16.** Contact	____________	**22.** Flux feed tube
____________	**17.** Solid slag	____________	**23.** Finished weld metal
____________	**18.** Welding wire	____________	**24.** Automatic wire feed
____________	**19.** Backing		

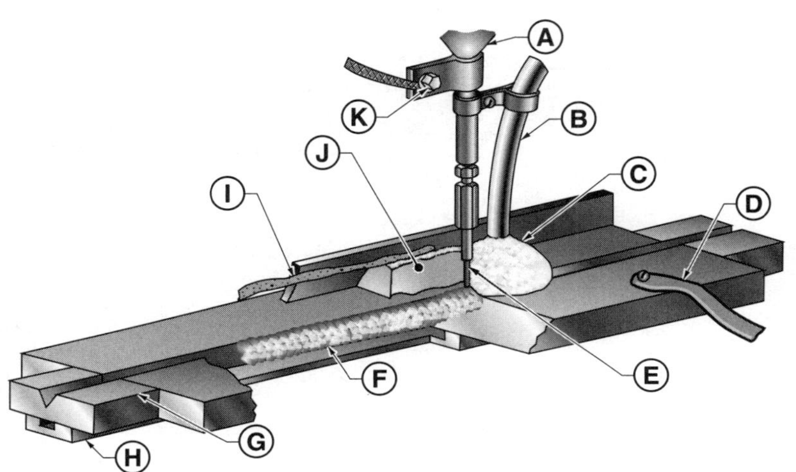

Identify the parts of the laser welding process shown.

____________ **25.** Lens

____________ **26.** Optical system

____________ **27.** Pumping source

____________ **28.** Ruby rod

____________ **29.** Concentrated beam

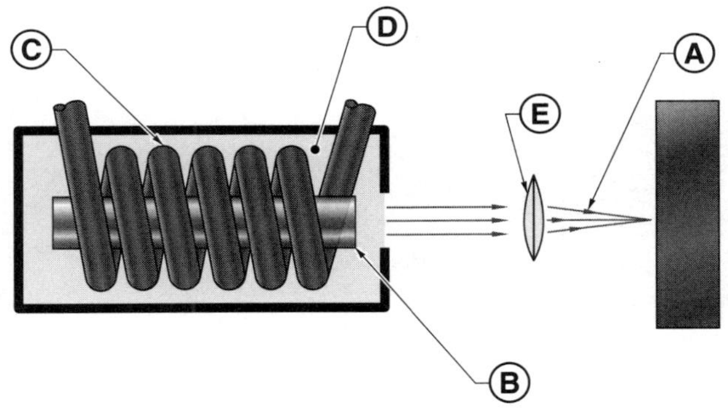

Identify the parts of the ultrasonic welding process shown.

_______________ **30.** Tip

_______________ **31.** Workpiece

_______________ **32.** Coupler

_______________ **33.** Transducer

_______________ **34.** Frequency coverter

_______________ **35.** Anvil

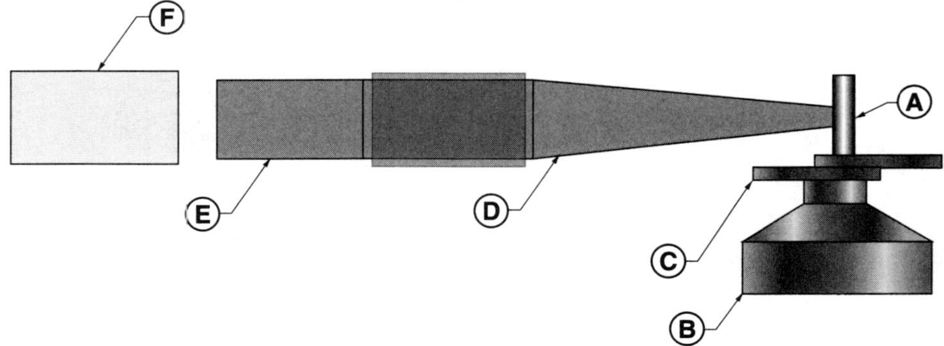

Name ______________________________ **Date** ______________

True-False

T F **1.** A programmable logic controller or some other automated controller is typically used with fixed automation systems to direct a weld sequence.

T F **2.** A robot controller directs the starting and stopping of servomotors as well as the rate of speed and acceleration of each servomotor.

T F **3.** A flexible automation system requires an operator to direct the movements of a torch and workpiece manually.

T F **4.** Fixed automation equipment uses mechanical and electrical means to guide a torch and workpiece.

T F **5.** A workpiece positioner may rotate or tilt by means of an electric motor that allows easier access to a weld seam.

T F **6.** A fixed automation system uses machines designed for multiple production functions.

T F **7.** Robot welding systems commonly use the oxyacetylene welding process for joining metals.

T F **8.** A robot is a programmed path device used to position a torch and at times a workpiece.

T F **9.** A six-axis robot is the most common configuration for a welding robot.

T F **10.** The biggest advantage that fixed automation has over flexible automation is the reprogrammability of the robot movement, allowing for varied movement of the robot.

T F **11.** The addition of servomotors allows a robot cell to better position weld joints for higher speeds and better weld quality.

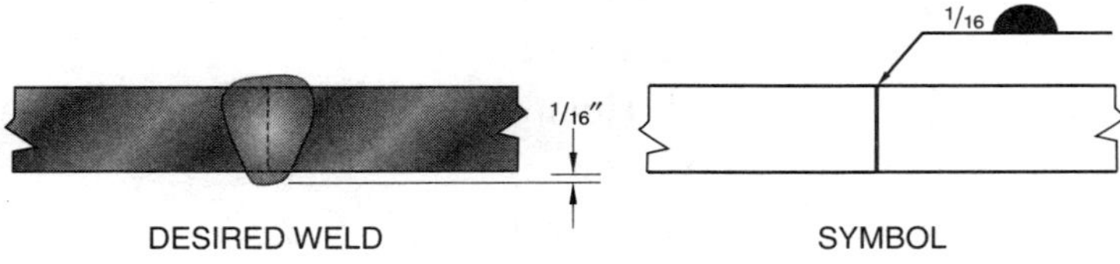

Multiple Choice

______________ **1.** A ___ is the device that a robot programmer uses to create robot movement programs.
- A. teach pendant
- B. program manipulator
- C. servomotor
- D. tool center point

______________ **2.** A ___ is a fixed-path mechanical apparatus that moves the torch in a specified path.
- A. workpiece positioner
- B. teach pendant
- C. torch positioner
- D. touch sensor

______________ **3.** ___ is generally the same equipment used for machine and semiautomatic applications.
- A. The positioner
- B. The jig
- C. The start station
- D. Automatic welding equipment

______________ **4.** Touch sensors use a ___ to identify the deviation from the programmed point.
- A. laser beam
- B. probe
- C. magnetic positioning device
- D. welding current voltage monitor

______________ **5.** A(n) ___ space is established to protect personnel from hazards.
- A. operating
- B. robot-limiting
- C. safeguarded
- D. hazardous

______________ **6.** The ___ is(are) used to start and stop the welding cycle.
- A. teach pendant
- B. operator controls
- C. servomotors
- D. robot interface

______________ **7.** ___ time is the time that a piece of equipment spends in the nonproductive activity of moving from one weld to another.
- A. Travel
- B. Holding
- C. Down
- D. Air cut

_______________ **8.** A laser location system uses a ___ to locate weld seam deviations.
A. welding wire
B. probe
C. laser beam
D. magnet

_______________ **9.** The ___ coordinate system is a system of locating points in space defined by perpendicular planes.
A. Cartesian
B. three-dimensional
C. Parisian
D. perpendicular

_______________ **10.** A ___ is an AC or DC motor with encoder feedback to indicate how far the motor has rotated.
A. servomotor
B. robomotor
C. manipulator
D. feedback motor

_______________ **11.** The most common configuration for a welding robot is a(n) ___-axis articulated robot manipulator driven by AC servomotors.
A. three
B. four
C. six
D. eight

_______________ **12.** In automatic welding, manufacturers generally strive for weld travel speeds from ___ inches per minute (ipm).
A. 5 to 15
B. 20 to 30
C. 30 to 40
D. 60 to 80

_______________ **13.** ___ welding controls adapt to the changing conditions of the weld by continually monitoring the weld voltage and adjusting the welding current to maintain a set arc length.
A. Pulsed
B. Adaptive
C. Flexible
D. Constant-voltage

_______________ **14.** Laser seam trackers use a laser beam to sweep across the weld path to detect ___.
A. the weld bead
B. weld seam deviations
C. a seamer
D. the weld pool

_______________ **15.** A ___ is designed to weld linear seams in rolled tubes or flat plates.
A. torch positioner
B. servomotor
C. teach pendant
D. seamer

_______________ **16.** What is the meaning of the dimension shown on the welding symbol?
A. scarf joint with 1⁄16″ root opening
B. square groove weld with 1⁄16″ root opening
C. surfacing weld with 1⁄16″ throat
D. square groove weld with 1⁄16″ root reinforcement

1/16

Name ______________________________ **Date** ____________________

True-False

T F **1.** Overheating the weld area results in a charred and discolored weld.

T F **2.** For best results, triangular-shaped filler rods are used for fillet welds.

T F **3.** Ventilation is not required for plastic welding.

T F **4.** When using the hand feed technique for welding plastic, a fanning motion should be used to distribute uniform heat over both the rod and the edges of the joint.

T F **5.** When welding plastic, the filler material should match the properties of the plastic being welded.

T F **6.** Thermosetting plastics can be softened repeatedly for different manufacturing processes.

T F **7.** The types of joints used in plastic welding are the same as those used in metal welding.

T F **8.** Different welding tips are required for specific joint applications.

T F **9.** Nitrogen gas is best used when welding PVC plastics.

T F **10.** Welds made by the induction process are not as strong as those obtained by other heating methods.

T F **11.** Filler materials come in flat, round, and triangular shapes.

T F **12.** A tacker tip is used with all types of joints to be welded.

Know Your Welding Symbols

Surfacing Weld Dimension

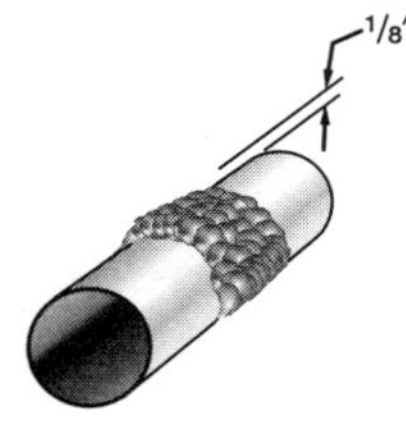

DESIRED WELD

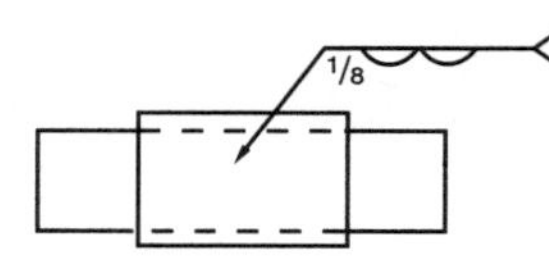

SYMBOL

T F **13.** Friction welding is useful for joining dissimilar metals, plastics, and composites in manufacturing and repair operations.

T F **14.** Friction welding is sometimes called spin welding.

T F **15.** When using the high-speed welding technique, an insufficient speed causes filler material to stretch due to a buildup of excessive heat.

T F **16.** Plastics are poor heat conductors and do not readily melt and flow.

Multiple Choice

______________ **1.** The technique for hand feed welding of plastic is similar to ___ welding of metals.
- A. resistance
- B. submerged arc
- C. shielded metal arc
- D. oxyacetylene

______________ **2.** The ___ group of plastics will soften only once when exposed to heat.
- A. thermoplastic
- B. thermosetting
- C. phenolic
- D. acrylic

______________ **3.** The ___ welding technique is commonly used for joining sections of pipe and tubing and for the assembly of many molded articles.
- A. friction
- B. hot gas
- C. adhesive bonding
- D. heated-tool

______________ **4.** Hot gas welding guns supply a welding temperature of approximately ___.
- A. 95°F to 175°F
- B. 204°F to 316°F
- C. 400°F to 925°F
- D. 1200°F to 1400°F

______________ **5.** ___ are not in the thermosetting group of plastics.
- A. Polyesters
- B. Ureas
- C. Acrylics
- D. Epoxies

_______________ **6.** What is the meaning of the welding symbol shown?
- A. surfacing weld ⅛″ in length
- B. melt-through ⅛″
- C. surfacing weld ⅛″ in thickness
- D. edge weld ⅛″ in thickness

_______________ **7.** To achieve a ___, heat the base material and filler material to the fusing point.
- A. liquidus temperature
- B. permanent bond
- C. molten plastic
- D. none of the above

_______________ **8.** During welding, exerting excessive pressure on the filler material can cause ___.
- A. excessive penetration
- B. excessive stretching
- C. excessive bonding
- D. air bubbles

_______________ **9.** To start a high-speed weld, hold the tool at a ___° angle.
- A. 30
- B. 45
- C. 60
- D. 90

Matching

Identify the parts of the high-speed welding process shown.

_______________ **1.** Base material

_______________ **2.** High-speed tool

_______________ **3.** Plasticized strip

_______________ **4.** Position of welder

_______________ **5.** Flowlines

_______________ **6.** Broad shoe

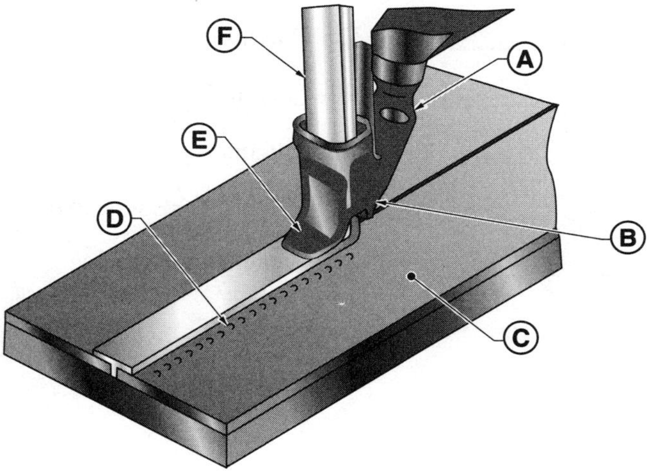

__________ **7.** ___ occurs when the heat for fusing is caused by rubbing the surfaces of the parts to be joined.

__________ **8.** ___ occurs when the edges to be joined are heated to fusing temperature and brought in contact.

__________ **9.** ___ is used to join parts with an adhesive placed between the faying (mating) surfaces.

__________ **10.** ___ is accomplished with a specially designed gun containing an electrical heating unit.

A. Heated-tool welding
B. Friction welding
C. Hot gas welding
D. Adhesive bonding

Identify each adhesive used in adhesive bonding.

__________ **11.** Acrylic

__________ **12.** Anaerobic adhesive

__________ **13.** Cyanoacrylate adhesive

__________ **14.** Epoxy

__________ **15.** Hot melt adhesive

__________ **16.** Polyurethane

__________ **17.** Polysulfide adhesive

__________ **18.** Silicone

__________ **19.** Solvent-base adhesive

__________ **20.** Water-base adhesive

A. Cures due to the absence of air that has been displaced between mated parts
B. Commonly used in the aerospace and building materials industry
C. Primarily used for wood and paper products
D. Has high temperature resistance and excellent sealing characteristics
E. A thermoplastic material that is applied in a molten state
F. Cures by evaporation, catalyst, or heat
G. Commonly used as contact cement for bonding large surface areas
H. Two-part adhesive that cures when resin and hardener are combined
I. Has a fast setting time and excellent flexibility
J. Cures instantly by reacting to trace surface moisture

Name ______________________________ **Date** ____________________

True-False

T F **1.** Destructive testing involves subjecting weld samples to loads until they fail.

T F **2.** Before a specimen is placed in a tensile testing machine, an accurate measurement of gauge length should be taken.

T F **3.** In tensile testing, a weld piece is pulled until it breaks.

T F **4.** In a root bend test, the test specimen is placed in a jig with the root down, such that the root face is in tension.

T F **5.** The Vickers test produces a square (pyramid-shaped) indentation.

T F **6.** The Rockwell hardness test is the most commonly used and versatile hardness test.

T F **7.** The guided bend and wraparound guided bend tests are types of impact tests used to determine soundness.

T F **8.** The drop weight test is a more reliable method than the Charpy V-notch test for measuring NDT.

T F **9.** The percent of elongation of a weld specimen is found by fitting the broken ends of the tested piece and measuring the new gauge length.

T F **10.** A metal bar that breaks after repeated bending is an example of fatigue failure.

T F **11.** The Knoop test is used to check the tensile strength of a weld specimen.

T F **12.** A Rockwell testing machine uses a variety of loads and indenters, requiring different scales.

T F **13.** The Brinell test is particularly suitable for very thin, case-hardened, or hard-faced components.

Flush Weld Contour

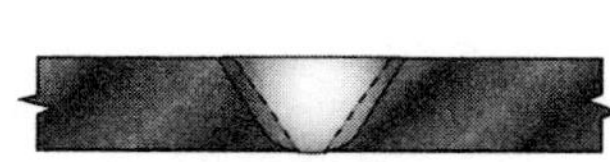

DESIRED WELD

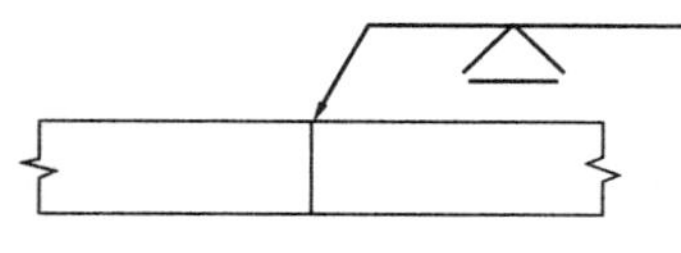

SYMBOL

Multiple Choice

_______________ **1.** A ___ test is a test in which a specimen is gripped in a vise and then bent and peeled apart with pincers to reveal the weld.

A. tensile
B. macroetch
C. guided bend
D. peel

_______________ **2.** The Knoop test produces a ___ indentation.

A. triangular
B. circular
C. hexagonal
D. rectangular

_______________ **3.** The guided bend test requires ___ to determine the degree of fusion and weld penetration.

A. a 10 mm ball
B. gauge marks
C. two specimens
D. a pin end

_______________ **4.** Yield point is the location on the load-extension curve where ___.

A. a part fails
B. an increase in strain occurs without an increase in stress
C. an increase in stress is applied without creating strain
D. stress is directly proportional to strain

_______________ **5.** A Bourdon tube is used to measure ___.

A. hydraulic load applied to a test specimen
B. tensile strength of a test specimen
C. hardness
D. density of a weld

_______________ **6.** ___ strain is strain that remains permanent after the stress is removed.

A. Proportional
B. Yield
C. Ultimate tensile
D. Plastic

_______________ **7.** An acceptable shear strength is usually at least ___% of the minimum specified tensile strength of the base metal.

A. 30
B. 40
C. 50
D. 60

_______________ **8.** A ___ uses the elastic deformation of a spring or diaphragm to indicate the mechanical load applied to a test specimen.
A. Bourdon tube
B. pin end
C. load cell
D. tension meter

_______________ **9.** When measuring ductility, ___ specimens must be used for calculating percent reduction of area.
A. rectangular
B. round
C. triangular
D. hexagonal

_______________ **10.** ___ is the maximum stress at which stress is directly proportional to strain.
A. Proportional balance
B. Median tensile strength
C. Stress limit
D. Proportional limit

_______________ **11.** Auto-refrigeration is ___.
A. cooling that occurs when gas expands
B. cooling that occurs when the weld torch is removed
C. forced cooling of a specimen
D. a way to prepare a test specimen

_______________ **12.** A ___ test specimen is prepared by sawing a slot in either end of the weld.
A. Rockwell hardness
B. fatigue strength
C. Charpy V-notch
D. nick-break

_______________ **13.** ___ tests are less commonly used for testing welds because the HAZ of a weld requires special preparation.
A. Toughness
B. Soundness
C. Strength
D. Hardness

_______________ **14.** A ___ test is a type of indentation hardness test that uses light loads of less than 200 g.
A. Brinell hardness
B. Rodgers hardness
C. microhardness
D. Rockwell hardness

_______________ **15.** What is the meaning of the welding symbol shown?
A. single-bevel-groove weld other side, flush contour
B. single-V-groove weld arrow side, flush contour
C. double-bevel-groove weld arrow side, concave contour
D. bevel-groove arrow side, concave contour

Matching

Identify the following destructive testing methods.

_______________ **1.** The ___ test involves testing a specimen with the weld root in tension.

_______________ **2.** The ___ test forces a steel ball into the surface of the metal.

_______________ **3.** The ___ test bends a piece of welded metal around a U-shaped die.

_______________ **4.** The ___ test uses hammer blows, stretching, or bending to break a small, notched specimen.

_______________ **5.** The ___ test uses a dynamic load to measure the energy needed to break a small, machine-notched specimen.

A. Brinell hardness
B. nick-break
C. fillet weld break
D. guided bend
E. Charpy V-notch

Identify the following weld tests.

_______________ **6.** Charpy V-notch

_______________ **7.** Guided bend

_______________ **8.** Drop weight

_______________ **9.** Rockwell

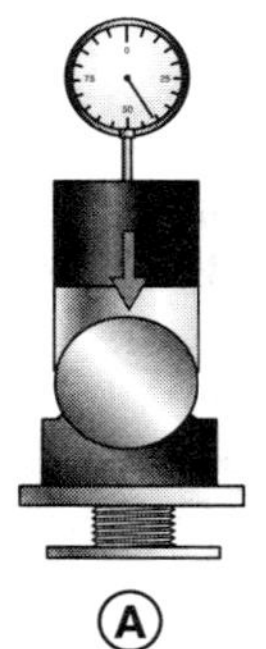

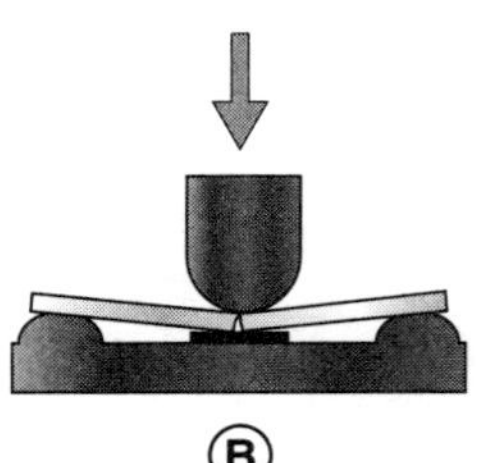

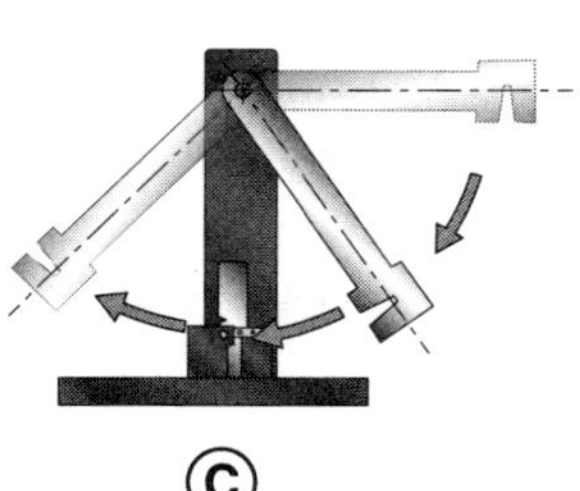

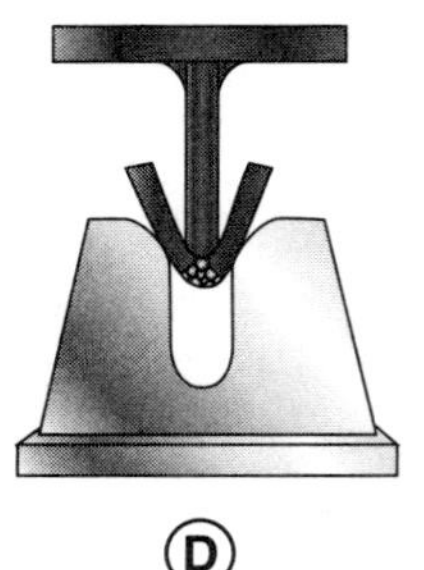

Name ______________________________ Date ______________________

True-False

T F **1.** Nondestructive testing requires a part to be permanently removed from service.

T F **2.** Cure time is the total time a penetrant is in contact with a component surface.

T F **3.** A black light is used with the fluorescent liquid penetrant examination method.

T F **4.** The liquid penetrant examination method is used to find defects in a weld by outlining surface defects.

T F **5.** Visual examination is a form of nondestructive testing.

T F **6.** Demagnetization is the ability of a material to retain a portion of an applied magnetic field after the magnetizing force has been removed.

T F **7.** A ferromagnetic material is a material that cannot be magnetized.

T F **8.** Internal weld defects can be detected with radiographic testing equipment.

T F **9.** The weld piece must be magnetized when using the magnetic particle examination method.

T F **10.** Prods used for the prod method of magnetic particle examination must never be crisscrossed.

T F **11.** When performing the magnetic particle examination procedure, demagnetization is not necessary for engine and machine parts, as they typically contain residual magnetization resulting from normal operation.

T F **12.** A couplant is a liquid substance used between a search unit and a test surface to permit or improve the transmission of ultrasonic energy.

Convex Weld Contour

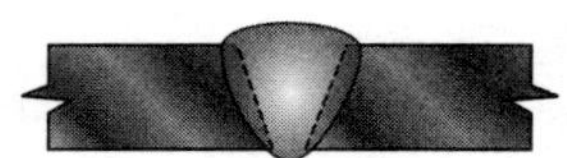

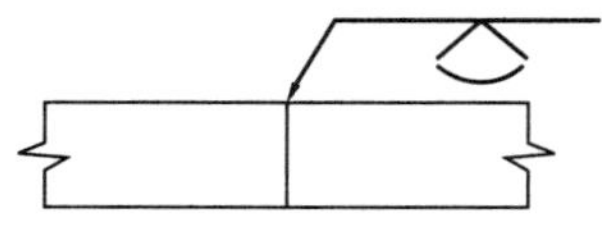

SYMBOL

Multiple Choice

_______________ **1.** The ___ test is a nondestructive testing procedure.
A. guided bend
B. peel
C. ultrasonic
D. shear

_______________ **2.** ___ time is the elapsed time between the application of a developer and the examination of a part.
A. Dwell
B. Cure
C. Developing
D. Resting

_______________ **3.** Visual examination is used to ___.
A. check for surface condition
B. locate internal weld defects
C. detect grain growth
D. measure tensile strength

_______________ **4.** ___ examination does not reveal very narrow discontinuities that are not closely aligned (parallel) to the weld, such as laminations.
A. Ultrasonic
B. Radiographic
C. Electromagnetic
D. Magnetic particle

_______________ **5.** ___ examination uses X rays and gamma rays to detect discontinuities.
A. Ultrasonic
B. Radiographic
C. Liquid penetration
D. Magnetic particle

_______________ **6.** A(n) ___ is a material that is applied to a test surface to accelerate bleedout and enhance the contrast of indications.
A. developer
B. penetrant
C. accelerator
D. cleaner

_______________ **7.** ___ magnetization is a concentric magnetic field produced by a straight conductor, such as a piece of wire, carrying an electrical current.
A. Longitudinal
B. Conducted
C. Circular
D. Transverse

_______________ **8.** A(n) ___ point is a point during a fabrication process where inspection and acceptance are required before fabrication can proceed.
A. dwell
B. pulse-echo
C. artifact
D. hold

_______________ **9.** A ___ wave is a transverse wave that represents wave motion in which the particle oscillation is perpendicular to the direction of wave propagation.
A. shear
B. longitudinal
C. perpendicular
D. horizontal

_______________ **10.** In ultrasonic welding, a ___ is one complete reflection of an ultrasonic beam.
A. half skip
B. full skip
C. transverse wave
D. longitudinal wave

_______________ **11.** What is the meaning of the welding symbol shown?
A. single-bevel-groove weld other side, concave contour
B. single-V-groove weld arrow side, convex contour
C. single-V-groove weld arrow side, concave contour
D. double-V-groove weld arrow side, convex contour

Matching

Identify the examination methods used for nondestructive examination.

_______________ **1.** The ___ examination method uses dyes to locate surface defects.

_______________ **2.** The ___ examination method uses high-frequency vibrations to locate defects.

_______________ **3.** The ___ examination method uses photographic film or sensitized paper.

_______________ **4.** The ___ examination method uses a strong magnetizing current and a finely divided powder to detect defects.

_______________ **5.** The ___ examination method produces an eddy current in the test piece.

A. liquid penetrant
B. magnetic particle
C. ultrasonic
D. radiographic
E. electromagnetic

Identify the components used with nondestructive examination methods.

______	**6.** A(n) ___ is a permanent, visible image on a recording medium produced by penetrating radiation passing through a material being tested.	**A.** A-scan presentation
______	**7.** A(n) ___ is a nonrelevant indication that appears on a radiograph.	**B.** artifact
______	**8.** A(n) ___ is a device or combination of devices whose demonstrated image determines radiographic quality and sensitivity.	**C.** equipment calibration standard
______	**9.** A(n) ___ uses a horizontal base line that indicates distance or time, and a vertical deflection from the base line that indicates relative amplitude of the returning signal.	**D.** radiograph
______	**10.** A(n) ___ is a test piece that contains typical discontinuities that demonstrate that calibration equipment is detecting the discontinuities for which the part is being inspected.	**E.** image quality indicator

Identify the proof testing method.

______	**11.** In a(n) ___ test, air is pressurized in a closed vessel to reveal leaks.	**A.** acoustic
______	**12.** A(n) ___ test consists of detecting acoustic signals produced by plastic deformation or crack formation during mechanical loading or thermal stressing of metals.	**B.** hydrostatic
______	**13.** In a(n) ___ test, closed containers are filled with water and a predetermined test pressure is applied.	**C.** pneumatic

Chapter 33 Review

Metallography

Name ______________________________ **Date** ______________________

True-False

T F **1.** The purpose of microscopic examination is to look for clues as to how a metal was made.

T F **2.** Cutting is the least common method of obtaining specimens from a component.

T F **3.** When a diamond-tipped cutting wheel is used to make a final cut, rough grinding is required.

T F **4.** Specimen mounting is usually permanent.

T F **5.** Cold mounting is performed in a vacuum to remove air bubbles from the mount.

T F **6.** Fine grinding is the last stage before polishing the mount.

T F **7.** Etching is usually performed by immersion.

T F **8.** Macroscopic examination is never performed using magnification.

T F **9.** Different etchants are used to reveal specific types of microstructural details.

T F **10.** Lighting has the greatest overall effect on the appearance of a surface.

T F **11.** The most common specimen orientation for microscopic examination is a longitudinal orientation.

T F **12.** Etching is the first stage of metallographic preparation before microstructural examination.

Concave Weld Contour

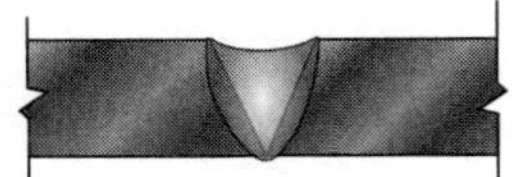

DESIRED WELD

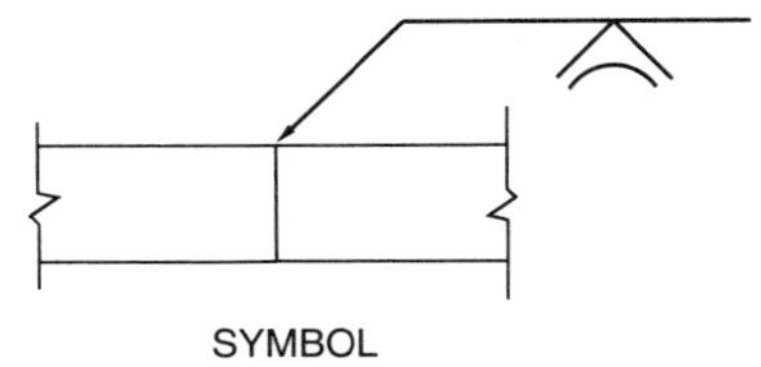

SYMBOL

Multiple Choice

________________ **1.** Microscopic examination is conducted at ___ magnification.

A. zero
B. low
C. medium
D. high

________________ **2.** To prevent overheating, flame or plasma cutting must be performed at a minimum distance of ___″ away from the area to be examined.

A. ¼
B. ½
C. 1
D. 2

________________ **3.** ___ are caused by a polishing operation that removes tiny nonmetallic particles such as carbides from the metal surface.

A. Pits
B. Bumps
C. Divots
D. Blooms

________________ **4.** ___ is a lighting method that uses a small region of a brighter light to increase detail on a dark area of a subject.

A. Buildup lighting
B. Main lighting
C. Backlighting
D. Fill lighting

________________ **5.** ___ polishing is a polishing process that is performed on a series of rotating wheels covered with a low-nap cloth.

A. Rough
B. Final
C. Electrolytic
D. Chemical

________________ **6.** What is the meaning of the welding symbol shown?

A. single-V-groove weld other side, concave contour
B. single-V-groove weld arrow side, concave contour
C. single-bevel groove weld, flush contour
D. single-bevel groove weld, concave contour

Matching

______________	**1.** ___ is the controlled selective attack on a metal surface.	**A.** Mounting **B.** Polishing **C.** Etching **D.** Bloom
______________	**2.** ___ is a slight haze that appears on the surface of a specimen.	
______________	**3.** ___ prevents rounding of the edges of specimens.	
______________	**4.** ___ is used to produce a scratch-free mirror finish on a specimen.	

______________	**5.** ___ illumination uses polarized light that is separated into two beams by a biprism.	**A.** Brightfield **B.** Darkfield **C.** Polarized **D.** Nomarski
______________	**6.** ___ illumination illuminates features perpendicular to the optical axis of a microscope.	
______________	**7.** ___ illumination reveals microstructural features in optically anisotropic metals.	
______________	**8.** ___ illumination illuminates the specimen at an oblique angle.	

______________	**9.** ___ is an intense lighting source that uses a single bulb in a reflector.	**A.** Reflected light **B.** A flashlight **C.** A spotlight **D.** Diffused light
______________	**10.** ___ is a lighting source that provides a pulse of very intense light.	
______________	**11.** ___ is a lighting source that bounces light off a wall or ceiling.	
______________	**12.** ___ is a lighting source that uses a semi-opaque screen.	

Identify the lighting methods for macroscopic examination.

_________________ **13.** Buildup

_________________ **14.** Fill

_________________ **15.** Main

_________________ **16.** Back

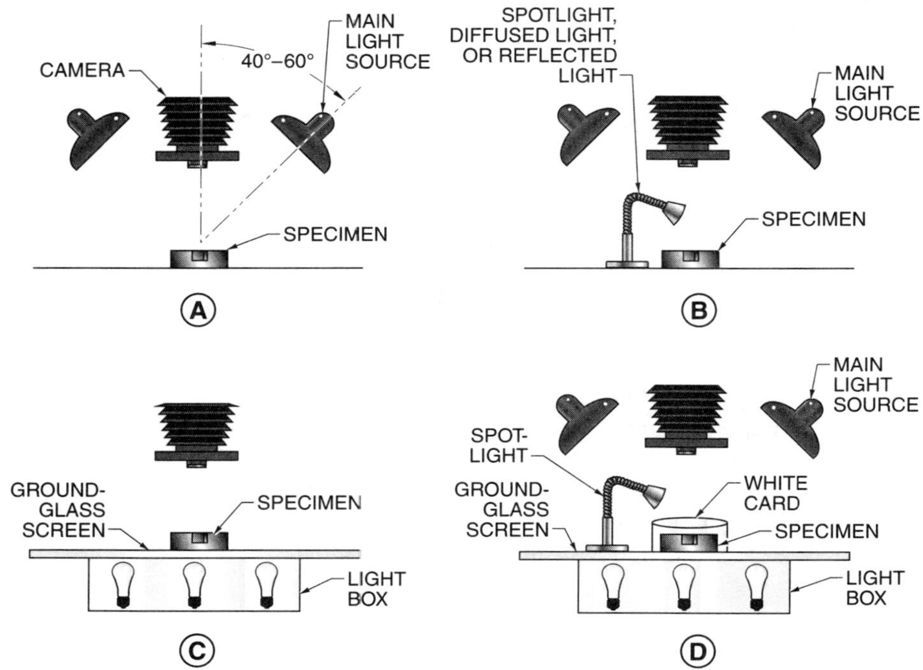

Chapter 34 Review
Weld Discontinuities and Failures

Name ______________________________ **Date** ____________________

True-False

T F **1.** A weld discontinuity is an interruption in the typical structure of a weld.

T F **2.** The average stress on a weld is in direct proportion to the reduction of the load-bearing cross-sectional area caused by a discontinuity.

T F **3.** The higher the load-bearing cross-sectional area, the higher the stress.

T F **4.** Cracks are the most serious discontinuities in weldments.

T F **5.** Throat cracks are cold cracks that are confined to the edges of a weld.

T F **6.** Crater cracks are star-shaped cracks that extend from the crater of a weld to the edge of the weld.

T F **7.** Underbead cracks are hydrogen cracks that occur in steels susceptible to hydrogen embrittlement during welding.

T F **8.** Root cracks are usually easy to detect.

T F **9.** Cavities only occur on the surface of a weld.

T F **10.** Porosity consists of cavity-type discontinuities formed by gas entrapment during solidification.

T F **11.** The primary causes of porosity are dirt, rust, and moisture on the surface of the base metal.

T F **12.** Wormholes can be prevented by using methods similar to those used to prevent porosity.

Consumable Insert

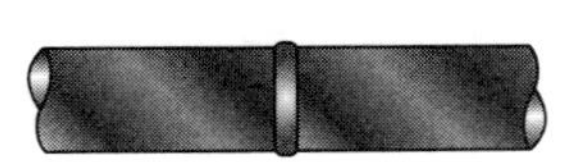

DESIRED WELD

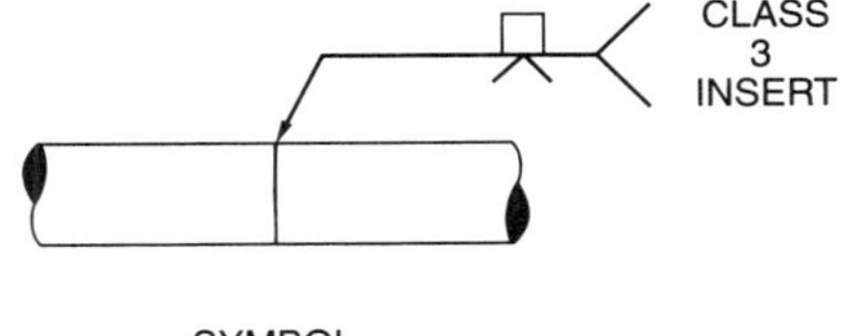

SYMBOL

Matching

Identify the crack types shown.

______________________ **1.** Underbead

______________________ **2.** Root

______________________ **3.** Crater

______________________ **4.** Longitudinal

______________________ **5.** Toe

______________________ **6.** Transverse

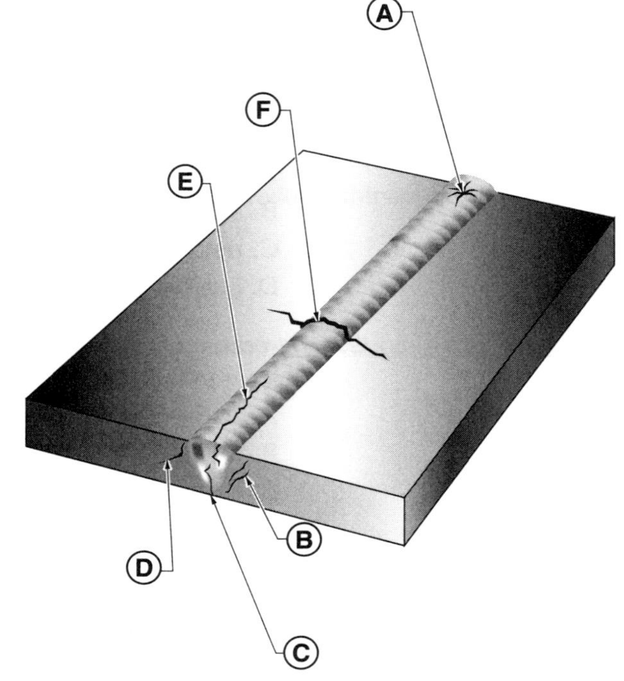

Name ______________________________ Date ____________________

True-False

T	F	**1.** Developing and qualifying WPSs are the responsibility of the manufacturer or contractor.
T	F	**2.** An essential variable is a welding variable that affects the mechanical properties of a weld.
T	F	**3.** Joint design may be affected by weld type, edge preparation, and backgouging.
T	F	**4.** Welding procedures are qualified by grouping base metals according to their P-number.
T	F	**5.** Filler metals do not require new WPSs if they share the same usability classification but have different A-numbers.
T	F	**6.** Shielding gas and flow rate are nonessential variables.
T	F	**7.** A minimum temperature must be specified for both preheat and interpass temperature controls.
T	F	**8.** Essential variables for brazing are the same as for arc welding.
T	F	**9.** Peening can be performed as soon as the weld cools completely.
T	F	**10.** It is necessary to specify nonessential variables in a WPS.
T	F	**11.** Mock-up tests are the only way to evaluate the effects of restricted joint access or complicated weldment design on weld quality.

Liquid Penetrant Examination

Examine-All-Around

NONDESTRUCTIVE EXAMINATION METHOD

EXAMINE-ALL-AROUND SYMBOL

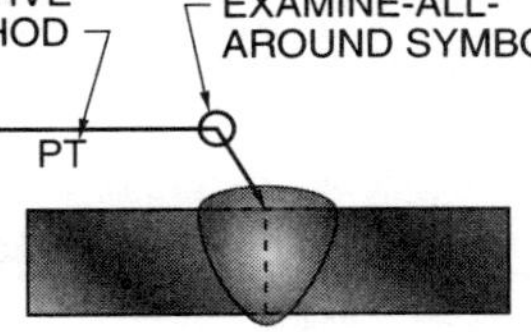

DESIRED WELD

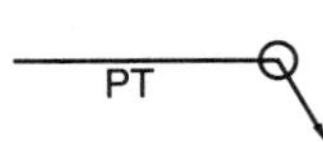

SYMBOL

T F **12.** P1 materials are specific chrome-moly steels that require preheat to approximately 300°F.

T F **13.** The welding procedure specification (WPS) provides formal documentation for all welding variables.

T F **14.** The type of weld bead is a variable that affects heat input and must be specified in the WPS and documented in the WPQR.

T F **15.** Prequalified welding procedures may be used as an alternative to testing.

T F **16.** Mock-up tests should not be used to simulate difficult or restricted welding conditions.

T F **17.** Destructive testing determines weld soundness, fusion characteristics, and depth of weld penetration.

Multiple Choice

______________ **1.** ___ is precise shaping to a desired profile using special tools to remove material.
A. Shearing
B. Grinding
C. Machining
D. Thermal cutting

______________ **2.** A ___ is the chemical composition or industry specification of a base metal.
A. base metal material specification
B. base metal weldability classification
C. base metal thickness range
D. filler metal specification

______________ **3.** Variables unique to ___ welding include joint design, electrode type and size, weld size and strength, and surface appearance.
A. oxyacetylene
B. submerged arc
C. gas tungsten arc
D. resistance

______________ **4.** A ___ complies with all the relevant stipulated conditions of a particular welding code or standard and is therefore acceptable for use under that code or standard without additional qualification testing.
A. welding procedure variable sheet
B. prequalified WPS
C. welding procedure specification
D. welding procedure qualification record

________________ **5.** A ___ is a document in which the actual values for the welding variables used to produce an acceptable test weldment as well as the results of tests conducted on the weldment are recorded.
- A. welding procedure qualification record
- B. welding process report
- C. standard welding procedure specification
- D. prequalified WPS

________________ **6.** ___ is the removal of weld metal from the root side of a welded joint so a back weld can be deposited to ensure complete fusion and joint penetration.
- A. Machining
- B. Edge preparation
- C. Backgouging
- D. Grinding

________________ **7.** ___ is parting of material when one blade forces the material past an opposing blade.
- A. Peeling
- B. Shearing
- C. Straining
- D. Shelling

________________ **8.** ___ is a welding variable that reduces residual stresses in a finished weld that may cause distortion or cracking.
- A. Peening
- B. Shearing
- C. Backgouging
- D. Machining

________________ **9.** ___ materials are low-carbon steels that generally do not require preheat.
- A. P1
- B. P2
- C. P3
- D. P4

________________ **10.** What is the meaning of the nondestructive examination symbol shown?
- A. proof testing, in the field
- B. proof testing, examine-all-around
- C. liquid penetrant examination, examine-all-around
- D. liquid penetrant examination, in the field

PT

Matching

______________	**1.** A(n) ___ requires a new welding procedure specification for metals where impact testing is required.	**A.** essential variable
______________	**2.** A(n) ___ may be changed in a WPS without requalification of the WPS.	**B.** supplementary essential variable
______________	**3.** A(n) ___ affects the mechanical properties of a weld.	**C.** nonessential variable

Identify the equipment used for the backgouging process shown.

______________ **4.** Electrode holder

______________ **5.** Base metal

______________ **6.** Carbon electrode

______________ **7.** Weld

______________ **8.** Air stream

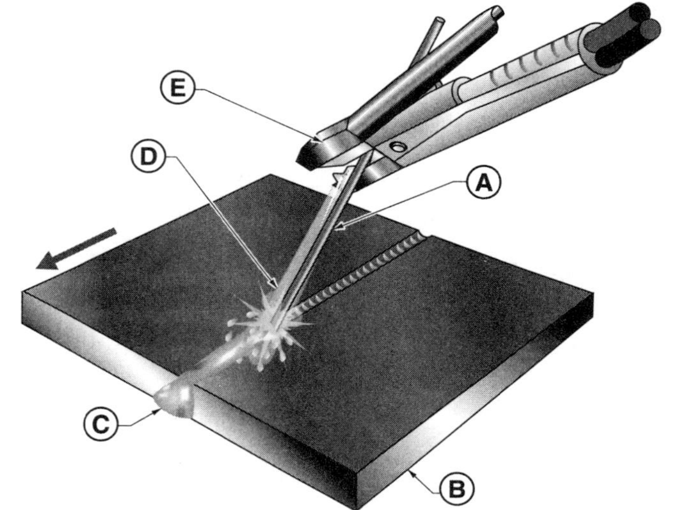

Chapter 36 Review
Welder Performance Qualification

Name ______________________________ Date ____________________

True-False

T F 1. Welder qualification in roll welding qualifies the welder for position welding.

T F 2. Welder certification is a test that demonstrates a welder's ability to produce welds that meet required standards.

T F 3. Qualification under one fabrication code or standard does not necessarily qualify a welder to weld under another, though the tests appear to be identical.

T F 4. When testing to ASME Section IX requirements, RT may be substituted for mechanical tests when GMAW with short circuiting transfer is used.

T F 5. When testing under AWS D1.3, *Structural Welding Code–Sheet Steel,* qualification is required for each position used.

T F 6. Qualification in the vertical and overhead positions qualifies the welder for all positions.

T F 7. Qualification on a product-specific groove weld test qualifies a welder for fillet welds in the same position.

T F 8. An open root joint is an unwelded joint that uses backing or consumable inserts.

Know Your Welding Symbols

Radiographic Examination with Direction of Radiation

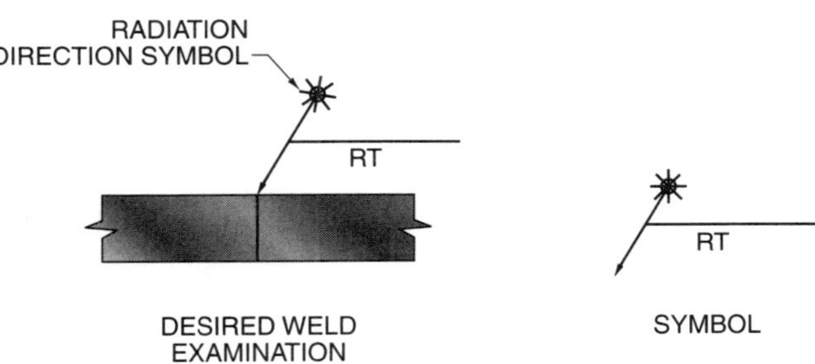

Multiple Choice

_______________ **1.** ___ can be used to relieve stress in a weld.
A. Backup welding
B. Inclusion welding
C. Peening
D. Rapid cooling

_______________ **2.** ___ is the amount of stress required for a metal to begin to deform plastically.
A. Tensile strength
B. Fatigue
C. Toughness
D. Yield strength

_______________ **3.** Approximately 90% of all failures in engineering components are related to ___.
A. embrittlement
B. ductility
C. hardness
D. fatigue

_______________ **4.** A(n) ___ load is a load that is applied suddenly or intermittently.
A. cyclical
B. impact
C. static
D. variable

_______________ **5.** The ___ is the lowest temperature at which an alloy is completely molten.
A. liquidus
B. melting point
C. solidus
D. thermal limit

_______________ **6.** ___ conductivity is the rate at which current flows through a metal.
A. Thermal
B. Magnetic
C. Electrical
D. Hydraulic

_______________ **7.** What is the meaning of the nondestructive examination symbol shown?
A. visual examination arrow side, in the field
B. visual examination other side, in the field
C. visual examination arrow side, examine-all-around
D. visual examination other side, examine-all-around

VT

Matching

______	**1.** ___ strength is a measure of the maximum stress that a material can resist under tensile stress.	**A.** Shear
______	**2.** ___ strength is the ability of a metal to resist being crushed.	**B.** Torsional
______	**3.** ___ strength is the quality of a metal that resists forces that cause a member to bend or deflect in the direction of the applied load.	**C.** Tensile
______	**4.** ___ strength is the ability of a metal to withstand forces that cause a member to twist.	**D.** Bending
______	**5.** ___ strength is the ability of a metal to withstand two equal forces acting in opposite directions.	**E.** Compressive

______	**6.** ___ is the failure of a material operating under alternating (cyclic) stresses at a value below the tensile strength of the material.	**A.** Ductility
______	**7.** ___ is the ability of a metal to absorb energy, such as impact loads, by deforming instead of cracking.	**B.** Toughness
______	**8.** ___ is a measure of the ability of a metal to yield plastically under load instead of fracturing.	**C.** Embrittlement
______	**9.** ___ is the ability of a material to resist indentation or scratching.	**D.** Hardness
______	**10.** ___ is the complete loss of ductility and toughness of a metal, so that it fractures when a small load is applied.	**E.** Fatigue

______	**11.** ___ is a measure of the stiffness of an object under tension or compression.	**A.** Creep
______	**12.** ___ is the internal resistance of a material.	**B.** Strain
______	**13.** ___ is the ability of a metal to be deformed by compression forces without developing defects.	**C.** Stress
______	**14.** ___ is slow plastic elongation that occurs during extended service at high temperatures.	**D.** Modulus of elasticity
______	**15.** ___ is the change in dimension that results when a load is applied.	**E.** Malleability

Identify the crystal structure patterns.

__________ **16.** Hexagonal close packed

__________ **17.** Body-centered cubic

__________ **18.** Face-centered cubic

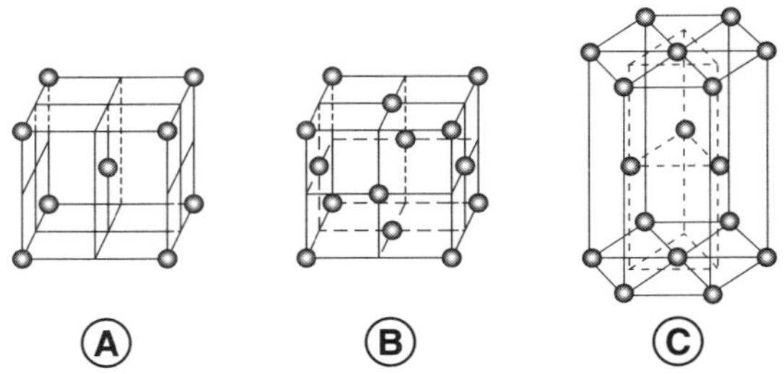

Ⓐ Ⓑ Ⓒ

Identify the regions of metallurgical structures created from the heat of welding.

__________ **19.** Heat-affected zone

__________ **20.** Base metal

__________ **21.** Weld metal

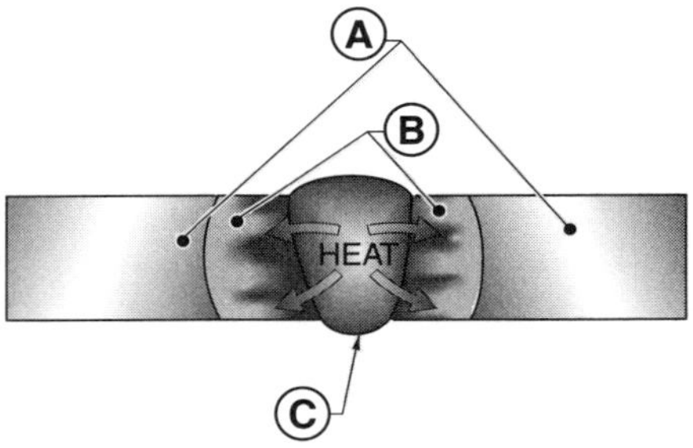

Ⓐ Ⓑ HEAT Ⓒ

Identify the mechanical force applied to each object to cause dimensional change.

__________ **22.** Shear

__________ **23.** Torsion

__________ **24.** Flexing

__________ **25.** Compression

__________ **26.** Tension

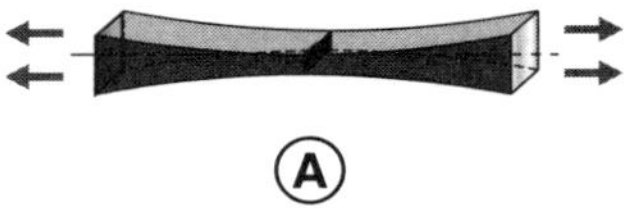

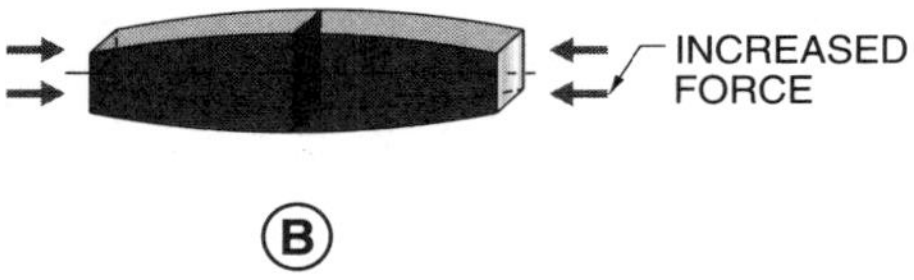

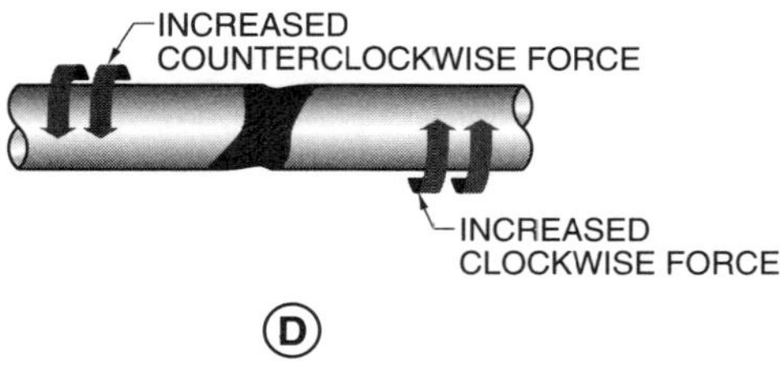

Ⓐ

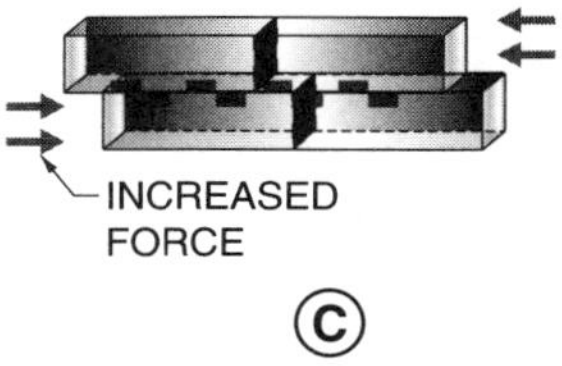

INCREASED FORCE
Ⓑ

INCREASED FORCE
Ⓒ

INCREASED COUNTERCLOCKWISE FORCE
INCREASED CLOCKWISE FORCE
Ⓓ

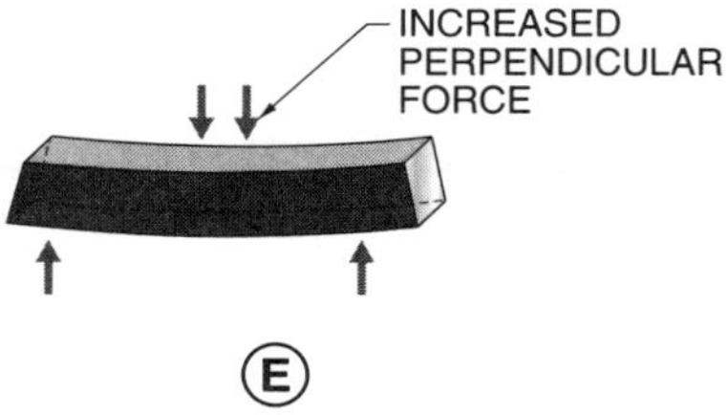

INCREASED PERPENDICULAR FORCE
Ⓔ

Name ______________________________ **Date** ________________

True-False

T F **1.** A materials test report is also called a certificate of analysis (COA).

T F **2.** A certificate of compliance (COC) is a statement by a manufacturer, without supporting documentation, that a supplied metal meets specifications.

T F **3.** If improper metal substitutions are made, significant damage to equipment or injury to workers may result.

T F **4.** Heat tint testing can be used to expose a metal's true color.

T F **5.** Nameplates identify the design, the pressure and temperature rating, the test pressure, and the materials of construction of fabricated equipment.

T F **6.** Qualitative identification is metal identification by a qualified person to confirm the identity of an unknown metal.

T F **7.** Magnetic particle testing is a qualitative identification method in which a magnet is laid on the surface of an unknown metal to test for a magnetic force.

T F **8.** Curie temperature is the temperature of magnetic transformation, above which a metal is nonmagnetic, and below which it is magnetic.

T F **9.** When chisel testing, long and curled chips result from mild steel and soft metals such as aluminum.

T F **10.** If particles of metal from previous tests are not removed prior to spark testing, they will contaminate the spark stream of the specimen being examined.

Spot Welds with Magnetic Examination of All Four Welds

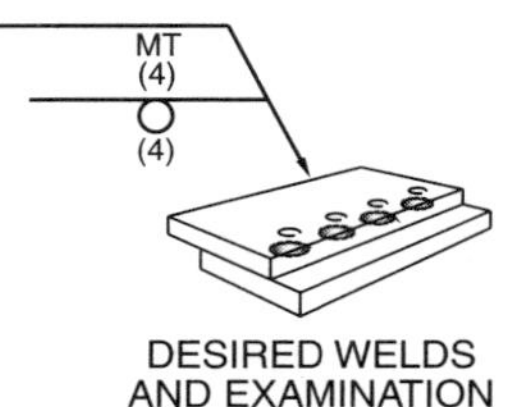

DESIRED WELDS AND EXAMINATION

MAGNETIC PARTICLE EXAMINATION SYMBOL
MT
(4)
(4)
NUMBER OF EXAMINATIONS
NUMBER OF SPOT WELDS
SYMBOL

T F **11.** Chemical spot testing is used to identify metals by the color changes that occur to a metal when it is contacted with specific chemical reagents.

T F **12.** The null point method is an alternative method of thermoelectric potential sorting used for identifying an unknown metal.

T F **13.** Diffraction is a modification of light in which the rays appear to be deflected to produce fringes of parallel light and dark colored bands.

T F **14.** Quantitative identification methods are used to analyze metals for every chemical element that may be present in the metals.

T F **15.** All quantitative examination methods are nondestructive.

T F **16.** Filler metals are identified by markings that are attached to or stamped on them.

Multiple Choice

______________ **1.** ___ is physical certification or documentation provided by a product manufacturer or supplier.
- A. Paperwork
- B. Product analysis
- C. Materials conformance report
- D. Nameplate marking

______________ **2.** A ___ is incorporated into a casting mold.
- A. color-coding system
- B. stencil mark
- C. nameplate
- D. foundry mark

______________ **3.** ___ is an identification marking that consists of colored stripes painted on one end of a metal.
- A. Color coding
- B. Stencil marking
- C. Foundry marking
- D. Plate marking

______________ **4.** A(n) ___ is the termination of a carrier line.
- A. spark line
- B. fork
- C. arrowhead
- D. burst

______________ **5.** ___ is an identification marking that consists of continuous or repeated ink markings on a metal.
- A. Color coding
- B. Stencil marking
- C. Line marking
- D. Plate marking

______________ **6.** ___ testing is a qualitative identification method that is used to identify a metal by the melting rate, the appearance of the metal when heat is applied, and the action of the molten metal.
A. Torch
B. Chisel
C. Heat tint
D. Magnetic response

______________ **7.** A(n) ___ is an incandescent (glowing) streak that traces the trajectory (path) of a particle (spark).
A. carrier line
B. fork
C. arrowhead
D. burst

______________ **8.** ___ identification is a method of metal identification that applies a physical stimulus to an unknown metal to produce a signal that is interpreted against a set of standards.
A. Qualitative
B. Semi-quantitative
C. Quantitative
D. Electrical

______________ **9.** ___ testing uses a small specimen, fine wire, and a beaker of distilled water.
A. Density
B. File
C. Magnetic
D. Distilled

______________ **10.** The ___ of an unknown metal influences the form of the spark stream produced during spark testing.
A. density
B. magnetic force
C. chemical composition
D. color

______________ **11.** A(n) ___ is a simple branching of a carrier line.
A. spark line
B. fork
C. arrowhead
D. burst

______________ **12.** ___ is an identification method involving the measurement of the electric potential generated when two metals are heated.
A. Electrical resistivity
B. Thermoelectric potential sorting
C. Chemical spot testing
D. Electrographic testing

__________ **13.** ___ is a nondestructive quantitative identification method that involves using a gamma ray beam to identify an unknown metal.
- A. Radiographic spectroscopy
- B. Gamma beam testing
- C. X-ray fluorescence spectrography
- D. Chemical analysis

__________ **14.** A(n) ___ is a complex branching of the carrier line.
- A. spark line
- B. fork
- C. arrowhead
- D. burst

__________ **15.** What is the meaning of the nondestructive examination symbol shown?
- A. four resistance spot welds
- B. magnetic particle examination of all four spot welds
- C. magnetic particle examination of four spot welds, examine-all-around
- D. magnetic particle examination of all four spot welds in the field

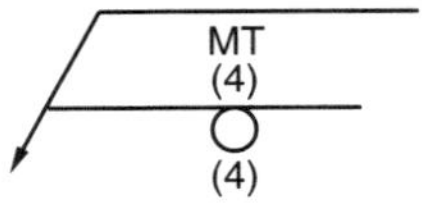

Matching

Match the metal with the type of spark stream it produces.

__________ **1.** Stainless steel (Type 40)

__________ **2.** Wrought iron

__________ **3.** Austenitic manganese steel

__________ **4.** Cemented tungsten carbide

__________ **5.** Carbon tool steel

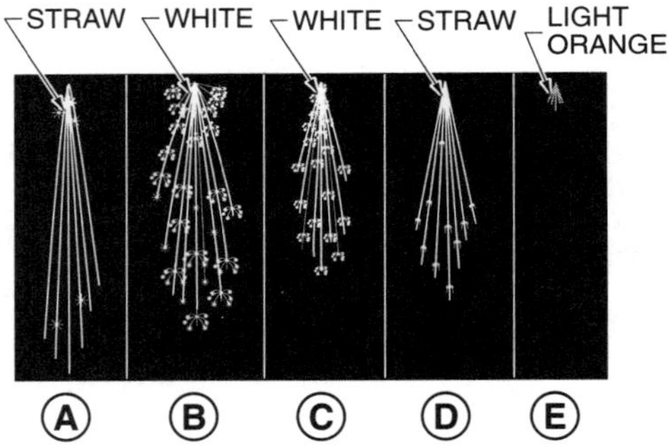

Identify the spark stream features and equipment shown.

__________ **6.** Carrier line

__________ **7.** Burst

__________ **8.** Grinder

__________ **9.** Fork

__________ **10.** Specimen

__________ **11.** Arrowhead

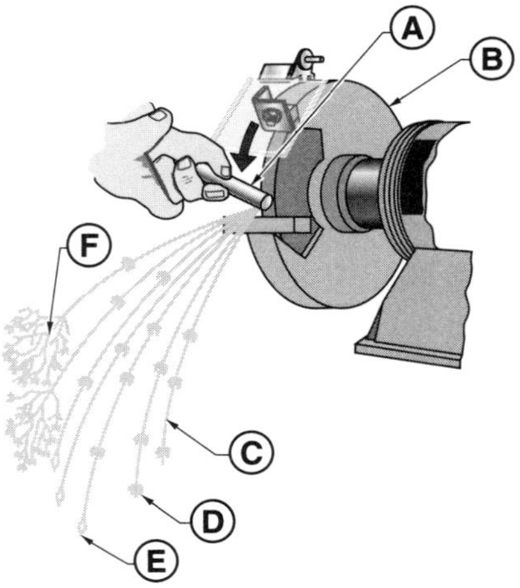

Name ______________________________ Date ________________

True-False

T F **1.** Carbon steels are alloys of iron, carbon, and manganese.

T F **2.** Medium-carbon steels are stronger than low-carbon steels.

T F **3.** Rimmed steel is steel with little or no deoxidizers added during steel production.

T F **4.** Martensite that forms in low-carbon steels is generally too soft and ductile to cause embrittlement or cracking.

T F **5.** Bainite is more brittle than martensite and forms at a faster cooling rate.

T F **6.** Carbon is the alloying element that has the greatest effect on the mechanical properties of steel.

T F **7.** The interpass temperature is the temperature of a weld area between passes of a multiple-pass weld.

T F **8.** Postheating is a stress-relief treatment for welding medium-carbon and high-carbon steels.

T F **9.** High-carbon steels are significantly easier to weld than other carbon steels.

T F **10.** Postheating temperatures for chrome-moly steels are typically lower than for carbon and low-alloy steels.

T F **11.** The recommended preheat temperature for low-alloy steels is about 50°F (28°C) above the temperature at which martensite begins to form on cooling.

T F **12.** The ductility of austenitic manganese steels increases when the steels are reheated.

Multiple Reference Lines Showing Sequence of Operations

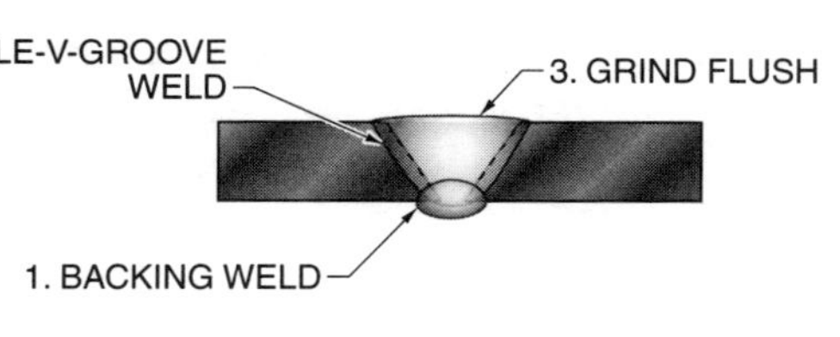

DESIRED WELD

THIRD OPERATION
SECOND OPERATION
FIRST OPERATION
G

SYMBOL

T F **13.** Graphitization is the formation of iron carbide that results in loss of ductility.

T F **14.** Carbon steel with a carbon equivalent less than 0.4% is weldable without preheat or postheating, depending on joint member thickness.

T F **15.** Free-machining steels are not usually welded unless absolutely necessary.

T F **16.** Killed steel is completely deoxidized during steel production by adding silicon or aluminum.

T F **17.** Hydrogen cracking is commonly detected by common nondestructive examination techniques.

T F **18.** In semikilled steel, no deoxidizers are added during production.

Multiple Choice

______________ **1.** ___ is not normally used to weld free-machining steels.
- A. SMAW
- B. FCAW
- C. GMAW
- D. GTAW

______________ **2.** ___ is the process of removing a controlled amount of oxygen from steel during steelmaking.
- A. Iron homogenization
- B. Steel deoxidation
- C. Oxygen reduction
- D. Carburization

______________ **3.** Postheating temperatures for stress relief should be in the range of ___.
- A. 600°F to 1000°F
- B. 750°F to 1500°F
- C. 900°F to 1250°F
- D. 1250°F to 1750°F

______________ **4.** ___ in free-machining steels can melt during welding, emitting weld fumes and creating a health hazard.
- A. Lead
- B. Oxygen
- C. Air
- D. Molybdenum

______________ **5.** ___ steel is suitable for cold bending and cold forming because of its high ductility.
- A. Semikilled
- B. Capped
- C. Killed
- D. Rimmed

______________ **6.** A ___ must be used when welding chrome-moly steels.
A. low current and slow speed
B. low current and rapid speed
C. high current and slow speed
D. high current and rapid speed

______________ **7.** Preheat temperatures for carbon steel range from ___, depending on carbon content.
A. 100°F to 400°F
B. 150°F to 500°F
C. 200°F to 500°F
D. 200°F to 700°F

______________ **8.** ___ steels are the easiest to weld since no special welding preparations are necessary.
A. Low-carbon
B. Medium-carbon
C. High-carbon
D. Free-machining

______________ **9.** The clip test is not applicable to thin steels but produces good results on sections up to ___″ thick.
A. ⅛
B. ¼
C. ⅓
D. ⅜

______________ **10.** What is the meaning of the multiple welding symbols shown?
A. Apply a backing weld. Then finish with a flat contour by grinding.
B. Apply a backing weld first. Then apply a single-V-groove weld. Then finish with a flat contour by grinding.
C. Apply a single-V-groove weld first. Then apply a backing weld. Then finish the back weld by grinding.
D. Apply a single-V-groove weld. Then finish with a flat contour by grinding.

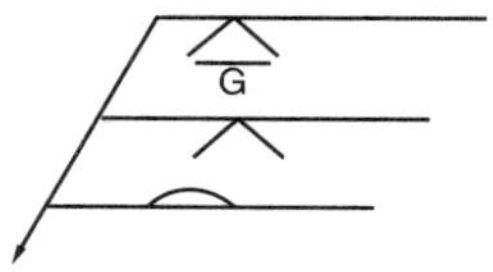

G

Matching

Identify the deoxidized steel shown.

_______________ **1.** Rimmed

_______________ **2.** Semikilled

_______________ **3.** Capped

_______________ **4.** Killed

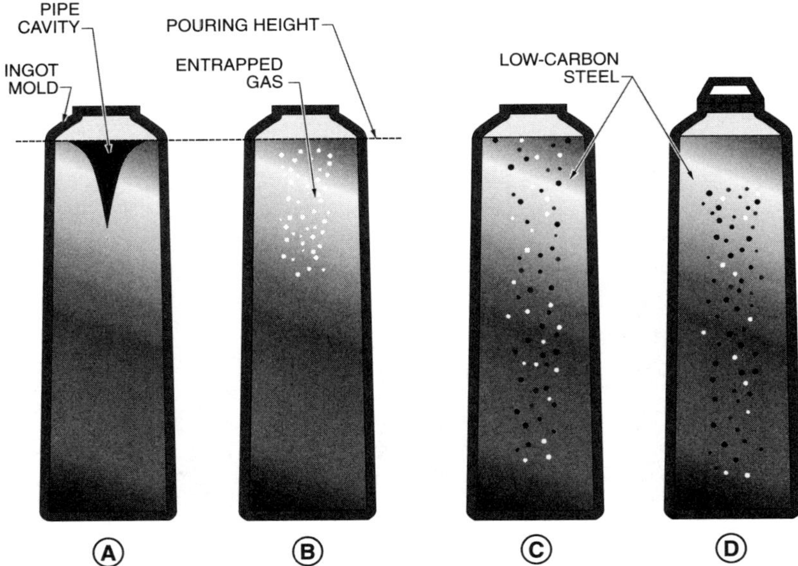

Chapter 40 Review

Weldability of Tool Steels and Cast Irons

Name ______________________________ **Date** ____________________

True-False

T F **1.** Tool steels are generally the hardest and strongest steels available.

T F **2.** Tool steels are highly resistant to hydrogen cracking in the HAZ when rapidly cooled.

T F **3.** Tool steels are always preheated for welding.

T F **4.** Filler metals must exactly match the tool steel composition.

T F **5.** Cast irons are easy to weld.

T F **6.** Nickel alloy filler metals are specially designed for welding cast irons.

T F **7.** Nickel alloy filler metals are nonmachinable.

T F **8.** Carbon steel filler metals are machinable.

T F **9.** Cast iron filler metals are nonmachinable.

T F **10.** Copper alloy filler metals are used for braze welding cast iron.

T F **11.** Water hardening tool steels, Group W, are high-carbon steels.

T F **12.** Shock resistant tool steels, Group S, have a high-carbon content.

T F **13.** Mold steels exhibit low hardness and low resistance.

T F **14.** Cold work tool steels have high wear resistance and poor-to-fair toughness.

Know Your Welding Symbols

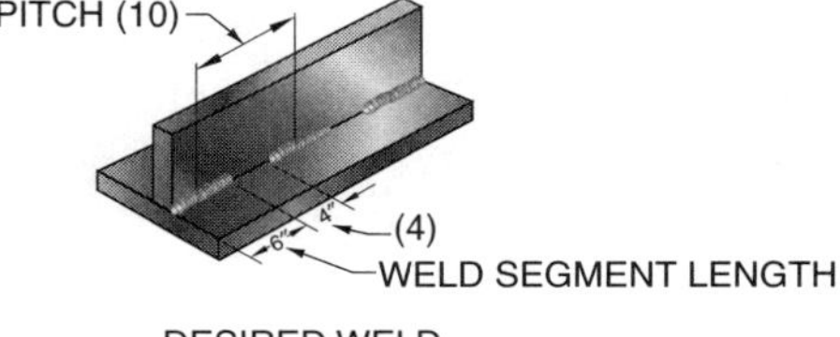

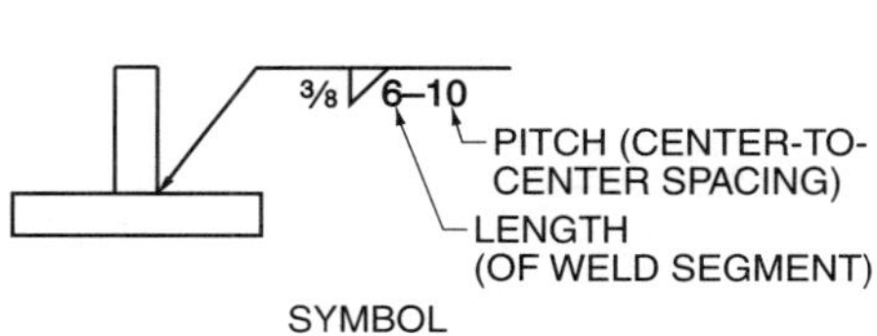

Multiple Choice

__________ **1.** ___ should not be used for welding tool steels.
- A. SMAW
- B. FCAW
- C. GMAW
- D. OFW

__________ **2.** ___ tool steels are typically the least costly tool steels.
- A. Cold work (Groups O, A, and D)
- B. Hot work (Group H)
- C. Water hardening (Group W)
- D. Mold and special purpose (Groups L, P, and F)

__________ **3.** High-speed tool steels are resistant to softening at temperatures up to ___°F.
- A. 500
- B. 750
- C. 1000
- D. 1500

__________ **4.** When repairing the cutting edge of a tool or die, the edge to be welded should be grooved approximately ___° for sufficient depth.
- A. 30
- B. 45
- C. 60
- D. 90

__________ **5.** ___ irons are produced by heat treating white iron.
- A. Gray
- B. Ductile
- C. Malleable
- D. Compacted graphite

__________ **6.** When welding cast iron ⅜″ thick or more, ___ is necessary.
- A. no V-joint
- B. a single-V joint with a 45° groove angle
- C. a single-V joint with a 60° groove angle
- D. a double-V joint with a 60° groove angle

__________ **7.** What is the meaning of the welding symbol shown?
- A. intermittent weld, 6″ length and 10″ pitch
- B. intermittent weld, between 6″ and 10″ in length
- C. back-step weld, between 6″ and 10″ in length
- D. back-step weld, 10″ leg size and 6″ width

6–10

Chapter **41** Review

Weldability of Stainless Steels

Name ______________________________ **Date** ____________________

True-False

T F **1.** Stainless steels are the least versatile family of metals.

T F **2.** Ferritic stainless steels contain less chromium than martensitic stainless steels.

T F **3.** Duplex stainless steels have higher strength than austenitic stainless steels.

T F **4.** Wrought and cast stainless steels can be welded using arc welding or OFW processes.

T F **5.** Austenitic stainless steels are less prone to distortion during welding than carbon steels.

T F **6.** A chill plate is a metal plate used as a heat sink during welding.

T F **7.** Before welding stainless steels, the surface of the weld area must be completely cleaned of all hydrocarbon contaminants.

T F **8.** Stainless steels obtain their corrosion resistance from a surface film composed largely of cobalt and nitrogen.

T F **9.** Stainless steel cannot be welded using the GTAW process.

T F **10.** Silicon increases weldability of stainless steels because it reduces the risk of hot cracking.

T F **11.** Only AC current should be used for arc welding stainless steel.

T F **12.** Austenitic stainless steels are susceptible to hot cracking.

Single-Bevel-Groove Weld

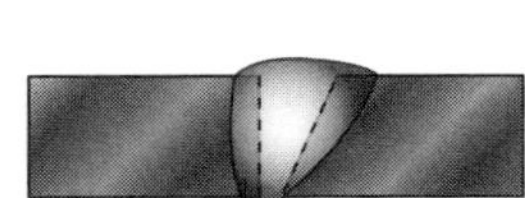

DESIRED WELD

BREAK INDICATES MEMBER TO BE BEVELED

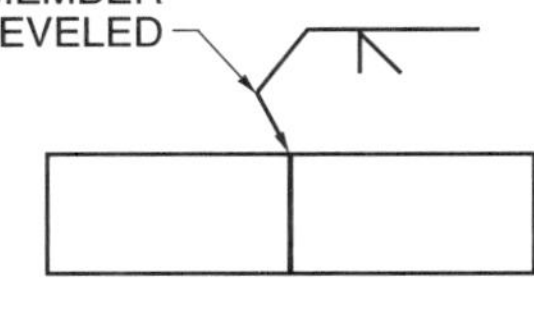

SYMBOL

Matching

______	**1.** The basic austenitic stainless steel is type ___.	**A.** 302
______	**2.** The basic ferritic stainless steel is type ___.	**B.** 329
______	**3.** The basic duplex stainless steel is type ___.	**C.** 410
______	**4.** The basic martensitic stainless steel is type ___.	**D.** 430

Name ______________________________ **Date** ________________

True-False

T F **1.** Nickel is incorporated as a major or minor constituent in approximately 3000 alloys.

T F **2.** Nickel alloys should not be heat-treated.

T F **3.** Joint cleanliness is the single most important requirement for welding nickel alloys.

T F **4.** Low heat input should be used when welding nickel alloys.

T F **5.** GMAW and GTAW are typically used to weld nickel and nickel alloys.

T F **6.** Wrought commercially pure coppers contain at least 99.9% copper.

T F **7.** Brasses are the least popular and most expensive of the copper alloys.

T F **8.** Copper and copper alloys are easy to weld.

T F **9.** Special coated electrodes are used for welding sheet copper with SMAW.

T F **10.** Aluminum alloys can be wrought or cast.

T F **11.** The aluminum oxide film that forms on the surface of aluminum alloys need not be removed prior to welding.

T F **12.** Aluminum alloys conduct heat at the same rate as steel.

T F **13.** All aluminum alloys may be welded using resistance welding.

T F **14.** Magnesium is one of the heaviest commercial metals.

T F **15.** Thick magnesium alloy workpieces generally require preheat to prevent weld cracking.

Double-Bevel-Groove Weld

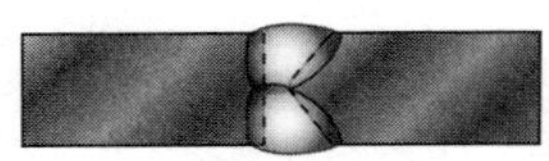

DESIRED WELD

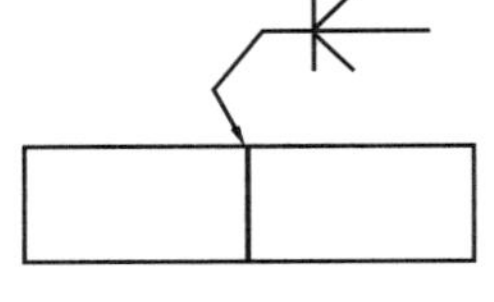

SYMBOL

T F **16.** Alpha titanium alloys are generally the lowest strength titanium alloys.

T F **17.** Beta titanium alloys have exceptional work-hardening characteristics.

T F **18.** Preheat is typically not required for titanium alloys.

Multiple Choice

______________ **1.** ___ hardening is hardening that occurs immediately by heating an alloy to an intermediate temperature to develop maximum strength.

A. Strain
B. Precipitation
C. Natural age
D. Artificial age

______________ **2.** Commercially pure coppers are ___.

A. hard
B. strong
C. very ductile
D. used for their low electrical conductivity

______________ **3.** Commercially pure coppers require preheat from ___, depending on joint thickness.

A. 200°F to 500°F
B. 250°F to 600°F
C. 250°F to 1000°F
D. 300°F to 1250°F

______________ **4.** Tin bronzes are susceptible to ___.

A. peeling
B. hot cracking
C. bonding
D. pickling

______________ **5.** Aluminum alloys have ___.

A. high density
B. poor corrosion resistance
C. poor weldability
D. low-temperature toughness

______________ **6.** ___ is commonly used for shielding when welding aluminum.

A. Nitrogen
B. Argon
C. Hydrogen
D. Helium

________________ **7.** Low-zinc brasses have higher thermal conductivity than high-zinc brasses and require preheat in the range 200°F to ___°F.

A. 600

B. 700

C. 800

D. 900

________________ **8.** When welding aluminum alloys, preheating of thick joints must not exceed ___°F to prevent distortion effects.

A. 200

B. 300

C. 400

D. 500

________________ **9.** When brazing titanium alloys, the brazing temperature must be below ___°F to prevent a reduction of mechanical properties.

A. 500

B. 750

C. 1000

D. 1650

________________ **10.** What is the meaning of the welding symbol shown?

A. weld arrow side of joint

B. double-bevel-groove weld

C. single-bevel-groove weld

D. weld other side of joint

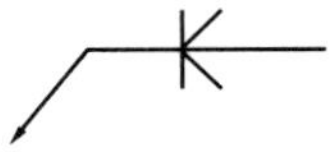

________________ **11.** Weld cracking in heat-treatable alloys can be minimized by using ___ series filler metal because it mixes with the molten base metal to produce a slight gain in strength.

A. 1XXX

B. 2XXX

C. 3XXX

D. 4XXX

________________ **12.** ___ is a chemical cleaning process that involves the use of strong acids to remove surface impurities, such as scales or contaminants.

A. Erosion-corrosion

B. Bonding

C. Hot cracking

D. Pickling

Matching

Identify the parts of the ASTM Temper Designation shown.

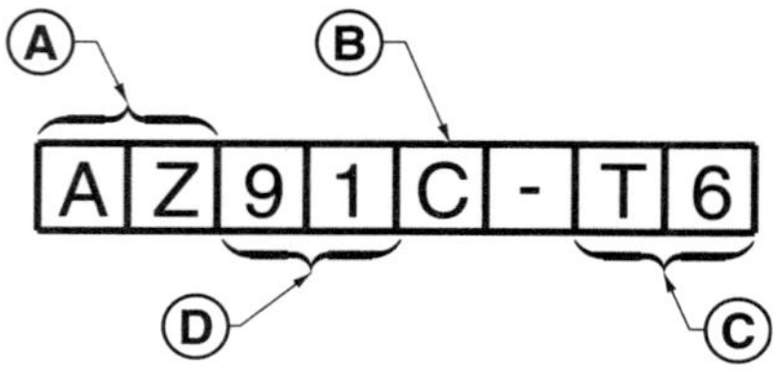

____________________ **1.** Indicates the amounts of the two alloying elements

____________________ **2.** Indicates temper condition

____________________ **3.** Indicates the two principal alloying elements

____________________ **4.** Distinguishes between different alloys with the same percentages of the two principal alloying elements

Name ______________________________ **Date** ____________________

True-False

T F **1.** Zinc and stainless steels are mutually insoluble base metals.

T F **2.** Welding is not limited to a certain combination of metals.

T F **3.** Galvanic corrosion is the acceleration of corrosion that occurs due to an electrochemical reaction between two dissimilar metals.

T F **4.** Solubility is the mutual intermixing of atoms of one metal or alloy with another with the formation of separate phases.

T F **5.** The differences between base metals may be significant if the metals belong to two different alloy families.

T F **6.** During cooling, the metal with the higher melting temperature solidifies first, and while the metal with the lower melting temperature is still molten, it is mechanically stronger and unsusceptible to hot cracking from thermal stress.

T F **7.** Buttering is a surfacing weld technique that involves applying weld metal on one or more joint surfaces to provide compatible base metal for subsequent completion of the weld.

T F **8.** For ferritic stainless steels, preheating highly restrained joints to between 400°F to 600°F will prevent cracking.

T F **9.** Dissimilar metal welding is the application of a welding process to join three different base metals.

T F **10.** Intermetallic phase is a chemical compound formed between metallic chemical elements that is nonmetallic (brittle).

Single Fillet Weld Unequal Leg Size

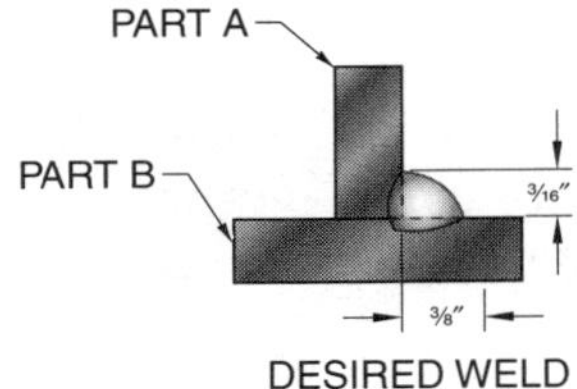

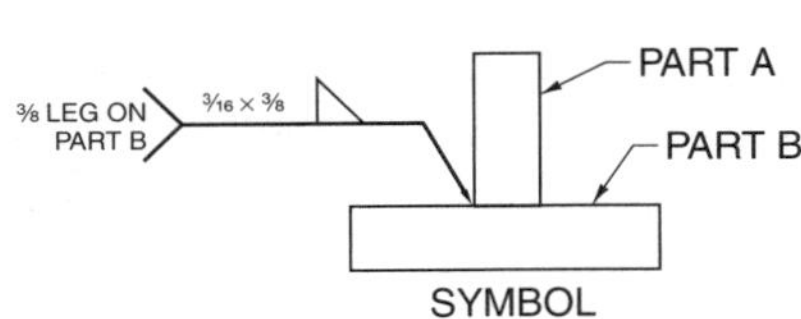

T F **11.** Brazing and soldiering are performed above the melting temperatures of the base metals being joined.

T F **12.** Dissimilar metals to be welded may have very different mechanical properties, such as strength, ductility, and toughness.

Multiple Choice

______________ **1.** The melting temperature of a filler metal must be between the melting temperatures of the base metals. When welding 304 stainless steel to 90-10 copper-nickel, the large difference in the melting temperature may lead to ___ on the copper-nickel side of the joint.

A. buttering
B. preheating
C. hot cracking
D. corrosion

______________ **2.** ___ alloys can be welded to carbon steel, stainless steel, and copper alloys using any arc welding process.

A. Iron
B. Cobalt
C. Aluminum
D. Nickel

______________ **3.** To avoid intermetallic phases, filler metal must be selected that permits mutual ___.

A. thermal expansion
B. corrosion resistance
C. buttering
D. solubility

______________ **4.** ___ are most typically used for dissimilar metal welding.

A. SMAW and GTAW
B. SMAW and GMAW
C. OAW and FCAW
D. FCAW and GMAW

______________ **5.** The difference between the melting temperatures of aluminum and carbon steel is 1200°F and ___°F.

A. 2400
B. 2800
C. 2850
D. 2900

______________ **6.** Hot cracking may be prevented when welding 304 stainless steel to 90-10 copper-nickel by buttering the copper-nickel with ___ Cu-Ni.

A. 50-50
B. 60-40
C. 70-30
D. 80-20

______________ **7.** For a proper weld, the filler metal must alloy readily with the base metals and be capable of being diluted by the base metals without the formation of ___ intermetallic phases.
- A. brittle
- B. fragile
- C. cracked
- D. soluble

______________ **8.** ___ is a surfacing weld technique where a base metal surface is overlaid with a material that has properties between those of the two base metals being joined.
- A. Dilution
- B. Brazing
- C. Buttering
- D. Soldering

______________ **9.** When welding copper to carbon steel, the copper should be preheated to between ___°F and 1000°F.
- A. 400
- B. 450
- C. 500
- D. 550

______________ **10.** When welding copper-nickel to austenitic stainless steel, heat input should be kept low with the preheating and interpass temperatures controlled to a maximum of ___°F.
- A. 75
- B. 150
- C. 300
- D. 600

______________ **11.** If there is a difference between the coefficients of ___ of base metals, the metal with the lower coefficient is subject to tensile stress when heated.
- A. thermal conductivity
- B. thermal expansion
- C. dilution
- D. buttering

______________ **12.** Buttering is particularly important for welding base metals with very different coefficients of thermal expansion or ___.
- A. corrosion resistance
- B. solubility
- C. thermal conductivity
- D. melting temperatures

______________ **13.** What is the meaning of the welding symbol shown?
- A. single fillet weld with equal legs, arrow side
- B. single fillet weld with unequal legs, arrow side
- C. 3/16″ fillet weld, 3/8″ depth of penetration, arrow side
- D. double fillet weld with equal legs, arrow side

3/16 × 3/8

Matching

Identify the steps of buttering.

_______________ **1.** Buttered face prepared for welding

_______________ **2.** Joint welded with filler metal

_______________ **3.** Face buttered with filler metal

_______________ **4.** Face prepared for buttering

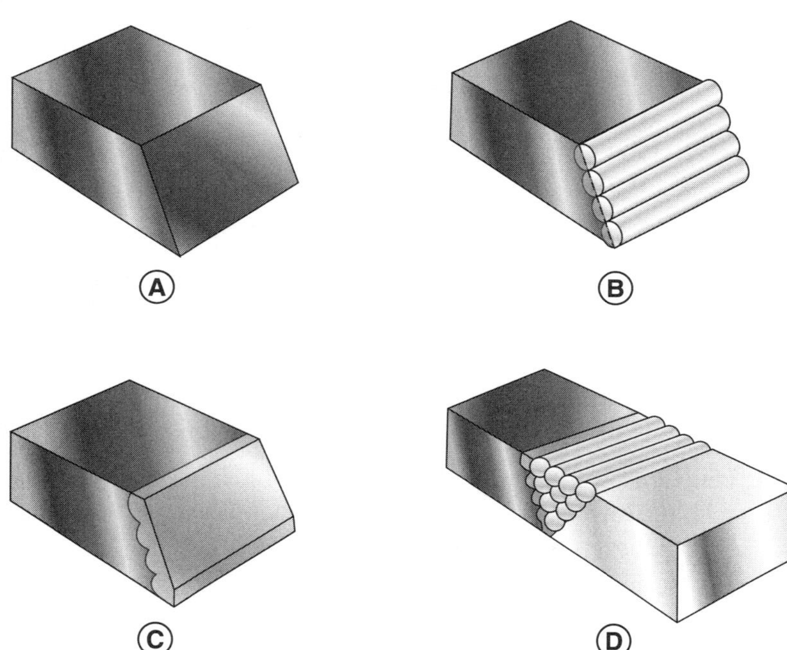

Name ____________________ Date ____________

True-False

T F 1. Distortion is an undesirable dimensional change of a part during fabrication.

T F 2. Distortion in welding arises from weld metal shrinkage and base metal shrinkage during cooling.

T F 3. Longitudinal shrinkage is a greater problem in groove welds than transverse shrinkage.

T F 4. During welding, the base metal temperature is substantially lower than the HAZ temperature because the base metal is less affected by the heat of welding.

T F 5. Distortion control is necessary to overcome poor fit-up and undesirable stresses.

T F 6. Heat shaping is the application of localized heating to cause movement of a distorted part and restore its dimensions.

T F 7. The V-heating pattern concentrates heat in one area in a circular motion and is applied with little, if any, forward motion.

T F 8. Heat shaping is applied using an oxyacetylene flame.

Multiple Choice

__________ 1. ___ shrinkage is shrinkage that occurs perpendicular to the weld axis.
A. Transverse
B. Perpendicular
C. Oblique
D. Longitudinal

Know Your Welding Symbols

Double Fillet Equal Leg Size

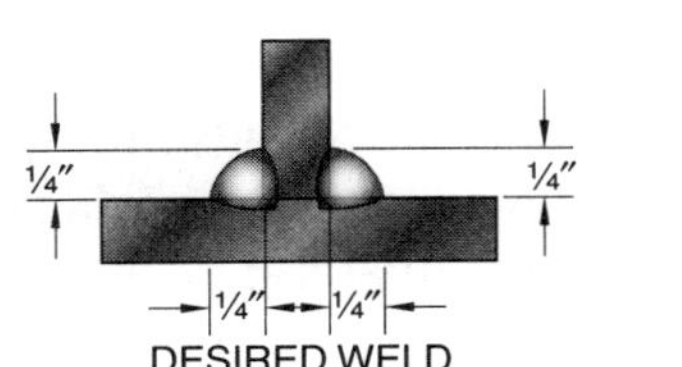

DESIRED WELD

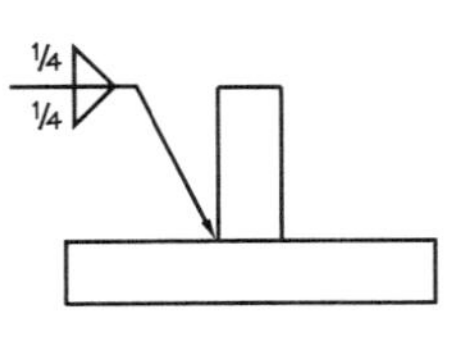

SYMBOL

_______________ **2.** The size of a plug weld is shown ___ the weld symbol.
- A. to the left of
- B. to the right of
- C. above
- D. below

_______________ **3.** The root opening dimension is located ___ the weld symbol.
- A. next to
- B. above
- C. below
- D. inside

_______________ **4.** When a weld is to extend completely around a joint, a small ___ is placed where the arrow connects to the reference line.
- A. square
- B. rectangle
- C. circle
- D. triangle

_______________ **5.** ___ welds are made on the other side of a joint before a groove weld is deposited to prevent excessive penetration.
- A. Back
- B. Backing
- C. Melt-through
- D. Surfacing

_______________ **6.** What is the meaning of the welding symbol shown?
- A. single-V-groove weld with backing weld
- B. single-V-groove weld with front weld
- C. single-V-groove weld with complete penetration
- D. single-V-groove weld with melt-through

Matching

Identify the weld type symbols.

_______________ **1.** Seam

_______________ **2.** J-groove

_______________ **3.** U-groove

_______________ **4.** Flare-V-groove

_______________ **5.** Flare-bevel groove

_______________ **6.** Plug

_______________ **7.** Fillet

_______________ **8.** Square groove

_______________ **9.** V-groove

_______________ **10.** Bevel-groove

_______________ **11.** Spot or projection

_______________ **12.** Slot

Ⓐ Ⓑ Ⓒ Ⓓ

Ⓔ Ⓕ Ⓖ Ⓗ

Ⓘ Ⓙ Ⓚ Ⓛ

section eight | Welding Technology

Chapter **46** Review

Materials and Fabrication Standards and Codes

Name ______________________________ **Date** ____________________

True-False

T F **1.** Materials standards are classified according to the kind of information they contain.

T F **2.** Standards are developed by standards committees.

T F **3.** A user enquiry is a formal procedure developed by standards committees and code-creating organizations to help users interpret issues and offer suggestions.

T F **4.** The International Standards Organization (ISO) is the largest source of materials standards.

T F **5.** The ASTM International standards designation uses a letter-number combination to refer to a particular type of metal.

T F **6.** The unified lettering system (ULS) is a common embedded designation that unifies all families of metals and alloys.

T F **7.** Aerospace Material Specifications (AMS) standards generally contain the most stringent quality requirements of any standards.

T F **8.** The American National Standards Institute adopts standards written and approved by member organizations.

T F **9.** Variations between materials standards allow them to cover many industrial applications and meet a wide range of quality requirements.

T F **10.** Many large cities publish their own specific welding codes.

Know Your Welding Symbols

Depth of Bevel

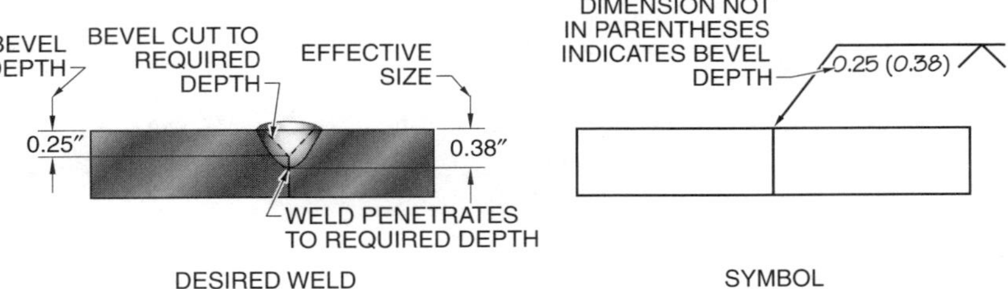

T F **11.** A manufacturing data report (MDR) certifies that all details of design, material, construction, and workmanship conform to the ASME International Boiler and Pressure Vessel Code.

T F **12.** Piping systems should never be used to transport flammable, toxic, or corrosive liquids.

T F **13.** Some states require welders to be certified by the state to work on bridges.

T F **14.** Ship and barge welding requirements are typically covered by jurisdictions or insurance companies.

T F **15.** Underwater welding can only be performed in a dry environment.

T F **16.** No restrictions are placed on the use of any welding process or procedure used on earthmoving and construction equipment, provided the weld produced meets the qualification requirements of the specification.

Multiple Choice

______________ **1.** A ___ is a type of standard that provides instructions for performing one or more repetitive technical functions.
- A. designation
- B. specification
- C. recommended practice
- D. code

______________ **2.** A ___ is a type of standard that indicates the technical and commercial requirements for a product.
- A. division
- B. specification
- C. recommended practice
- D. code

______________ **3.** A ___ is a document that serves as a model for the measurement of a property or the establishment of a procedure.
- A. standard
- B. specification
- C. recommended practice
- D. code

______________ **4.** A(n) ___ is a standard that is mandatory and is used by a jurisdictional body.
- A. MSDS
- B. description
- C. recommended practice
- D. code

______________ **5.** The ___ states the intent of the user enquiry.
- A. scope
- B. purpose
- C. content
- D. proposed reply

______________ **6.** Whenever possible, welding consumables should be referred to by ___ designations rather than commercial names.
A. ASTM International (ASTM)
B. American National Standards Institute (ANSI)
C. ASME International (ASME)
D. American Welding Society (AWS)

______________ **7.** ASME International Boiler and Pressure Vessel Code-approved materials and welding consumables are assigned the prefix letter ___ to indicate approval.
A. A
B. F
C. P
D. S

______________ **8.** European standards are produced by the ___.
A. International Organization for Standardization (ISO)
B. ASME International (ASME)
C. European Standards Council (CEN)
D. ASTM International (ASTM)

______________ **9.** A ___ is certification issued by the primary manufacturer verifying the chemical analysis and mechanical test properties of stock obtained from a starting ingot or billet of metal.
A. manufacturing data report
B. certificate of compliance
C. mill test report
D. filler metal approval

______________ **10.** To verify that a filler metal meets a specification, the ___ and accompanying paperwork are checked.
A. invoice
B. filler metal box
C. purchase order
D. mill test report

______________ **11.** ___ is any repair that does not restore a mechanical component to its original design.
A. Restoration
B. Alteration
C. Modification
D. Reconstruction

______________ **12.** ___ piping is carbon-steel, standard-size pipe of small diameter that conveys products from intermediate facilities to consumers.
A. Transportation
B. Delivery
C. Distribution
D. Pressure

______________ **13.** ___ is used for its lightness coupled with its strength and atmospheric corrosion resistance.
- A. Sheet metal
- B. Structural steel
- C. Reinforcing steel
- D. Structural aluminum

______________ **14.** In the welding symbol shown, what does the number 0.25 indicate?
- A. depth of groove
- B. complete penetration
- C. V-groove width
- D. fillet weld leg size

0.25 (0.38)

Matching

Identify the parts of the ASTM Standards Designation shown.

______________ **1.** Standard title

______________ **2.** Version revised in same year

______________ **3.** General classification

______________ **4.** Year of adoption

______________ **5.** Indicates metric version

______________ **6.** One to four digit number

A B C D E F

B 187 - 97 COPPER ROD, BAR, AND SHAPES

Identify the parts of the underwater welding equipment shown.

______________ **7.** Power source

______________ **8.** Electrode lead

______________ **9.** Gas-filled enclosure

______________ **10.** Workpiece connection

______________ **11.** Waterproof electrode holder

______________ **12.** 400 A knife switch

______________ **13.** Air supply

______________ **14.** Workpiece lead

______________ **15.** Work

______________ **16.** Communication link to diver

______________ **17.** Shielding gas

Section 1 Activities

Introduction to Welding

Name ______________________________ **Date** ____________________

1. List and define three major welding processes used in industry.

2. List five principal job titles of welders.

3. List five basic rules contributing to oxyacetylene welding safety.

4. List five methods used to clean a container before it is welded or cut.

5. Sketch the following weld joints in the space provided.

A. Butt B. T C. Lap D. Edge E. Corner

6. Sketch the following types of butt joints in the space provided.

A. Square B. Single bevel C. Single-V

D. Double-V E. Single-U F. Double-U

7. Sketch the following types of T-joints in the space provided.

A. Square B. Single bevel C. Double bevel D. Single-J E. Double-J

8. Sketch the following types of lap joints in the space provided.

A. Single fillet B. Double fillet

9. Sketch the following types of corner joints in the space provided.

A. Flush B. Half-open C. Full-open

10. Sketch the following types of edge joints in the space provided.

A. Square B. Single bevel C. Double bevel

11. Sketch the following welding positions for a square-groove joint and a T-joint in the space provided.

Square-groove joint

A. Flat B. Horizontal C. Vertical D. Overhead

T-Joint

E. Flat F. Horizontal G. Vertical H. Overhead

12. Sketch a fillet weld in a T-joint in the space provided. Identify the weld legs, actual throat, weld toes, weld face, and weld root.

13. Sketch a single-V-groove weld in a butt joint with a root face in the space provided. Identify the root opening, root face, groove face, depth of bevel, face reinforcement, and root reinforcement.

14. List five categories of chemical hazard information that must be contained on an MSDS.

Name ______________________________ **Date** ________________

exercise 2-1

Lighting the Oxyacetylene Torch

Conditions

Number 0 welding tip
Welding torch
Sparklighter
Oxyacetylene welding equipment

Performance

The welder will demonstrate the correct procedure for lighting an oxyacetylene welding torch.

Criteria

Correct sequence of procedures will be evaluated by the instructor.

Procedure

1. Mount a number 0 welding tip on the welding torch. Turn the regulator adjusting screws all the way out.
2. Stand to one side and open the oxygen and acetylene cylinder valves slowly. The oxygen cylinder valve should be opened all the way and the acetylene cylinder valve should be opened approximately one-quarter of a turn to one complete turn.
3. Stand to the side and turn the regulator adjusting screws to obtain the required working pressures.
4. Open the acetylene needle valve on the torch approximately one-quarter of a turn.
5. With the welding tip pointing down, position the sparklighter approximately 1″ from the end of the tip. Ignite the flame quickly to prevent wasting acetylene.
6. Adjust the flame so that it jumps slightly away from the torch tip. If the flame produces a great amount of black smoke, increase the amount of acetylene by opening the acetylene needle valve.
7. With the acetylene burning, slowly open the oxygen needle valve to produce a neutral flame.

8. An excess amount of oxygen will produce an oxidizing flame. An excess amount of acetylene will produce a carburizing flame.

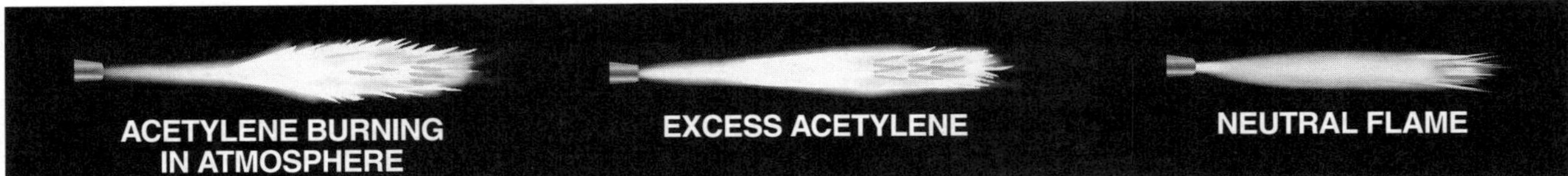

9. To shut off the torch, close the oxygen needle valve first.
10. Promptly close the acetylene valve.
11. Close the oxygen and acetylene cylinder valves to shut down the entire welding apparatus.
12. To remove pressure on the regulators, open the oxygen needle valve on the torch, then open the acetylene needle valve. Close the needle valves when no pressure is indicated on the regulators.
13. Turn out the adjusting screws on the regulators to release pressure on the regulator.

exercise 2-2

Testing the Flames

Conditions

Refer to Exercise 2-1

Performance

The welder will demonstrate the correct procedure for adjusting the welding torch to obtain carburizing, oxidizing, and neutral flames.

Criteria

Correct flame characteristics will be evaluated by the instructor.

Procedure

1. Refer to Exercise 2-1, Steps 1 through 8.

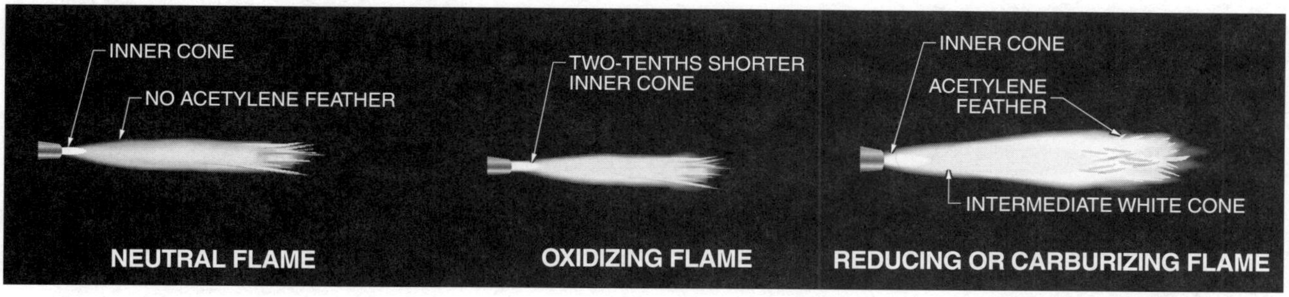

INNER CONE
NO ACETYLENE FEATHER
NEUTRAL FLAME
TWO-TENTHS SHORTER INNER CONE
OXIDIZING FLAME
INNER CONE
ACETYLENE FEATHER
INTERMEDIATE WHITE CONE
REDUCING OR CARBURIZING FLAME

2. Obtain a neutral, carburizing, and oxidizing flame by increasing or decreasing the flow of oxygen and/or acetylene.
3. Identify the characteristics of the flame.
4. Apply each flame to metal and observe the results of each.

exercise 2-3

Carrying a Weld Pool without Filler Metal Using OAW

Conditions

Flat position
Refer to Exercise 2-1
1⁄16″ × 3″ × 5″ mild steel

Performance

The welder will demonstrate the correct procedure for carrying a weld pool without filler metal.

Criteria

Weld beads should be consistent in width, ripple formation, and penetration, and run parallel to the length of the workpiece.

Procedure

1. Be sure the surface of the workpiece is free of oil, dirt, and scale.
2. Light the torch and adjust for a neutral flame.
3. Hold the torch at a 45° angle with the inner cone of the flame 1⁄8″ from the workpiece.
4. Use a circular motion with the torch to distribute the heat evenly. Start from the right side of the piece, moving to the left. (If left-handed, start from the left side of the piece, moving to the right.)
5. Maintain a consistent travel speed to prevent melt-through in the workpiece.
6. Practice depositing beads approximately 3⁄8″ apart until properly formed beads are produced consistently.

exercise 2-4

Depositing Beads with Filler Metal Using OAW

Conditions

Flat position
Refer to Exercise 2-3, Steps 1 through 5.

Performance

The welder will demonstrate the correct procedure for depositing beads with filler metal.

Criteria

Weld beads should be consistent in width, ripple formation, and penetration, and run parallel to the length of the workpiece.

Procedure

1. Refer to Exercise 2-3, Steps 1 through 5.
2. Maintain a 45° angle to the workpiece with the torch and the filler metal.
3. Dip the filler metal into the weld pool with an in-and-out motion while manipulating the torch.
4. At the end of the weld, withdraw the torch slightly from the workpiece and fill the crater by dipping the filler metal into the weld pool.
5. Practice depositing beads approximately ⅜″ apart until properly formed beads are produced consistently.

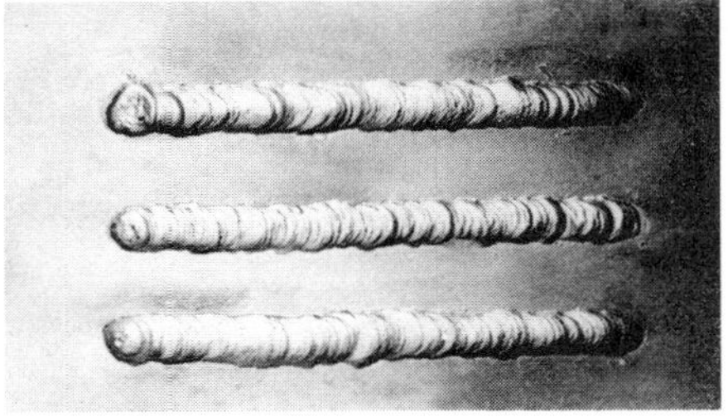

exercise 2-5

Welding a Butt Joint in Flat Position Using OAW

Conditions

Flat position
Refer to Exercise 2-1
Two pieces of 1⁄16″ × 1½″ × 5″ mild steel

Performance

The welder will demonstrate the correct procedure for welding a butt joint in flat position.

Criteria

Welding technique, weld appearance, and weld strength as evaluated by the instructor.

Procedure

1. Tack weld the workpieces to form a butt joint. Allow a gap of 1⁄16″.
2. Use the same motion with the torch and filler metal as practiced when depositing beads with filler metal. Add sufficient filler metal to build up the weld bead approximately 1⁄16″ above the surface of the workpiece.
3. Maintain a weld pool approximately 1⁄4″ to 3⁄8″ wide.
4. Advance the torch approximately 1⁄16″ with each rotation of the torch movement to maintain consistent bead width.
5. Fill the crater at the end of the weld to prevent a weak spot in the weld.

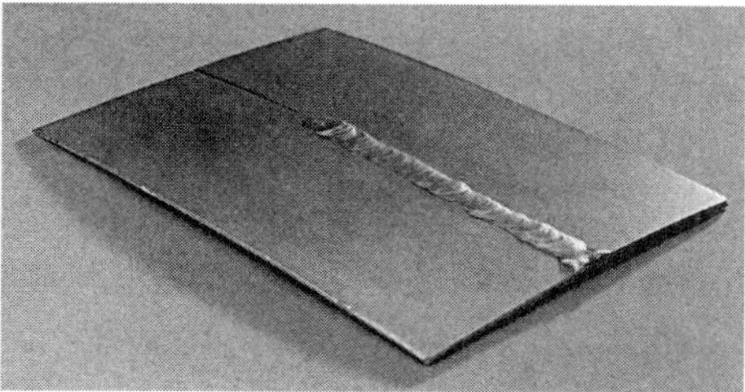

exercise 2-6

Welding a Flange Joint in Flat Position Using OAW

Conditions

Flat position
Refer to Exercise 2-5

Performance

The welder will demonstrate the correct procedure for welding a flange joint in flat position.

Criteria

Welding technique, weld appearance, and weld strength as evaluated by the instructor.

Procedure

1. Prepare the joint by bending the workpieces to obtain a flange that extends above the surface approximately the thickness of the workpiece.
2. Butt the two flanged edges and tack weld.
3. Hold the torch on the starting end until a weld pool is formed.
4. Carefully manipulate the torch to maintain a consistent weld pool across the joint.
5. Withdraw the torch at the end of the joint to prevent burning a hole in the joint.

exercise 2-7

Welding a Corner Joint in Flat Position Using OAW

Conditions

Flat position
Refer to Exercise 2-6, Steps 3 through 5

Performance

The welder will demonstrate the correct procedure for welding a corner joint in flat position.

Criteria

Welding technique, weld appearance, and weld strength as evaluated by the instructor.

Procedure

1. Tack weld the two workpieces to form a corner joint.
2. Refer to Exercise 2-6, Steps 3 through 5.
3. If additional build-up is required, filler metal may be added as the weld pool is carried across the joint.

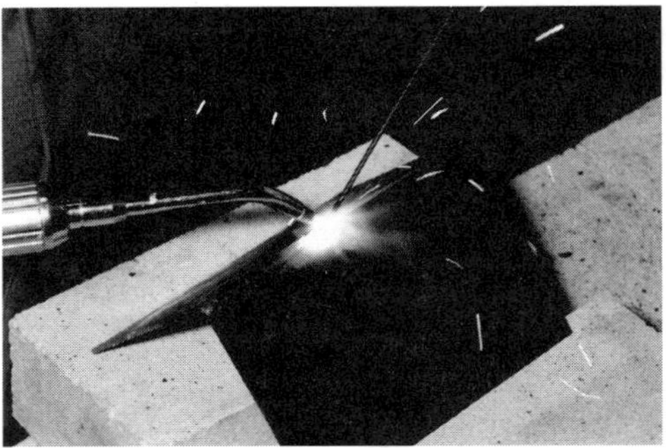

exercise 2-8

Welding a Lap Joint in Flat Position Using OAW

Conditions

Flat position
Refer to Exercise 2-5

Performance

The welder will demonstrate the correct procedure for welding a lap joint in flat position.

Criteria

Welding technique, weld appearance, and weld strength as evaluated by the instructor.

Procedure

1. Tack weld the two workpieces to form a lap joint.
2. Place a firebrick under the bottom workpiece to obtain flat position.
3. Use a semicircular motion with the torch. Direct more of the welding heat to the bottom workpiece by increasing the duration of the torch movement on the bottom workpiece. This will prevent overheating of the top workpiece.
4. Weld one side of the joint and repeat the weld on the other side.

exercise 2-9

Welding a T-Joint in Flat Position Using OAW

Conditions

Flat position
Refer to Exercise 2-5

Performance

The welder will demonstrate the correct procedure for welding a T-joint in flat position.

Criteria

Welding technique, weld appearance, and weld strength as evaluated by the instructor.

Procedure

1. Tack weld the two workpieces to form a T-joint.
2. Place a firebrick under one side to obtain flat position.
3. Hold the torch at a 45° angle to the bottom workpiece.
4. Use the same technique used for welding a lap joint, but direct the heat from the torch equally to both workpieces.
5. Use a semicircular torch movement to distribute the weld metal and prevent undercutting.

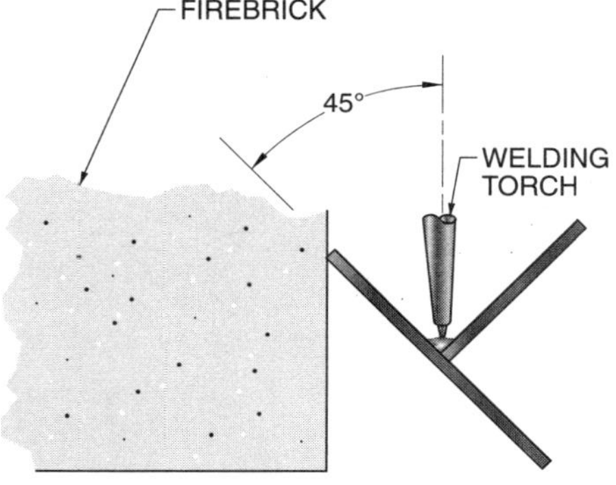

exercise 2-10

Welding a Butt Joint in Horizontal Position Using OAW

Conditions

Horizontal position
Refer to Exercise 2-5, Steps 2 through 5

Performance

The welder will demonstrate the correct procedure for welding a butt joint in horizontal position.

Criteria

Welding technique, weld appearance, and weld strength as evaluated by the instructor.

Procedure

1. Tack weld the two workpieces to form a butt joint. Position the workpiece so the weld joint is in horizontal position.
2. Refer to Exercise 2-5, Steps 2 through 5.
3. To prevent overheating, direct more heat to the bottom workpiece.
4. Dip the filler metal above the center of the weld pool to fill the crater and to prevent undercutting.

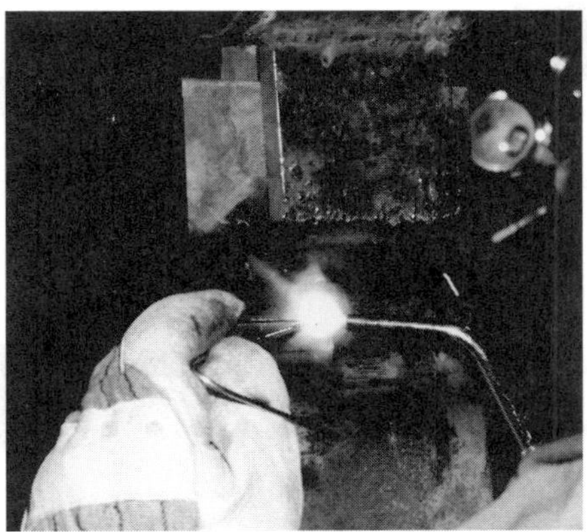

exercise 2-14

Welding a T-Joint in Vertical Position Using OAW

Conditions

Vertical position (uphill)
Refer to Exercise 2-5

Performance

The welder will demonstrate the correct procedure for welding a T-joint in vertical position.

Criteria

Welding technique, weld appearance, and weld strength as evaluated by the instructor.

Procedure

1. Tack weld the two workpieces to form a T-joint. Position the workpiece so the weld joint is in vertical position.
2. Start the weld at the bottom edge and work upward.
3. Hold the torch and filler metal at the same angle as in flat position.
4. Use a crescent motion with the torch to properly fill the joint. Direct the flame more toward the filler metal to control the weld pool and filler metal deposited.

exercise 2-15

Welding a Butt Joint in Overhead Position Using OAW

Conditions

Overhead position
Refer to Exercise 2-5

Performance

The welder will demonstrate the correct procedure for welding a butt joint in overhead position.

Criteria

Welding technique, weld appearance, and weld strength as evaluated by the instructor.

Procedure

1. Tack weld the two workpieces to form a butt joint. Position the workpiece so the weld joint is in overhead position.

2. Use the same technique as for flat position, except use a circular motion with the filler metal. This motion helps distribute the deposited weld and prevents drops of metal from falling from the weld area.

3. Watch the flame closely. If the weld pool begins to run, withdraw the torch slightly.

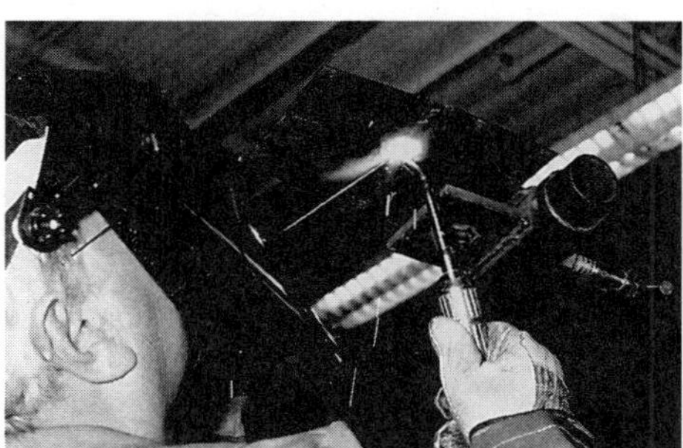

exercises 2-16 through 2-19

Welding Heavy Steel in Flat Position Using OAW

Conditions

Flat position
¼″ × 2″ × 5″ for single-V butt joint and T-joint using the forehand technique
½″ × 2″ × 5″ for double-V butt joint

Performance

The welder will demonstrate the correct procedure for welding the following joints:
Exercise 2-16. Welding a Single-V Butt Joint Using the Forehand Technique
Exercise 2-17. Welding a Single-V Butt Joint Using the Backhand Technique
Exercise 2-18. Welding a Double-V Butt Joint Using the Forehand Technique
Exercise 2-19. Welding a T-Joint Using the Forehand Technique

Criteria

Welding technique, weld appearance, and weld strength as evaluated by the instructor.

Procedure

1. Bevel the edges of the workpieces as required.

2. Tack weld the two workpieces, allowing a 1⁄16″ space.

3. Hold the torch at a 60° angle from vertical rather than the 45° angle used on thinner metals.

4. Complete the welds using the specified technique.

exercise 2-20

Welding Gray Cast Iron Using OAW

Conditions

Flat position
Gray cast iron

Performance

The welder will demonstrate the correct procedure for welding gray cast iron.

Criteria

Welding technique, weld appearance, and weld strength as evaluated by the instructor.

Procedure

1. Prepare the edges to be welded.
2. Preheat the entire workpiece to a dull red.
3. Concentrate the flame near the starting point of welding. When the metal begins to melt, move the flame from side to side to form a pool in the V.
4. Heat the filler metal and dip it into the flux. Insert the fluxed end of the filler metal into the weld pool. Do not dip the filler metal in and out of the weld pool.
5. Move the weld pool as the V is entirely filled. Repeat the operation until the joint is filled across the workpiece.

exercise 2-21

Welding Aluminum in Flat Position Using OAW

Conditions

Flat position
Refer to Exercise 2-1
3⁄16″ × 2″ × 5″ aluminum

Performance

The welder will demonstrate the correct procedure for welding a single-V butt joint and a T-joint.

Criteria

Welding technique, weld appearance, and weld strength as evaluated by the instructor.

Procedure

1. Prepare the edges of the workpieces to be welded. Apply the recommended flux.
2. Pass the flame over the starting point until the flux melts.
3. Determine when welding should begin by scraping the filler metal over the surface of the workpiece.
4. Use a forehand technique, holding the torch 30° above horizontal.
5. The torch should be moved forward without any side-to-side motion.
6. Dip the filler metal in and out of the weld pool with a forward motion.
7. Maintain the same procedure across the weld joint.

Section 3 Activities

Shielded Metal Arc Welding (SMAW)

Name ______________________ **Date** ______________

1. Use arrows to show current flow in the following circuits:

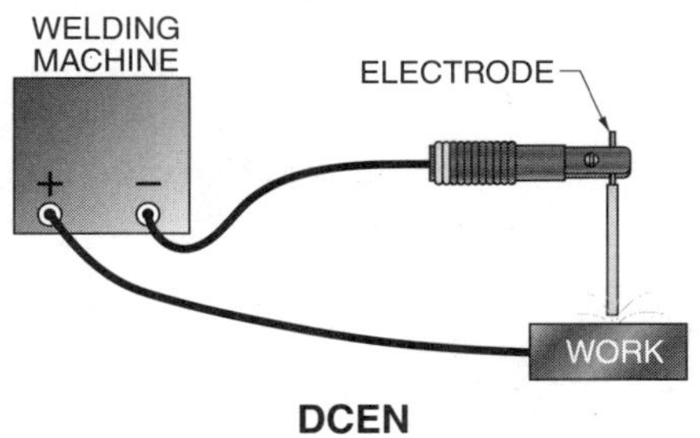

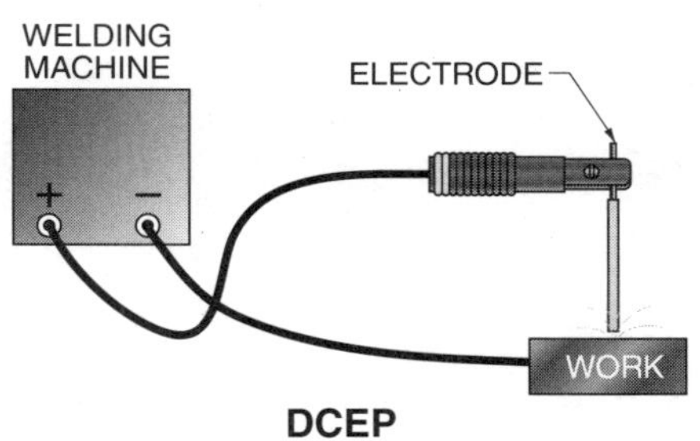

2. Define the following electrical principle terms:

Current: ______________________

Voltage: ______________________

Resistance: ______________________

3. Supply the information pertaining to the electrodes used in shielded metal arc welding (SMAW).

AWS Classification	Current Required (DCEN, DCEP, AC)	Welding Position (All, Flat, Horizontal)	Penetration Characteristics	F Group (F1, F2, F3, F4)
EXX10				
EXX20				
EXX11				
EXX12				
EXX13				
EXX14				
EXX24				
EXX15				
EXX16				
EXX27				
EXX18				
EXX28				

exercises 3-1 through 3-3

Striking the Arc in Shielded Metal Arc Welding (SMAW)

Conditions

Flat position
DCEN
Exercise 3-1. E6012 – ⅛″
Exercise 3-2. E6013 – ⅛″
Exercise 3-3. E7024 – ⅛″
¼″ mild steel

Performance

The welder will strike the arc using the scratching and/or tapping method in the desired location.

Criteria

The arc is struck in the correct location without excess spatter and without the electrode sticking to the plate.

Procedure

1. Practice striking (starting) the arc using the methods shown.
2. Continue practicing until this operation can be performed quickly and easily.

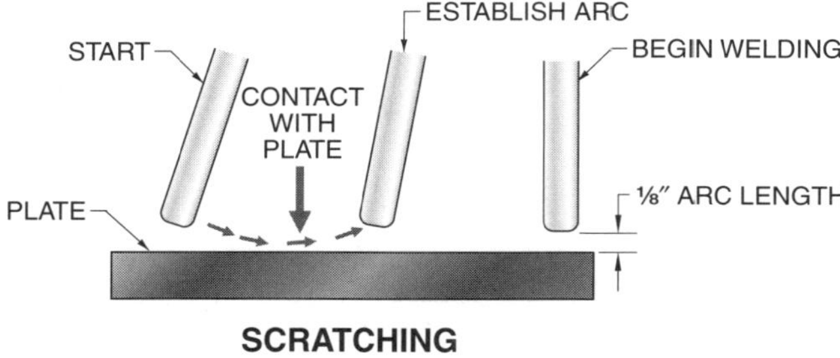

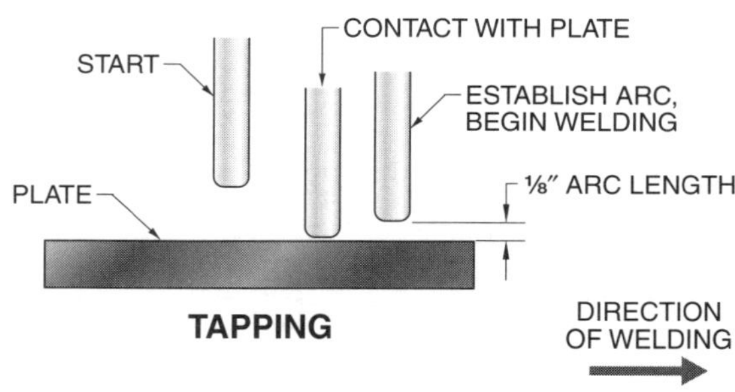

exercises 3-4 through 3-6

Depositing Short Beads Using SMAW

Conditions

Flat position
DCEN
Exercise 3-4. E6012 – ⅛″
Exercise 3-5. E6013 – ⅛″
Exercise 3-6. E7024 – ⅛″
¼″ mild steel

Performance

The welder will demonstrate the correct procedure for depositing short beads.

Criteria

The beads deposited should be approximately ⅜″ in width, and parallel to the length of the workpiece.

Procedure

1. Mark the workpiece with soapstone as shown.
2. Deposit beads using a 90° work angle and a 15° to 25° travel angle.

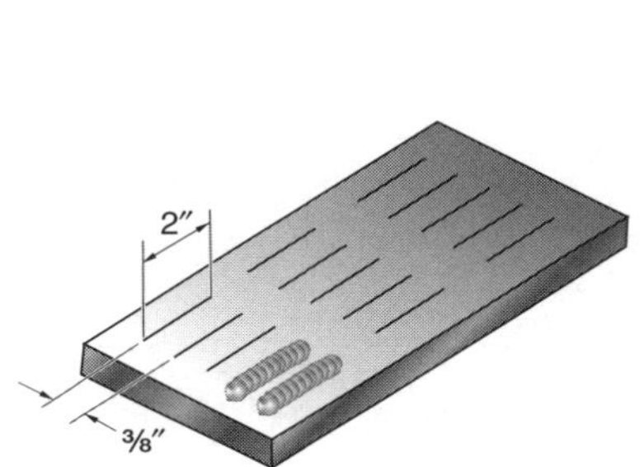

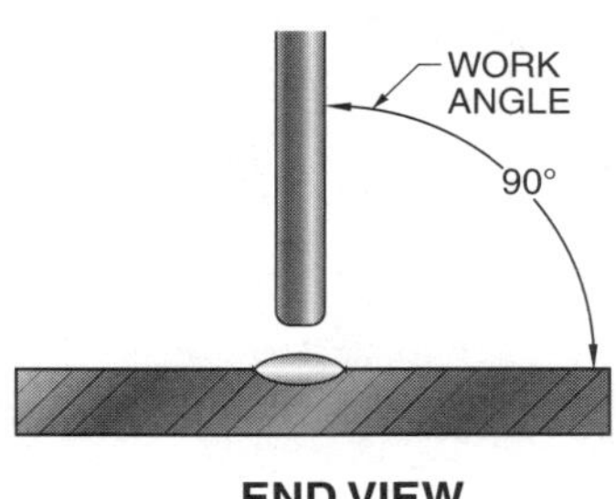

END VIEW

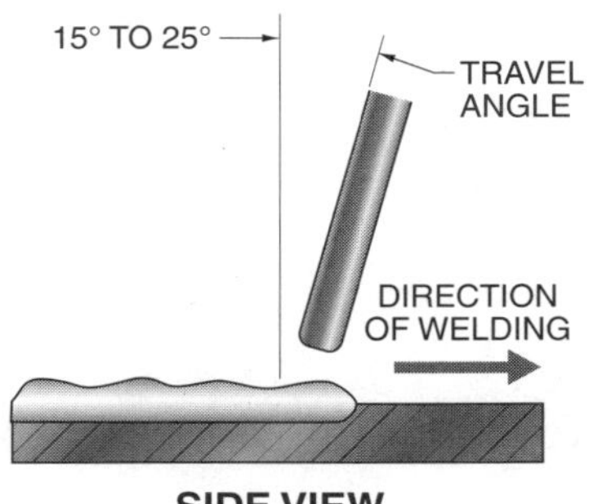

SIDE VIEW

exercises 3-7 through 3-11

Depositing Continuous Beads Using SMAW

Conditions

Flat position
Exercise 3-7. E6010 – ⅛″ DCEP
Exercise 3-8. E6011 – ⅛″ DCEP
Exercise 3-9. E6012 – ⅛″ DCEN
Exercise 3-10. E6013 – ⅛″ DCEN
Exercise 3-11. E7024 – ⅛″ DCEN
¼″ mild steel

Performance

The welder will deposit continuous beads.

Criteria

The beads should be consistent in width and parallel to the length of the workpiece.

Procedure

1. Use a soapstone to mark a series of lines ¾″ apart.
2. Deposit beads using the lines as a guide.
3. Remove the slag and examine the beads for consistent ripple formation.

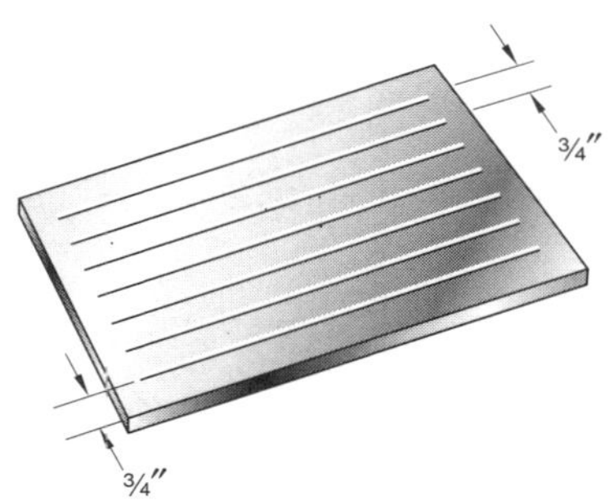

exercises 3-12 through 3-16

Moving the Electrode in Several Directions

Conditions

Flat position
Exercise 3-12. E6010 – ⅛″ DCEP
Exercise 3-13. E6011 – ⅛″ DCEP
Exercise 3-14. E6012 – ⅛″ DCEN
Exercise 3-15. E6013 – ⅛″ DCEN
Exercise 3-16. E7024 – ⅛″ DCEN
¼″ mild steel

Performance

The welder will deposit a continuous bead, moving the electrode in different directions.

Criteria

The bead deposited should show consistent bead formation resulting from correct arc length, work angle, travel angle, and speed of travel.

Procedure

1. Draw lines on the workpiece as shown.
2. Deposit a continuous bead as shown by the direction of the arrows.

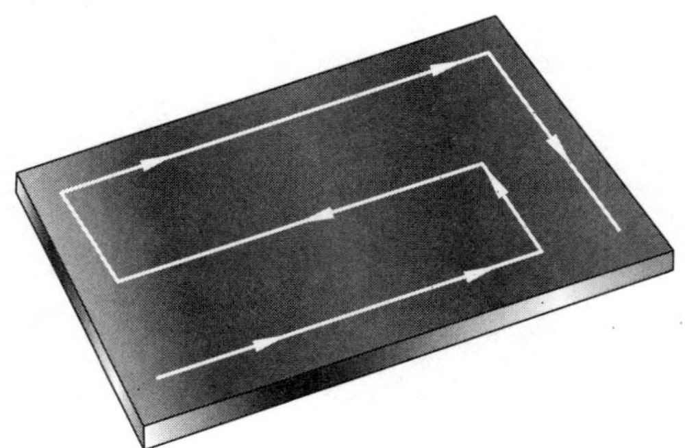

exercises 3-17 through 3-21

Restarting the Arc

Conditions

Flat position
Exercise 3-17. E6010 – ⅛″ DCEP
Exercise 3-18. E6011 – ⅛″ DCEP
Exercise 3-19. E6012 – ⅛″ DCEN
Exercise 3-20. E6013 – ⅛″ DCEN
Exercise 3-21. E7024 – ⅛″ DCEN
¼″ mild steel

Performance

The welder will demonstrate the correct procedure for restarting an arc.

Criteria

Consistent, uniform beads are deposited with the preceding craters properly filled when the arc is restarted.

Procedure

1. Draw a series of straight lines on the workpiece.
2. Divide the lines into 2″ sections.

3. Deposit a bead over the first 2″ section and break the arc.
4. Restart the arc and overlap the first 2″ section to completely fill the crater.

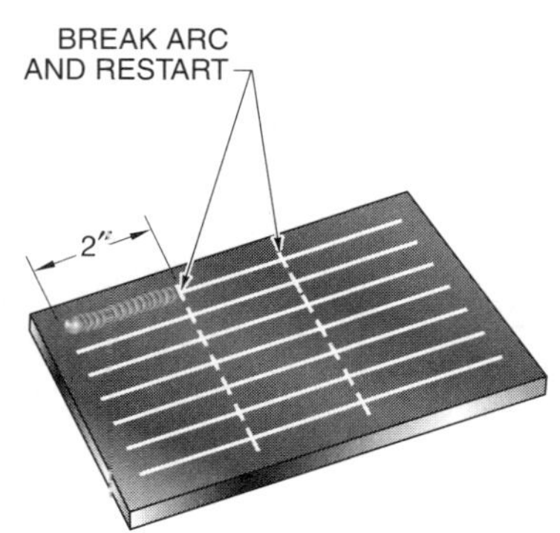

exercises 3-22 through 3-24

Weaving the Electrode

Conditions

Flat position
Exercise 3-22. E6010 – ⅛″ DCEP
Exercise 3-23. E6012 – ⅛″ DCEN
Exercise 3-24. E7024 – ⅛″ DCEN
¼″ workpiece

Performance

The welder will demonstrate the correct technique for weaving the electrode.

Criteria

The deposited bead is consistent in height, ripple shape, and penetration.

Procedure

1. Lay out a series of lines parallel to the length of the workpiece at 1″ intervals.
2. Deposit straight beads over the lines.

3. Use a weaving motion to deposit a bead between the straight beads on the workpiece.
4. Use a crescent, figure eight, or rotary weaving motion as necessary with the electrodes used.

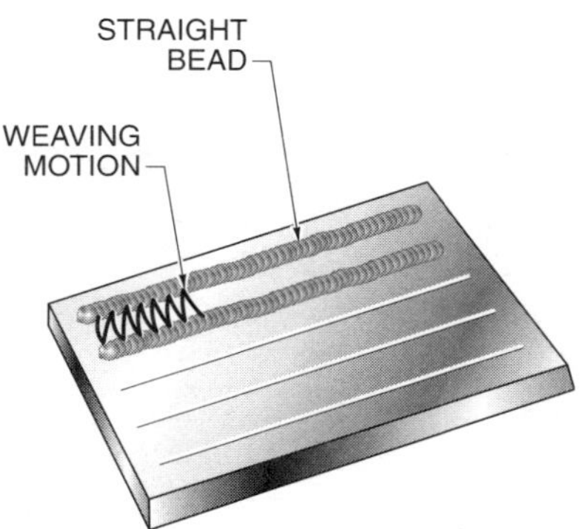

exercise **3-25**

Surfacing

Conditions

Flat position
E6010 – ⅛″ DCEP, layer 1
E6011 – ⅛″ DCEP, layer 2
E6012 – ⅛″ DCEN, layer 3
E6013 – ⅛″ DCEN, layer 4
E7024 – ⅛″ DCEN, layer 5
⅜″ or thicker mild steel

Performance

The welder will deposit successive overlapping beads to increase the thickness of the workpiece.

Criteria

Each bead should overlap the previous bead by half, and completely penetrate the previous bead and the base metal with no voids in the weld metal applied.

Procedure

1. Deposit a bead along the edge of the workpiece and parallel to the length.
2. Remove the slag and deposit the next bead, which should penetrate half of the width of the previous bead and the base metal.

3. Repeat the operation until the entire workpiece is filled.
4. Using the designated electrode, deposit beads as described in Steps 1 and 2, parallel to the width of the workpiece.
5. Complete layers two through five, alternating the deposited beads by 90° for each layer.

exercises 3-26 through 3-30

Welding a Butt Joint in Flat Position Using SMAW

Conditions

Flat position
Exercise 3-26. E6010 – ⅛″ DCEP
Exercise 3-27. E6011 – ⅛″ DCEP
Exercise 3-28. E6012 – ⅛″ DCEN
Exercise 3-29. E6013 – ⅛″ DCEN
Exercise 3-30. E7024 – ⅛″ DCEN
Two pieces of 3⁄16″ to ¼″ mild steel

Performance

The welder will demonstrate the correct procedure for welding a butt joint in flat position.

Criteria

Welding technique, weld appearance, and weld strength as evaluated by the instructor.

Procedure

1. Form a butt joint and tack weld together.
2. Position the workpiece so the weld joint is in flat position.
3. Using a 90° work angle and a 15° to 30° travel angle, deposit a bead along the joint.
4. Let cool and repeat the procedure on the reverse side.

WORK ANGLE

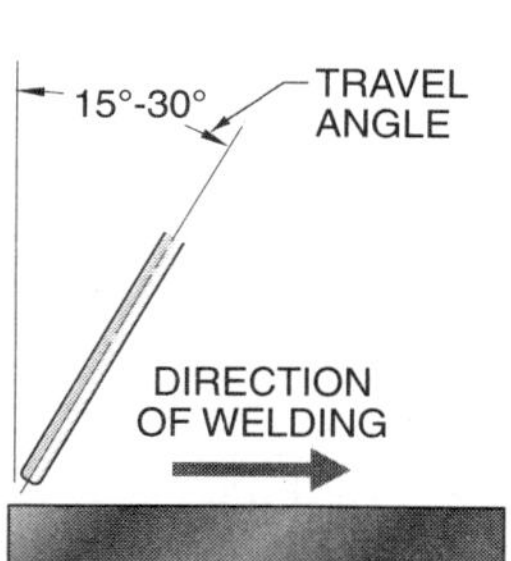

TRAVEL ANGLE

exercises 3-31 through 3-35

Welding an Open Butt Joint in Flat Position Using SMAW

Conditions

Flat position
Exercise 3-31. E6010 – ⅛″ DCEP
Exercise 3-32. E6011 – ⅛″ DCEP
Exercise 3-33. E6012 – ⅛″ DCEN
Exercise 3-34. E6013 – ⅛″ DCEN
Exercise 3-35. E7024 – ⅛″ DCEN
Two pieces of 3⁄16″ to ¼″ mild steel

Performance

The welder will demonstrate the correct procedure for welding an open butt joint in flat position.

Criteria

Welding technique, weld appearance, and weld strength as evaluated by the instructor.

Procedure

1. Tack weld the two pieces to form a butt joint with a ⅛″ root opening. (Use the bare end of a ⅛″ diameter electrode for consistent spacing.)
2. Refer to Exercises 3-26 through 3-30 for procedure.
3. Examine the butt weld. The weld bead should penetrate completely to the underside of the workpiece.

exercises 3-36 through 3-40

Welding a Lap Joint in Flat Position Using SMAW

Conditions

Flat position
Exercise 3-36. E6010 – ⅛″ DCEP
Exercise 3-37. E6011 – ⅛″ DCEP
Exercise 3-38. E6012 – ⅛″ DCEN
Exercise 3-39. E6013 – ⅛″ DCEN
Exercise 3-40. E7024 – ⅛″ DCEN
Two pieces of 3⁄16″ to ¼″ mild steel

Performance

The welder will demonstrate the correct procedure for welding a lap joint in flat position.

Criteria

Welding technique, weld appearance, and weld strength as evaluated by the instructor.

Procedure

1. Tack weld the two pieces to form a T-joint.
2. Position the workpiece so the joint is in flat position, or rest it against a firebrick to obtain flat position.
3. Use a 45° work angle and a 15° to 30° travel angle to deposit the bead.
4. Advance the electrode as the proper fillet contour is formed.

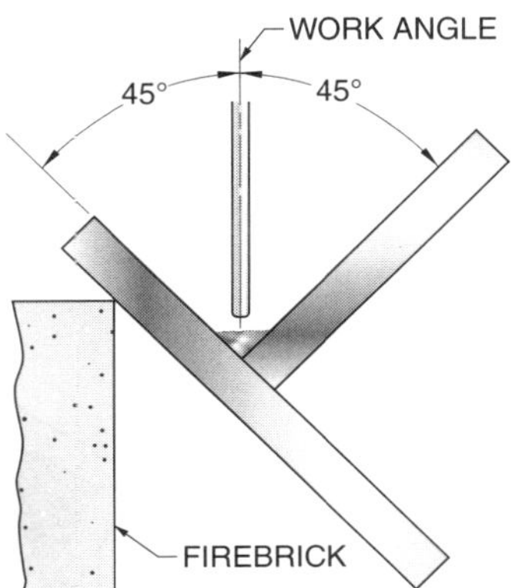

exercises 3-51 through 3-55

Welding a Multiple-Pass T-Joint in Flat Position Using SMAW

Conditions

Flat position
Exercise 3-51. E6010 – ⅛″ DCEP
Exercise 3-52. E6011 – ⅛″ DCEP
Exercise 3-53. E6012 – ⅛″ DCEN
Exercise 3-54. E6013 – ⅛″ DCEN
Exercise 3-55. E7024 – ⅛″ DCEN
Two pieces of 3⁄16″ to ¼″ mild steel

Performance

The welder will demonstrate the correct procedure for welding a multiple-pass T-joint.

Criteria

Welding technique, weld appearance, and weld strength as evaluated by the instructor.

Procedure

1. Refer to Exercises 3-46 through 3-50, and deposit a single pass on both sides of the T-joint.
2. Remove the slag.

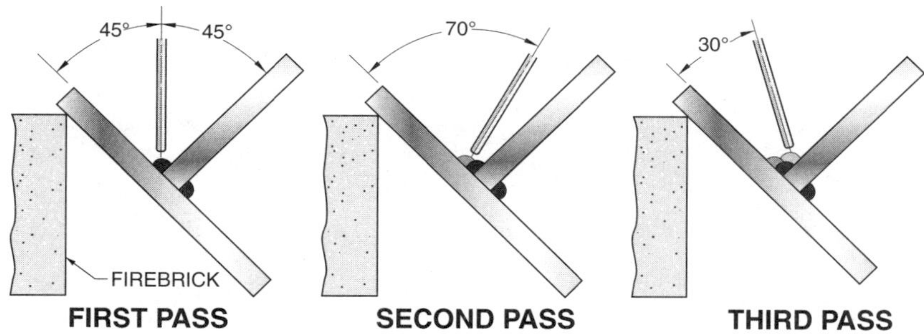

3. Deposit the second pass using a 70° work angle and a 15° to 30° travel angle. Cover the root pass by one-half to two-thirds and fuse into the base metal evenly along the toe. Remove the slag.
4. Deposit the third pass using a 30° work angle and a 15° to 30° travel angle. Cover the second pass by approximately one-third and fuse into the base metal evenly along the toe. Remove the slag.
5. Deposit welds on the other side of the joint.
6. If more weld metal is required, additional passes may be deposited.

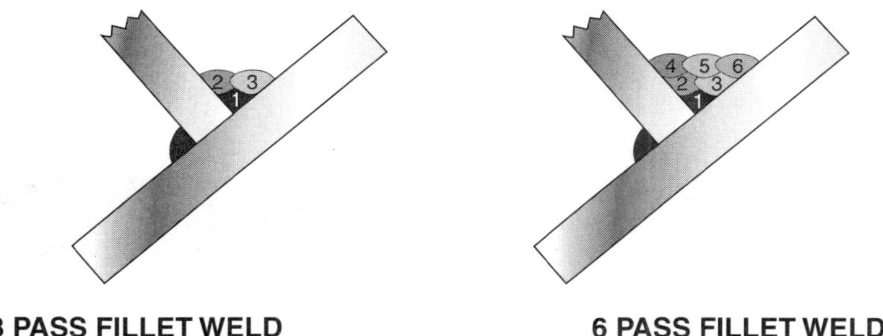

exercises **3-56** through **3-60**

Welding a Multiple-Pass T-Joint in Flat Position with a Cover Pass Using SMAW

Conditions

Flat position
Exercise 3-56. E6010 – ⅛″ DCEP
Exercise 3-57. E6011 – ⅛″ DCEP
Exercise 3-58. E6012 – ⅛″ DCEN

Exercise 3-59. E6013 – ⅛″ DCEN
Exercise 3-60. E7024 – ⅛″ DCEN
Two pieces of 3⁄16″ to ¼″ mild steel

Performance

The welder will demonstrate the correct procedure for welding a multiple-pass T-joint in flat position with a cover pass.

Criteria

Welding technique, weld appearance, and weld strength as evaluated by the instructor.

Procedure

1. Refer to Exercises 3-51 through 3-55 and deposit a single pass on both sides of the T-joint.
2. Remove the slag.
3. Deposit the second pass using a 70° work angle and a 15° to 30° travel angle. Cover the root pass by one-half to two-thirds and fuse into the base metal evenly along the toe. Remove the slag.
4. Deposit the third pass using a 30° work angle and a 15° to 30° travel angle. Cover the second pass by approximately one-third and fuse into the base metal evenly along the toe. Remove the slag.
5. Use a weaving motion to deposit the cover pass on the joint. Pause slightly at the toes of the weld to prevent undercutting.
6. Deposit welds on the other side of the joint.

exercises 3-61 through 3-65

Welding a Corner Joint in Flat Position Using SMAW

Conditions

Flat position
Exercise 3-61. E6010 – ⅛″ DCEP
Exercise 3-62. E6011 – ⅛″ DCEP
Exercise 3-63. E6012 – ⅛″ DCEN
Exercise 3-64. E6013 – ⅛″ DCEN
Exercise 3-65. E7024 – ⅛″ DCEN
Two pieces of 3⁄16″ to ¼″ mild steel

Performance

The welder will demonstrate the correct procedure for welding a corner joint in flat position.

Criteria

Welding technique, weld appearance, and weld strength as evaluated by the instructor.

Procedure

1. Tack weld two pieces to form an open corner joint.
2. Position the workpiece so the joint is in flat position.
3. Use a 45° work angle and a 15° to 30° travel angle to deposit a fillet weld in the joint.
4. If additional weld metal is required, deposit a second pass using a weaving motion.

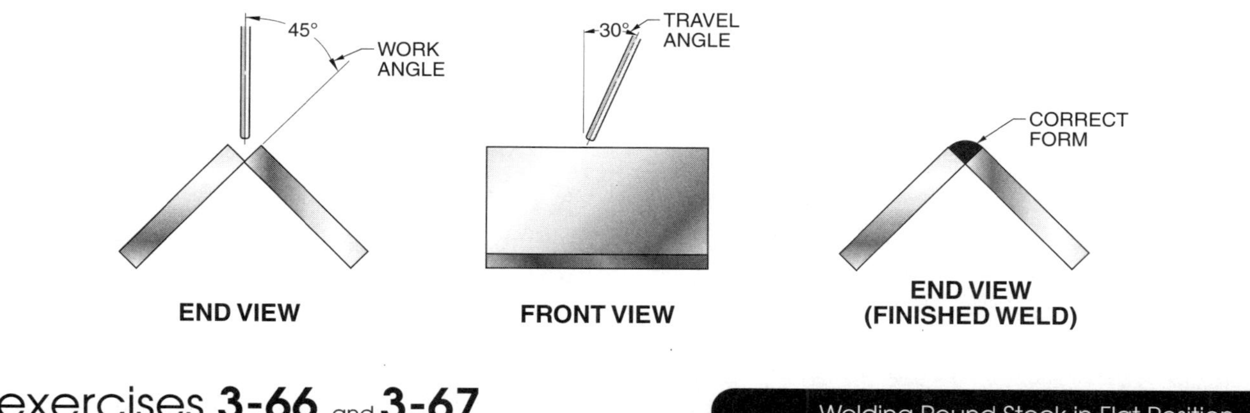

exercises 3-66 and 3-67

Welding Round Stock in Flat Position Using SMAW

Conditions

Flat position
Exercise 3-66. E6010 – ⅛″ DCEP
Exercise 3-67. E6013 – ⅛″ DCEN
Round stock

Performance

The welder will demonstrate the correct procedure for welding round stock in flat position.

Criteria

Welding technique, weld appearance, and weld strength as evaluated by the instructor.

Procedure

1. Bevel both ends of the round stock, leaving a root face in the center. Both sides should have the same groove angle.
2. Place the workpieces in a vise or a section of angle iron to maintain the required welding position.
3. Tack weld both sides.
4. Deposit a root pass on one side, then the other side.
5. Use a weaving motion to fill the groove completely.

exercises 3-68 through 3-70

Depositing Beads in Horizontal Position Using SMAW

Conditions

Horizontal position
Exercise 3-68. E6010 – ⅛″ DCEP
Exercise 3-69. E6012 – ⅛″ DCEN
Exercise 3-70. E7018 – ⅛″ DCEP
¼″ × 4″ × 6″ mild steel

Performance

The welder will deposit beads in horizontal position.

Criteria

The beads should be 5⁄16″ to ⅜″ wide and parallel to the edge of the workpiece.

Procedure

1. Draw a series of lines ½″ apart and parallel to the length of the workpiece.
2. Position the workpiece so the weld joint is in horizontal position.
3. Using the guidelines drawn, deposit straight, consistent beads with a 5° to 10° work angle and a 5° to 20° travel angle. Remove the slag from each bead.

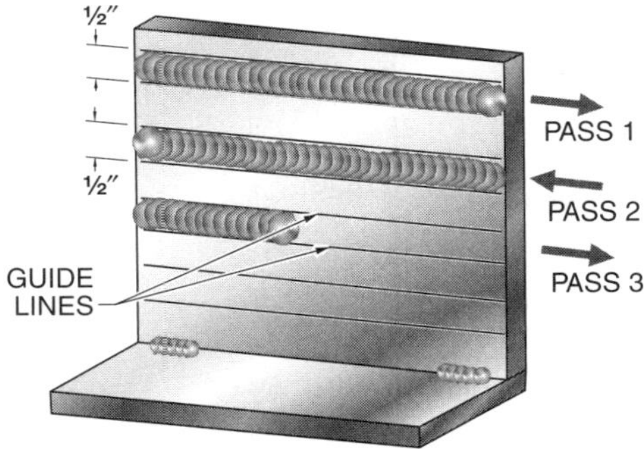

exercises 3-71 through 3-73

Surfacing in Horizontal Position

Conditions

Horizontal position
Exercise 3-71. E6010 – ⅛″ DCEP
Exercise 3-72. E6013 – ⅛″ DCEP
Exercise 3-73. E7018 – ⅛″ DCEP
¼″ × ⅜″ × 3″ or 4″ mild steel

Performance

The welder will deposit successive overlapping beads to increase the thickness of the workpiece.

Criteria

Each bead should overlap the previous bead by half and completely penetrate the previous bead and the base metal with no voids in the weld metal applied.

Procedure

1. Position the workpiece in horizontal position.
2. Deposit a bead along the edge of the workpiece and parallel to its length.
3. Remove the slag and deposit the next bead, which should penetrate half the width of the previous bead and the base metal.
4. Repeat the operation until the entire workpiece is filled.

exercise 3-74

Welding a Lap Joint in Horizontal Position Using SMAW

Conditions

Horizontal position
Adjust to correct current
E6010 – ⅛″ DCEP
Two pieces of ¼″ × 1½″ × 6″ mild steel

Performance

The welder will demonstrate the correct procedure for welding a lap joint in horizontal position.

Criteria

Welding technique, weld appearance, and weld strength as evaluated by the instructor.

exercises 3-77 through 3-81

Depositing Beads in Vertical Position (Downhill) Using SMAW

Conditions

Vertical position (downhill)
Exercise 3-77. E6010 – ⅛″ DCEP
Exercise 3-78. E6011 – ⅛″ DCEP
Exercise 3-79. E6012 – ⅛″ DCEP
Exercise 3-80. E6013 – ⅛″ DCEP
Exercise 3-81. E7048 – ⅛″ DCEP

Performance

The welder will deposit beads in vertical position using the downhill technique.

Criteria

The deposited bead should be consistent in height, ripple shape, and penetration.

Procedure

1. Draw a series of guidelines parallel to the length of the workpiece.
2. Start at the top using a 90° work angle and a 15° to 30° travel angle to deposit straight beads.
3. Maintain a short arc length.

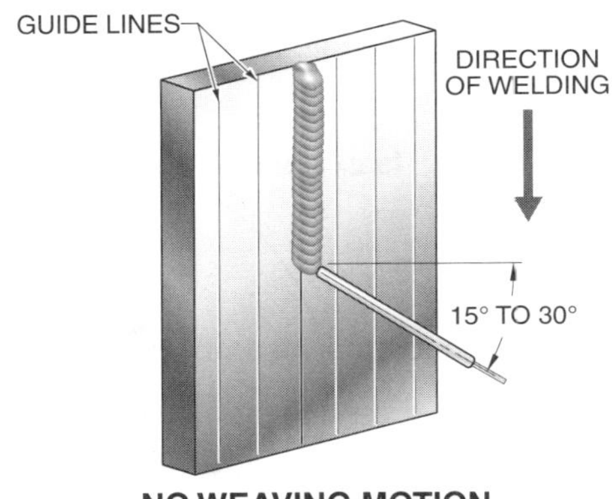

NO WEAVING MOTION

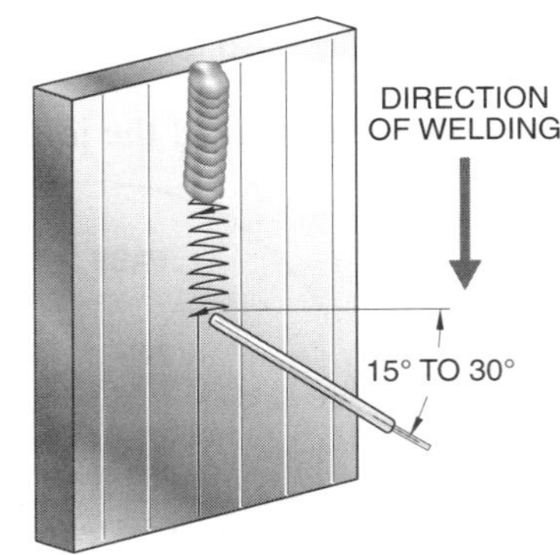

SLIGHT WEAVING MOTION

exercises 3-82 through 3-84

Depositing Beads in Vertical Position (Uphill) Using SMAW

Conditions

Vertical position (uphill)
Exercise 3-82. E6010 – ⅛″ DCEP

Exercise 3-83. E6011 – ⅛″ DCEP
Exercise 3-84. E7018 – ⅛″ DCEP

Performance

The welder will deposit beads in vertical position using the uphill technique.

Criteria

The deposited bead should be consistent in height, ripple shape, and penetration.

Procedure

1. Draw a series of guidelines parallel to the length of the workpiece.
2. Start at the bottom using a 90° work angle. To control the heat and bead formation, use a whipping motion with the E6010 and E6011 electrodes. Use a slight side-to-side weaving motion with the E7018 electrode.
3. Maintain a short arc length.

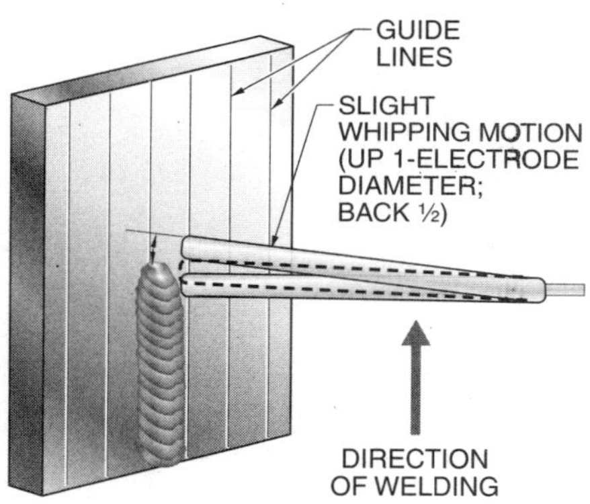

exercises **3-85** and **3-86**

Welding a Lap Joint in Vertical Position (Uphill) Using SMAW

Conditions

Vertical position (uphill)
Exercise 3-85. E6010 – ⅛″ DCEP
Exercise 3-86. E7018 – ⅛″ DCEP
Two pieces of ¼″ × 2″ × 4″ mild steel

Performance

The welder will demonstrate the correct procedure for welding a two-pass lap joint in vertical position (uphill).

Criteria

Welding technique, weld appearance, and weld strength as evaluated by the instructor.

Procedure

1. Tack weld the two pieces to form a lap joint. Position the workpiece so the joint is in vertical position.
2. Use a 45° work angle and a whipping motion (E6010), or a slight weaving motion (E7018), to deposit the root pass.
3. Remove the slag and deposit the second pass using a weaving motion. Pause at the toes of the weld to eliminate undercutting.

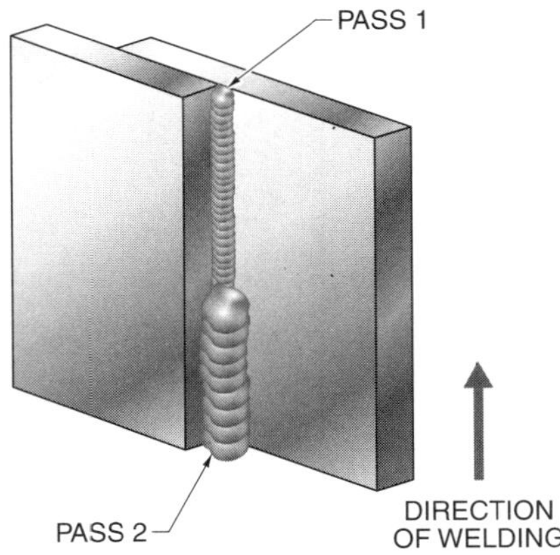

exercises 3-87 and 3-88

Welding a Single-V-Groove Butt Joint in Vertical Position (Uphill) Using SMAW

Conditions

Vertical position (uphill)
Exercise 3-87. E6010 – ⅛″ DCEP
Exercise 3-88. E7018 – ⅛″ DCEP
Two pieces of ¼″ × 4″ × 6″ mild steel

Performance

The welder will demonstrate the correct procedure for welding a single-V butt joint in vertical position (uphill).

Criteria

Welding technique, weld appearance, and weld strength as evaluated by the instructor.

Procedure

1. Bevel one side of each plate 30° along its length. Tack weld the two pieces to form a single-V-groove joint with a 60° groove angle. Allow a 1⁄16″ root opening.
2. Position the workpiece so the joint is in vertical position.
3. Deposit the root pass and penetrate completely to the root side. Remove the slag. Use a whipping motion (E6010) or a push motion (E7018).
4. Increase the current slightly and deposit the second pass, completely penetrating the root pass and the base metal. Remove the slag.
5. Reduce the current and deposit the third pass (cover pass) with a weaving motion.

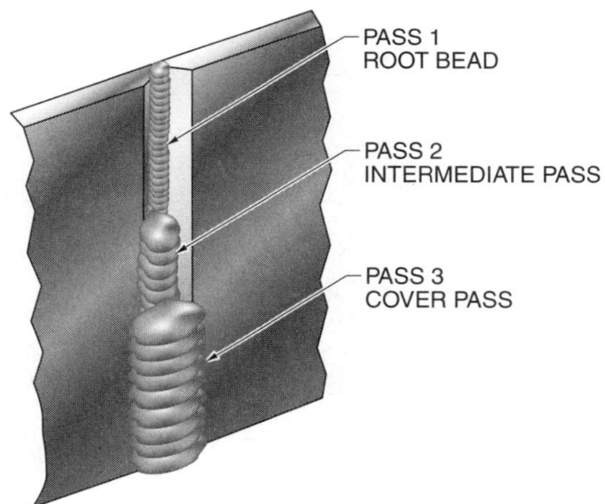

exercises 3-89 and 3-90

Welding a Multiple-Pass T-Joint in Vertical Position (Uphill) Using SMAW

Conditions

Vertical position (uphill)
Exercise 3-89. E6010 – 1⁄8″ DCEP
Exercise 3-90. E7018 – 1⁄8″ DCEP
Two pieces of 1⁄4″ × 4″ × 6″ mild steel

Performance

The welder will demonstrate the correct procedure for welding a multiple-pass T-joint in vertical position (uphill).

Criteria

Welding technique, weld appearance, and weld strength as evaluated by the instructor.

Procedure

1. Tack weld the two pieces to form a T-joint. Position the workpiece so the joint is in vertical position.
2. Deposit the root pass using a 45° work angle. Remove the slag.
3. Deposit the second pass using a 70° work angle. Remove the slag.
4. Deposit the third pass using a 30° work angle. Remove the slag.

exercises 3-91 through 3-93

Depositing Beads in Overhead Position Using SMAW

Conditions

Overhead position
Exercise 3-91. E6010 – ⅛″ DCEP
Exercise 3-92. E6011 – ⅛″ DCEP
Exercise 3-93. E7018 – ⅛″ DCEP
¼″ or ⅜″ mild steel

Performance

The welder will deposit beads in overhead position.

Criteria

The deposited bead should be consistent in height, ripple shape, and penetration.

Procedure

1. Draw a series of guidelines ½″ apart, parallel to the length of the workpiece.
2. Position the workpiece in overhead position.
3. Reduce the current as recommended and deposit straight beads over the guidelines. Reverse the direction.
4. After depositing beads across the workpiece, fill in the space between the beads using a weaving motion.

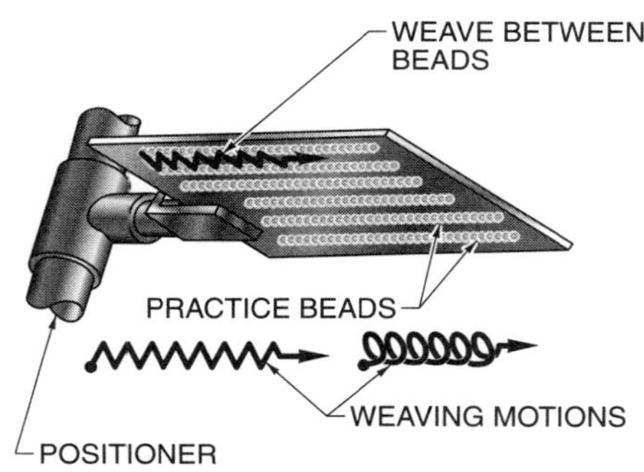

exercises 3-94 and 3-95

Welding a Multiple-Pass Lap Joint in Overhead Position Using SMAW

Conditions

Overhead position
Exercise 3-94. E6010 – ⅛″ DCEP
Exercise 3-95. E7018 – ⅛″ DCEP
Two pieces of ¼″ × 2″ × 4″ mild steel

Performance

The welder will demonstrate the correct procedure for welding a multiple-pass lap joint in overhead position.

Criteria

Welding technique, weld appearance, and weld strength as evaluated by the instructor.

Procedure

1. Tack weld two pieces to form a lap joint. Position the workpiece so the joint is in overhead position. Reduce the current as recommended.
2. Use a 45° work angle and a 15° travel angle to deposit the first pass on both sides of the joint. Remove the slag.
3. Deposit the second pass on the bottom toe of the root pass. Cover the root pass by one-half to two-thirds and fuse into the base metal evenly along the toe. Adjust the work angle as necessary. Remove the slag.
4. Deposit the third pass on the top toe of the root pass. Cover the second pass by approximately one-third and fuse into the top edge of the joint by about 1⁄16″. Adjust the work angle as necessary. Remove the slag.
5. Deposit welds on the other side of the joint.

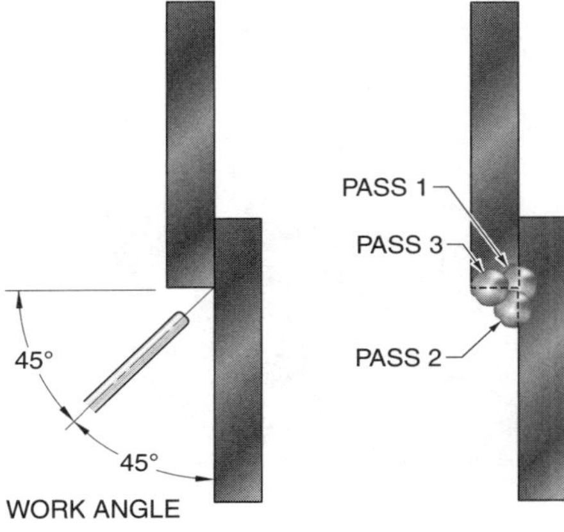

END VIEW

exercises 3-96 and 3-97

Welding a Multiple-Pass T-Joint in Overhead Position Using SMAW

Conditions

Overhead position
Exercise 3-96. E6010 – ⅛″ DCEP
Exercise 3-97. E7018 – ⅛″ DCEP
Two pieces of ¼″ × 2″ × 4″ mild steel

Performance

The welder will demonstrate the correct procedure for welding a multiple-pass T-joint in the overhead position.

Criteria

Welding technique, weld appearance, and weld strength as evaluated by the instructor.

Procedure

1. Tack weld two pieces to form a T-joint. Position the workpiece so the joint is in overhead position. Reduce the current as recommended.
2. Use a 45° work angle and a 5° to 15° travel angle to deposit the first pass on both sides of the joint. Remove the slag.
3. Deposit the second pass on the bottom toe of the root pass. Cover the root pass by one-half to two-thirds and fuse into the base metal evenly along the toe. Adjust the work angle as necessary. Remove the slag.
4. Deposit the third pass. Cover the second pass by approximately one-third and fuse into the base metal along the toe. Remove the slag.
5. Deposit welds on the other side of the joint.

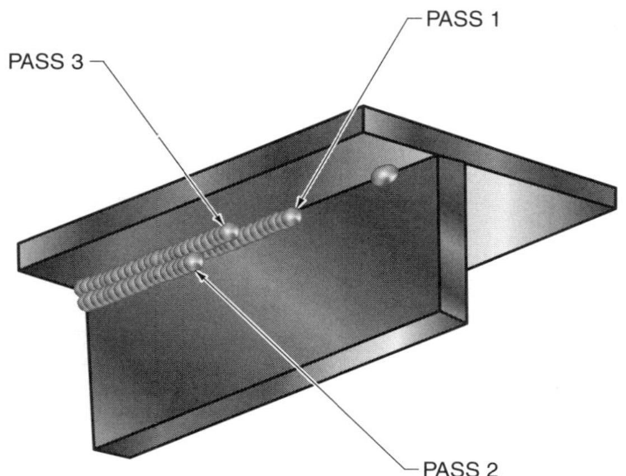

exercises 3-98 and 3-99

Welding a Single-V-Groove Butt Joint in Overhead Position Using SMAW

Conditions

Overhead position
Exercise 3-98. E6010 – ⅛″ DCEP
Exercise 3-99. E6011 – ⅛″ DCEP
Two pieces of ¼″ × 2″ × 4″ mild steel

Performance

The welder will demonstrate the correct procedure for welding a single-V butt joint in overhead position.

Criteria

Welding technique, weld appearance, and weld strength as evaluated by the instructor.

Procedure

1. Bevel one side of each plate 30° along its length. Tack weld the two pieces to form a single-V-groove butt joint with a 60° groove angle and a 1⁄16″ root opening.
2. Position the workpiece so the joint is in overhead position.
3. Deposit the root pass and obtain complete penetration to the root side. Remove the slag.
4. Deposit two additional passes to cover the groove faces of the joint. Each weld should fuse into the edges of the joint by approximately 1⁄16″. Remove the slag after depositing each weld.

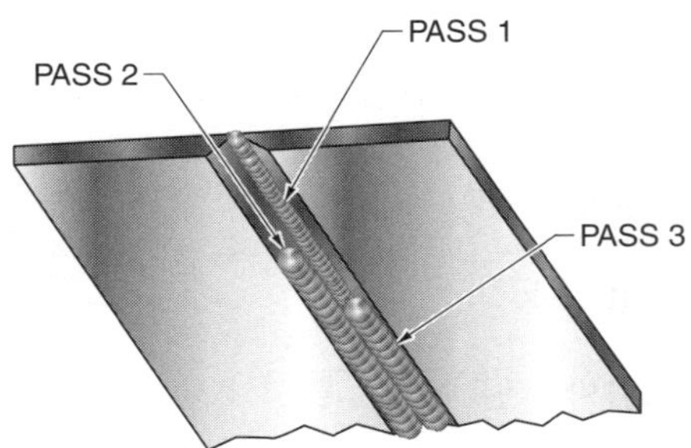

exercise 3-100

Welding a Multiple-Pass Single-V-Groove Joint with Backing in Horizontal Position

SENSE LEVEL I QUALIFICATION PRACTICE TEST

Conditions

Horizontal position
DCEP
E7018 – 3⁄32″
E7018 – 1⁄8″
Two pieces of 3⁄8″ × 6″ × 7″ mild steel
One piece of 1⁄4″ × 1½″ × 9″ mild steel for backing

Performance

The welder will demonstrate the correct procedure for welding a multiple-pass single-V-groove joint with a backing strip in the horizontal position.

Criteria

The weld passes visual inspection and a standard AWS Guided Bend Test.

Procedure

1. Bevel one side of each plate 22.5° along its length.
2. Obtain a 1⁄4″ × 1½″ × 9″ mild steel strip for backing. Tack the plates to the backing strip to form a single-V-groove joint with a 1⁄4″ root opening and a 45° groove angle. The backing strip should extend beyond the joint to form 1″ starting and runoff tabs.
3. Set the current between 80 A and 90 A for the 3⁄32″ E7018 electrode.
4. Position the workpiece so the joint is in horizontal position.
5. Deposit the first of two root passes on the lower plate with a 35° work angle above horizontal and a 5° to 10° drag angle. Deposit the second root pass with a 35° work angle below horizontal. Use a steady drag, and remove the slag after each weld.
6. Deposit a two-bead intermediate layer with a 3⁄32″ E7018 electrode using a steady drag. Remove slag after each pass.
7. Switch to 1⁄8″ E7018 and increase the current to between 105 A and 125 A.
8. Deposit a second two-bead intermediate layer. Remove slag after each pass.
9. Deposit a three-bead cover layer and remove slag after each pass. Visually inspect the weld.
10. Cut two 1½″ test straps from the workpiece, remove the backing, and prepare the straps for bending. Make one face bend and one root bend. Inspect specimens for acceptability.

exercise 3-101

Welding a Multiple-Pass Single-V-Groove Joint with Backing in Vertical Position (Uphill)

SENSE LEVEL I QUALIFICATION PRACTICE TEST

Conditions

Vertical position
Refer to Exercise 3-100

Performance

The welder will demonstrate the correct procedure for welding a multiple-pass single-V-groove joint with a backing strip in vertical position.

Criteria

The weld passes visual inspection and a standard AWS Guided Bend Test.

Procedure

1. Refer to Exercise 3-100 and prepare a single-V-groove joint with backing.
2. Position the workpiece so the joint is in the vertical position.
3. Start at the bottom of the joint and deposit a root pass and an intermediate pass with a 3⁄32″ E7018 electrode. Weave from side to side, moving up the joint in 1⁄8″ increments. Clean each weld thoroughly to prevent slag inclusions.
4. Deposit a second intermediate pass and a cover pass with a 1⁄8″ E7018 electrode.
5. Refer to Exercise 3-100, Step 10.

advanced exercise 3-102

Making a Multiple-Pass Fillet Weld, T-Joint, 2F Position, Using SMAW with Stainless Steel Electrodes

Conditions

Position: 2F
Materials: 1/4″ × 2″ × 6″ mild steel
Electrodes: 1/8″ E309-15 or -16
Polarity: DCEP
Amperage: 75 to 115
Voltage: 19 to 20

Performance

The welder will demonstrate the correct procedure and technique for depositing a weld in a T-joint in the horizontal 2F position.

Criteria

Welding technique, weld appearance, and weld strength will be evaluated by the instructor.

Procedure

1. Tack weld two pieces to form a T-joint. Position workpiece so joint is in 2F position.
2. Use a 45° work angle and a 5° to 10° drag travel angle to deposit root pass on both sides of joint. Remove slag.
3. Deposit second pass on bottom toe of root pass with a 50° to 55° work angle and a 5° to 10° drag travel angle. Cover root by one-half to two-thirds. Fuse into base metal evenly along toe. Remove slag.
4. Deposit third pass on top toe of root pass with a 30° to 35° work angle and a 5° to 10° drag travel angle. Cover second pass by one-third to one-half. Fuse into base metal evenly along toe. Remove slag.
5. Deposit welds on other side of joint.

Additional Practice

Repeat exercise using a Z-weave motion for second and third passes. Pause at each toe to ensure complete fusion into sides of joint.

advanced exercise **3-103**

Making a Multiple-Pass Fillet Weld, T-Joint, 3F Position, Using SMAW with Stainless Steel Electrodes

Conditions

Position: 3F
Materials: 1/4″ × 2″ × 4″ mild steel
Electrodes: 1/8″ E309-15 or -16
Polarity: DCEP
Amperage: 75 to 115
Voltage: 19 to 20

Performance

Refer to Exercise 3-102.

Criteria

Refer to Exercise 3-102.

Procedure

1. Tack weld two pieces to form a T-joint. Position workpiece so joint is in 3F position.
2. Use a 45° work angle and a 5° to 10° push travel angle to deposit root pass on both sides of joint. Remove slag.
3. Deposit second pass on either toe of root pass with a 50° to 55° work angle and a 5° to 10° push travel angle. Cover root by one-half to two-thirds. Fuse into base metal evenly along toe. Remove slag.
4. Deposit third pass on opposite toe of root pass with a 30° to 35° work angle and a 5° to 10° push travel angle. Cover second pass by one-third to one-half. Fuse into base metal evenly along toe. Remove slag.
5. Deposit welds on other side of joint.

Additional Practice

Repeat exercise using a Z-weave motion for second and third passes. Pause at each toe to ensure complete fusion into sides of joint.

advanced exercise **3-104**

Making a Multiple-Pass Fillet Weld, T-Joint, 4F Position, Using SMAW with Stainless Steel Electrodes

Conditions

Position: 4F
Materials: 1/4″ × 2″ × 6″ mild steel
Electrodes: 1/8″ E309-15 or -16
Polarity: DCEP
Amperage: 75 to 115
Voltage: 19 to 20

Performance

Refer to Exercise 3-102.

Criteria

Refer to Exercise 3-102.

Procedure

1. Tack weld two pieces to form a T-joint. Position workpiece so joint is in 4F position.
2. Use a 45° work angle and a 5° to 10° drag travel angle to deposit root pass on both sides of joint. Remove slag.
3. Deposit second pass on bottom toe of root pass with a 50° to 55° work angle and a 5° to 10° drag travel angle. Cover root by one-half to two-thirds. Fuse into base metal evenly along toe. Remove slag.
4. Deposit third pass on top toe of root pass with a 30° to 35° work angle and a 5° to 10° drag travel angle. Cover second pass by one-third to one-half. Fuse into base metal evenly along toe. Remove slag.
5. Deposit welds on other side of joint.

Additional Practice

Repeat exercise using a Z-weave motion for second and third passes. Pause at each toe to ensure complete fusion into sides of joint.

advanced exercise **3-105**

Making Multiple-Pass Groove Welds, 3G Position with and without Backing, Using SMAW with Stainless Steel Electrodes

Conditions

Position: 3G
Materials: 2 pieces 1″ × 3″ × 7″ mild steel

Backing: ¼″ × 1½″ × 9″ mild steel
Electrodes: 3⁄32″ E309-15 or -16
1⁄8″ E309-15 or -16
Polarity: DCEP
Amperage: 40 to 80 for 3⁄32″, 75 to 115 for 1⁄8″
Voltage: 17 to 19 for 3⁄32″, 19 to 20 for 1⁄8″

Performance

The welder will demonstrate the correct procedure and technique for depositing welds in single-V groove joints in the vertical 3G position.

Criteria

The welds will pass visual inspection and AWS side bend tests.

Procedure with Backing

1. Prepare a single-V groove joint with a 60° groove angle and backing strip. Tack assembly together with a ¼″ root opening. Position workpiece so joint is in 3G position.
2. Use a 90° work angle and a 5° to 10° push travel angle. Move up joint while pushing electrode into joint to deposit root pass. Remove slag.
3. Deposit second pass using same electrode angles and weaving motion. Pause briefly at each toe to fuse into edges of joint. Remove slag.
4. Change to 1⁄8″ electrodes and increase amperage between 75 and 115. Deposit remaining passes with same electrode angles and weaving motion. Remove slag after each pass.
5. Face reinforcement should be 1⁄8″ maximum.
6. Root reinforcement should be 1⁄16″ maximum.
7. Prepare best weld for a side bend test.

Procedure without Backing

1. Prepare a single-V groove open root joint with a 60° groove angle and a 3⁄32″ to 1⁄8″ root face. Tack assembly together with a 3⁄32″ to 1⁄8″ root opening. Position workpiece so joint is in 3G position.
2. Use a 90° work angle and a 5° to 10° push travel angle. Move up joint while pushing electrode into joint to deposit root pass. Remove slag.
3. Use a weaving motion to deposit remaining passes. Switch to 1⁄8″ electrodes after second pass, and increase amperage between 75 and 115. Remove slag after each weld.
4. Face reinforcement should be 1⁄8″ maximum.
5. Root reinforcement should be 1⁄16″ maximum.
6. Prepare best weld for a side bend test.

advanced exercise **3-106**

Making Multiple-Pass Groove Welds, 4G Position with and without Backing, Using SMAW with Stainless Steel Electrodes

Conditions

Position: 4G
Materials: 2 pieces 1″ × 3″ × 7″ mild steel
Backing: ¼″ × 1½″ × 9″ mild steel
Electrodes: ³⁄₃₂″ E309-15 or -16
⅛″ E309-15 or -16
Polarity: DCEP
Amperage: 40 to 80 for ³⁄₃₂″, 75 to 115 for ⅛″
Voltage: 17 to 19 for ³⁄₃₂″, 19 to 20 for ⅛″

Performance

The welder will demonstrate the correct procedure and technique for depositing welds in single-V groove joints in the overhead 4G position.

Criteria

The welds will pass visual inspection and AWS side bend tests.

Procedure with Backing

1. Prepare a single-V groove joint with a 60° groove angle and backing strip. Tack assembly together with a ¼″ root opening. Position workpiece so joint is in 4G position.
2. Use a 90° work angle and a 5° to 10° drag travel angle to deposit root pass. Move along joint with a steady drag pushing electrode into joint. Remove slag.
3. Deposit second pass using same electrode angles and a steady drag travel.
4. Change to ⅛″ electrodes and increase amperage between 75 and 115. Deposit remaining passes with same electrode angles and drag motion. Remove slag after each pass.
5. Face reinforcement should be ⅛″ maximum.
6. Root reinforcement should be ¹⁄₁₆″ maximum.
7. Prepare best weld for a side bend test.

Procedure without Backing

1. Prepare a single-V groove open root joint with a 60° groove angle and a ³⁄₃₂″ to ⅛″ root face. Tack assembly together with a ³⁄₃₂″ to ⅛″ root opening. Position workpiece so joint is in 4G position.
2. Use a 90° work angle and a 5° to 10° drag travel angle to deposit root pass. Move along joint pushing electrode into joint to obtain complete penetration. Remove slag.

3. Deposit second pass using same electrode angles and drag travel. Use a steady drag travel along joint to obtain complete penetration.

4. Change to ⅛″ electrodes and increase amperage between 75 and 115. Deposit remaining passes with same electrode angles and drag motion. Remove slag after each pass.

5. Face reinforcement should be ⅛″ maximum.

6. Root reinforcement should be 1⁄16″ maximum.

7. Prepare best weld for a side bend test.

advanced exercise **3-107**

Making a Multiple-Pass Fillet Weld, 2F Position, Carbon Steel Pipe to Plate Using SMAW with Steel Electrodes

Conditions

Position: 2F
Materials: 4″ to 6″ schedule 40 to 80 carbon steel pipe
¼″ carbon steel plate to fit
Electrodes: ⅛″ E6010 or E6011
⅛″ E7018
Polarity: DCEP
Amperage: 75 to 125
Voltage: 19 to 20

Performance

The welder will demonstrate the correct procedure and technique for depositing a weld to weld pipe to plate in the 2F position.

Criteria

The weld will pass visual inspection criteria.

Procedure

1. Tack weld pipe to plate. Position workpiece so joint is in 2F position.

2. Use a 45° work angle and a 5° to 10° push travel angle to deposit root pass on both sides of joint. Remove slag.

3. Deposit second pass on bottom toe of root pass with a 50° to 55° work angle and a 5° to 10° push travel angle. Cover root by one-half to two-thirds. Fuse into base metal evenly along toe. Remove slag.

4. Deposit third pass on top toe of root pass with a 30° to 35° work angle and a 5° to 10° push travel angle. Cover second pass by one-third to one-half. Fuse into base metal evenly along toe. Remove slag.

Additional Practice

Repeat exercise using a Z-weave motion for second and third passes. Pause at each toe to ensure complete fusion into sides of joint.

advanced exercise **3-108**

Making a Multiple-Pass Fillet Weld, 4F Position, Carbon Steel Pipe to Plate, Using SMAW with Mild Steel Electrodes

Conditions

Position:	4F
Materials:	4″ to 6″ schedule 40 to 80 carbon steel pipe
	¼″ carbon steel plate to fit
Electrodes:	⅛″ E6010 or E6011
	⅛″ E7018
Polarity:	DCEP
Amperage:	75 to 125
Voltage:	19 to 20

Performance

The welder will demonstrate the correct procedure and technique for depositing a weld to weld pipe to plate in the 4F position.

Criteria

The weld will pass visual inspection criteria.

Procedure

1. Tack weld pipe to plate. Position workpiece so joint is in 4F position.
2. Use a 45° work angle and a 5° to 10° drag travel angle to deposit root pass on both sides of joint. Remove slag.
3. Deposit second pass on bottom toe of root pass with a 50° to 55° work angle and a 5° to 10° drag travel angle. Cover root by one-half to two-thirds. Fuse into base metal evenly along toe. Remove slag.
4. Deposit third pass on top toe of root pass with a 30° to 35° work angle and a 5° to 10° drag travel angle. Cover second pass by one-third to one-half. Fuse into base metal evenly along toe. Remove slag.

Additional Practice

Repeat exercise using a Z-weave motion for second and third passes. Pause at each toe to ensure complete fusion into sides of joint.

advanced exercise **3-109**

Making a Multiple-Pass Fillet Weld, 5F Position, Carbon Steel Pipe to Plate, Using SMAW with Mild Steel Electrodes

Conditions

Position: 5F
Materials: 4″ to 6″ schedule 40 to 80 carbon steel pipe
¼″ carbon steel plate to fit
Electrodes: ⅛″ E6010 or E6011
⅛″ E7018
Polarity: DCEP
Amperage: 75 to 125
Voltage: 19 to 20

Performance

The welder will demonstrate correct procedure and technique for depositing a weld to weld pipe to plate in the 5F position.

Criteria

The weld will pass visual inspection criteria.

Procedure

1. Tack weld pipe to plate. Position workpiece so joint is in 5F position.
2. Use a 45° work angle and a 5° to 10° push travel angle to deposit root pass on both sides of joint. Remove slag.
3. Deposit second pass on bottom toe of root pass with a 50° to 55° work angle and a 5° to 10° push travel angle. Cover root by one-half to two-thirds. Fuse into base metal evenly along toe. Remove slag.
4. Deposit third pass on top toe of root pass with a 30° to 35° work angle and a 5° to 10° push travel angle. Cover second pass by one-third to one-half. Fuse into base metal evenly along toe. Remove slag.

Additional Practice

Repeat exercise using a Z-weave motion for second and third passes. Pause at each toe to ensure complete fusion into sides of joint.

advanced exercise **3-110**

Making Multiple-Pass Groove Welds, 2G Position, on Carbon Steel Pipe with and without Backing, Using SMAW with Mild Steel Electrodes

Conditions

Position: 2G
Materials: 4″ to 6″ schedule 40 to 80 carbon steel pipe
Backing ring to fit
Electrodes: ⅛″ E6010 or E6011
⅛″ E7018
Polarity: DCEP
Amperage: 75 to 125
Voltage: 19 to 20

Performance

The welder will demonstrate the correct procedure and technique for depositing welds to weld pipe to plate in the 2G position.

Criteria

The welds will pass visual inspection and AWS side bend tests.

Procedure with Backing

1. Bevel two pieces of pipe with a 37.5° groove face and no root face. Insert backing ring and tack weld to form a single-V groove joint. Position workpiece so joint is in 2G position.
2. Use a 90° work angle and a 5° to 10° drag travel angle. Move around joint while pushing electrode into joint to deposit root pass. Remove slag.
3. Deposit second pass on bottom toe of root pass with a 50° to 55° work angle and a 5° to 10° drag travel angle. Cover root by one-half to two-thirds. Fuse into base metal evenly along toe. Remove slag.
4. Deposit third pass on top toe of root pass with a 30° to 35° work angle and a 5° to 10° drag travel angle. Cover second pass by one-third to one-half. Fuse into base metal evenly along toe. Remove slag.
5. Face reinforcement should be ⅛″ maximum.
6. Root reinforcement should be 1⁄16″ maximum.
7. Prepare best weld for a side bend test.

Procedure without Backing

1. Prepare a single-V groove open root joint, 3⁄32″ to ⅛″ root face and 3⁄32″ to ⅛″ root opening. Tack and position workpiece so joint is in 2G position.
2. Use a 90° work angle and a 5° to 10° drag travel angle. Move around joint with a slight whipping motion to deposit root pass. Remove slag.

3. Deposit second pass on bottom toe of root pass with a 50° to 55° work angle and a 5° to 10° drag travel angle. Cover root by one-half to two-thirds. Fuse into base metal evenly along toe. Remove slag.

4. Deposit third pass on top toe of root pass with a 30° to 35° work angle and a 5° to 10° drag travel angle. Cover second pass by one-third to one-half. Fuse into base metal evenly along toe. Remove slag.

5. Face reinforcement should be ⅛″ maximum.

6. Root reinforcement should be 1⁄16″ maximum.

7. Prepare best weld for a side bend test.

advanced exercise **3-111**

Making Multiple-Pass Groove Welds, 5G Position, on Carbon Steel Pipe with and without Backing, Using SMAW with Mild Steel Electrodes

Conditions

Position: 5G
Materials: 4″ to 6″ schedule 40 to 80 carbon steel pipe
Backing ring to fit
Electrodes: ⅛″ E6010 or E6011
⅛″ E7018
Polarity: DCEP
Amperage: 75 to 125
Voltage: 19 to 20

Performance

The welder will demonstrate the correct procedure and technique for depositing welds to weld pipe to plate in the 5G position.

Criteria

The welds will pass visual inspection and AWS side bend tests.

Procedure with Backing

1. Bevel two pieces of pipe with a 37.5° groove face and no root face. Insert backing ring and tack weld to form a single-V groove joint. Position workpiece so joint is in 5G position.

2. Use a 90° work angle and a 5° to 10° push travel angle to deposit root pass. Move up joint while pushing electrode into joint. Remove slag.

3. Deposit second pass on either toe of root pass with a 50° to 55° work angle and a 5° to 10° push travel angle. Cover root by one-half to two-thirds. Fuse into base metal evenly along toe. Remove slag.

4. Deposit third pass on opposite toe of root pass with a 30° to 35° work angle and a 5° to 10° push travel angle. Cover second pass by one-third to one-half. Fuse into base metal evenly along toe. Remove slag.

5. Face reinforcement should be ⅛″ maximum.
6. Root reinforcement should be 1⁄16″ maximum.
7. Prepare best weld for a side bend test.

Procedure without Backing

1. Prepare a single-V groove open root joint, 3⁄32″ to ⅛″ root face, and 3⁄32″ to ⅛″ root opening. Tack and position workpiece so joint is in 5G position.
2. Use a 90° work angle and a 5° to 10° push travel angle to deposit root pass. Move up joint with a slight whipping motion. Remove slag.
3. Deposit second pass on either toe of root pass with a 50° to 55° work angle and a 5° to 10° push travel angle. Cover root by one-half to two-thirds. Fuse into base metal evenly along toe. Remove slag.
4. Deposit third pass on opposite toe of root pass with a 30° to 35° work angle and a 5° to 10° drag travel angle. Cover second pass by one-third to one-half. Fuse into base metal evenly along toe. Remove slag.
5. Face reinforcement should be ⅛″ maximum.
6. Root reinforcement should be 1⁄16″ maximum.
7. Prepare best weld for a side bend test.

advanced exercise **3-112**

Making Multiple-Pass Groove Welds, 6G Position, on Carbon Steel Pipe with and without Backing, Using SMAW with Mild Steel Electrodes

Conditions

Position: 6G
Materials: 4″ to 6″ schedule 40 to 80 carbon steel pipe
Backing ring to fit
Electrodes: ⅛″ E6010 or E6011
⅛″ E7018
Polarity: DCEP
Amperage: 75 to 125
Voltage: 19 to 20

Performance

The welder will demonstrate the correct procedure and technique for depositing welds to weld pipe to plate in the 6G position.

Criteria

The welds will pass visual inspection and AWS side bend tests.

Procedure with Backing

1. Bevel two pieces of pipe with a 37.5° groove face and no root face. Insert backing ring and tack weld to form a single-V groove joint. Position workpiece so joint is in 6G position.
2. Use a 90° work angle and a 5° to 10° push travel angle to deposit root pass. Move up joint while pushing electrode into joint. Remove slag.
3. Deposit second pass on bottom toe of root pass with a 50° to 55° work angle and a 5° to 10° push travel angle. Cover root by one-half to two-thirds. Fuse into base metal evenly along toe. Remove slag.
4. Deposit third pass on top toe of root pass with a 30° to 35° work angle and a 5° to 10° push travel angle. Cover second pass by one-third to one-half. Fuse into base metal evenly along toe. Remove slag.
5. Face reinforcement should be ⅛″ maximum.
6. Root reinforcement should be 1⁄16″ maximum.
7. Prepare best weld for a side bend test.

Procedure without Backing

1. Prepare a single-V groove open root joint, 3⁄32″ to ⅛″ root face and 3⁄32″ to ⅛″ root opening. Tack and position workpiece so joint is in 6G position.
2. Use a 90° work angle and a 5° to 10° push travel angle to deposit root pass. Move up joint with a slight whipping motion. Remove slag.
3. Deposit second pass on bottom toe of root pass with a 50° to 55° work angle and a 5° to 10° push travel angle. Cover root by one-half to two-thirds. Fuse into base metal evenly along toe. Remove slag.
4. Deposit third pass on top toe of root pass with a 30° to 35° work angle and a 5° to 10° push travel angle. Cover second pass by one-third to one-half. Fuse into base metal evenly along toe. Remove slag.
5. Face reinforcement should be ⅛″ maximum.
6. Root reinforcement should be 1⁄16″ maximum.
7. Prepare best welds with and without backing for side bend tests

Name ____ **Date** ________________________

exercise 4-1

Depositing Beads on Mild Steel in Flat Position Using GTAW without Filler Metal

Conditions

Flat position
DCEN
50 A to 60 A
3/32″, EWTh-2, EWLa-1.5, or EWCe-2 tapered tungsten electrode with 1/8″ to 3/16″ electrode stickout
100% argon shielding gas, 15 cfh to 20 cfh with 15-second postpurge
High frequency set to start (automatic)
16-gauge, 4″ × 6″

Performance

The welder will deposit continuous beads without filler metal.

Criteria

The beads should be approximately 1/8″ wide and parallel to the length of the workpiece.

Procedure

1. To start the arc, activate the current flow by pushing the foot pedal or turning on the switch at the torch.
2. Position the torch at a 45° angle, with the electrode 1/8″ from the workpiece.
3. When the arc starts, raise the torch to a 90° work angle and a 20° push angle.
4. Maintain a short arc while depositing beads approximately 3/8″ apart. Hold the torch in position at the end of each bead until the postflow times out.

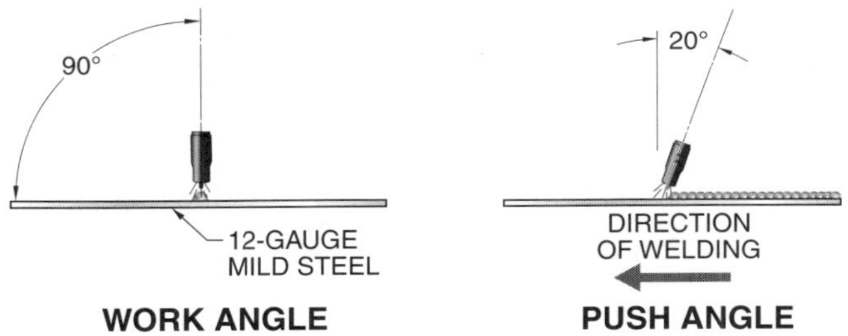

exercise 4-2

Depositing Beads on Mild Steel in Flat Position Using GTAW with Filler Metal

Conditions

Flat position
Refer to Exercise 4-1
1⁄16″ recommended filler metal

Performance

The welder will deposit continuous beads using filler metal.

Criteria

The beads should be approximately 3⁄16″ wide, convex in shape, and parallel to the length of the workpiece.

Procedure

1. Refer to Exercise 4-1.
2. While maintaining the arc, hold the filler metal at a 20° angle. Dip the filler metal into the leading edge of the weld pool using an in-and-out motion.
3. Use a small rotary motion to form a bead approximately 3⁄16″ wide.
4. Deposit a series of straight, consistent beads 3⁄8″ apart along the workpiece. Hold the torch in position at the end of each bead until the postflow times out.

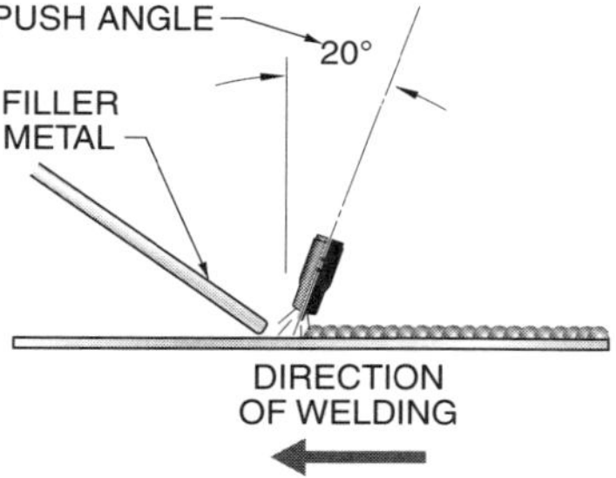

exercise 4-3

Welding a Butt Joint on Mild Steel in Flat Position Using GTAW

Conditions

Flat position
Refer to Exercise 4-1
Two pieces of 16-gauge, 1½″ × 6″
1⁄16″ recommended filler metal

Performance

The welder will demonstrate the correct procedure for welding a butt joint in flat position.

Criteria

Welding technique, weld appearance, and weld strength as evaluated by the instructor.

Procedure

1. Tack weld the two pieces of steel to form a butt joint with a 1⁄32″ root opening.
2. Use the same procedure for depositing beads with filler metal, using the joint as the center of the weld across the workpiece. Hold the torch in position at the end of the weld until the postflow times out.
3. Penetration should be complete, with a bead width of approximately 3⁄16″.

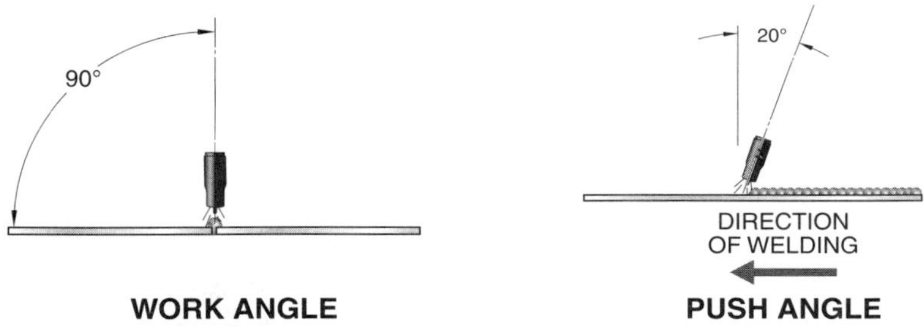

exercise 4-4

Welding an Open Corner Joint on Mild Steel in Flat Position Using GTAW

Conditions

Flat position
Refer to Exercise 4-1
35 A to 45 A
Two pieces of 16-gauge, 1½″ × 4″ mild steel
1⁄16″ filler metal

Performance

The welder will demonstrate the correct procedure for depositing a fillet weld in an outside corner joint with and without filler metal.

Criteria

The weld with filler metal should be about 1 to 1½ electrode diameters wide. The autogenous weld should be about one electrode diameter wide. Both welds should have complete penetration through to the root side of the joint.

Procedure

1. Tack the two pieces to form an open corner joint and place it in the flat position.
2. The work angle is 90° with a 15° to 20° push angle. The filler metal is raised 20° above the joint.
3. Activate the foot/hand remote control to start the arc.
4. Deposit the weld using the dip technique. Dip the filler metal into the leading edge of the weld pool and withdraw it while moving the torch with a steady push. Keep the end of the filler metal in the flow of shielding gas.
5. At the end of the joint, fill the crater and hold the torch in position until the postflow times out.
6. Repeat the exercise using the lay-wire technique. Hold the filler metal in the joint with a slight downward pressure and move the torch with a steady push.
7. Repeat the exercise without filler metal. Once the weld pool forms, move the torch with a steady push.
8. Repeat each exercise using pulsed current.

exercise 4-5

Welding a Lap Joint on Mild Steel in Horizontal Position Using GTAW

Conditions

Horizontal position
Refer to Exercise 4-1
Two pieces of 16-gauge, 1½″ × 6″
1⁄16″ recommended filler metal

Performance

The welder will demonstrate the correct procedure for welding a lap joint in horizontal position.

Criteria

Welding technique, weld appearance, and weld strength as evaluated by the instructor.

Procedure

1. Tack weld the two pieces of steel to form a lap joint.
2. Hold the torch at an 80° to 85° work angle and a 20° push angle.
3. Add filler metal using an in-and-out motion while directing the arc to the bottom workpiece.
4. Maintain a consistent bead along the workpiece. Hold the torch in position at the end of the weld until the postflow times out.

5. Repeat the exercise using the lay-wire technique.
6. Repeat the exercise without filler metal.
7. Repeat each exercise using pulsed current.

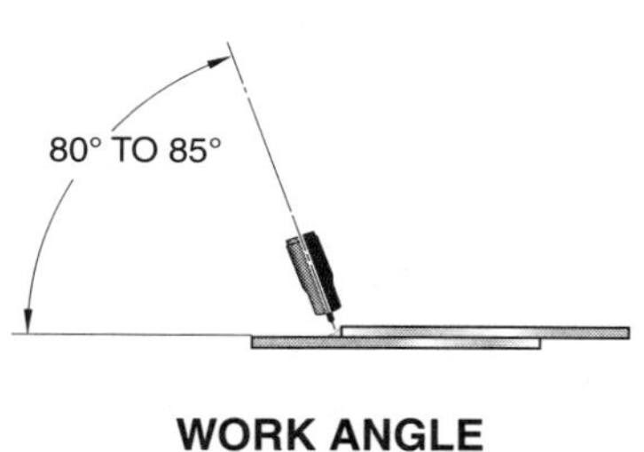

WORK ANGLE

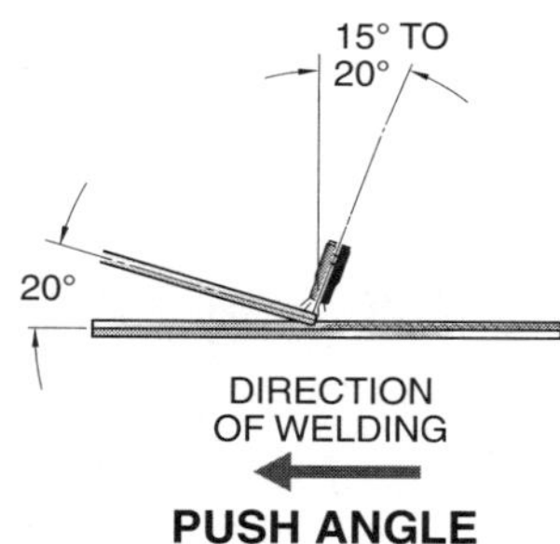

PUSH ANGLE

exercise 4-6

Welding a T-Joint on Mild Steel in Horizontal Position Using GTAW

Conditions

Horizontal position
Refer to Exercise 4-1
Two pieces of 16-gauge, 1½″ × 6″
1⁄16″ filler metal

Performance

The welder will demonstrate the correct procedure for welding a T-joint in horizontal position.

Criteria

Welding technique, weld appearance, and weld strength as evaluated by the instructor.

Procedure

1. Tack weld the two pieces to form a T-joint.
2. Hold the torch at a 45° work angle and a 10° to 15° push angle.
3. Add filler metal by holding it at a 20° angle from the bottom plate and a 20° angle from the vertical plate.
4. Weave the torch slightly while adding filler metal with an in-and-out motion.

5. Maintain a consistent bead along the workpiece. Hold the torch in position at the end of the weld until the postflow times out.
6. Repeat the exercise using the lay-wire technique.
7. Repeat each exercise using pulsed current.

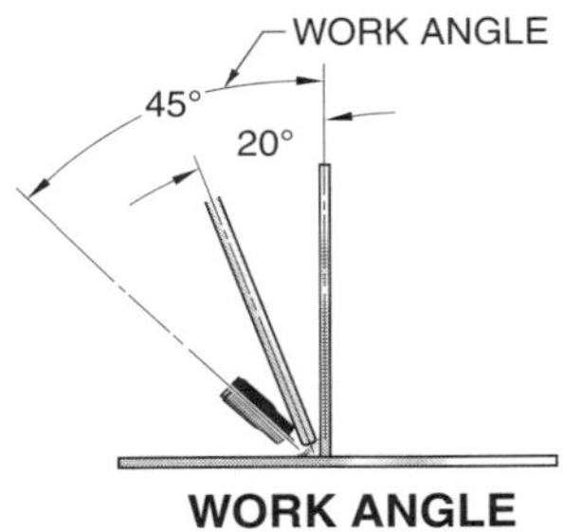

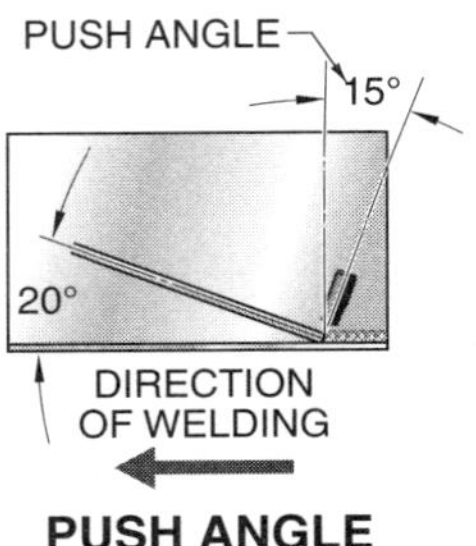

exercise 4-7

Welding a Butt Joint on Mild Steel in Horizontal Position Using GTAW

Conditions

Horizontal position
Refer to Exercise 4-1
Two pieces of 16-gauge, 1½″ × 6″
¹⁄₁₆″ filler metal

Performance

The welder will demonstrate the correct procedure for welding a butt joint in horizontal position.

Criteria

Welding technique, weld appearance, and weld strength as evaluated by the instructor.

Procedure

1. Tack weld the two pieces to form a butt joint. Position the workpiece so the joint is in horizontal position.
2. Hold the torch at a 90° work angle and a 5° to 15° push angle.
3. Position the filler rod at a 20° angle in line with the weld bead.

4. Weave the torch slightly while adding filler metal with an in-and-out motion on the top half of the leading edge of the weld pool.
5. Maintain a consistent bead along the workpiece. Hold the torch in position at the end of the weld until the postflow times out.
6. Repeat the exercise using the lay-wire technique.
7. Repeat each exercise using pulsed current.

exercise **4-8**

Welding a Butt Joint on Mild Steel in Vertical Position (Uphill) Using GTAW

Conditions

Vertical position (uphill)
Refer to Exercise 4-1
Two pieces of 16-gauge, 1½″ × 6″
1⁄16″ filler metal

Performance

The welder will demonstrate the correct procedure for welding a butt joint in vertical position (uphill).

Criteria

Welding technique, weld appearance, and weld strength as evaluated by the instructor.

Procedure

1. Tack weld the two pieces to form a butt joint. Position the workpiece so the joint is in vertical position.
2. Hold the torch at a 90° work angle and a 5° to 15° push angle.
3. Angle the filler metal 20° above the joint.
4. Start at the bottom of the workpiece and use a slight weaving motion while adding filler metal with an in-and-out motion.
5. Maintain a consistent bead along the workpiece. Hold the torch in position at the end of the weld until the postflow times out.
6. Repeat the exercise using the lay-wire technique.
7. Repeat each exercise using pulsed current.

exercise **4-9**

Welding a T-Joint on Mild Steel in Vertical Position (Uphill) Using GTAW

Conditions

Vertical position (uphill)
Refer to Exercise 4-1
Two pieces of 16-gauge, 1½″ × 6″
1⁄16″ filler metal

Performance

The welder will demonstrate the correct procedure for welding a T-joint in vertical position (uphill).

Criteria

Welding technique, weld appearance, and weld strength as evaluated by the instructor.

Procedure

1. Tack weld the two pieces to form a T-joint. Position the workpiece so the joint is in vertical position.
2. Hold the torch at a 45° work angle and a 20° push angle.
3. Position the filler metal at a 20° angle centered between the two members.
4. Start at the bottom of the workpiece and weave the torch slightly while adding filler metal with an in-and-out motion.
5. Maintain a consistent bead along the workpiece. Hold the torch in position at the end of the weld until the postflow times out.
6. Repeat the exercise using the lay-wire technique.
7. Repeat each exercise using pulsed current.

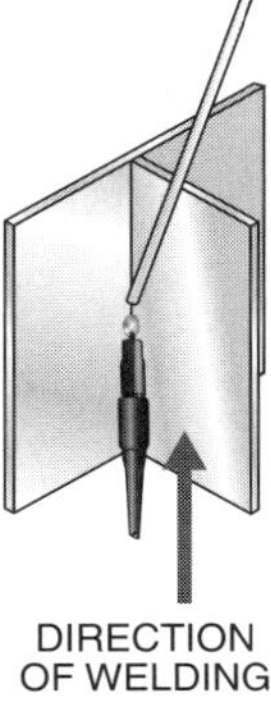

exercise **4-10**

Welding a Butt Joint on Mild Steel in Overhead Position Using GTAW

Conditions

Overhead position
Refer to Exercise 4-1
Two pieces of 16-gauge, 1½″ × 6″
¹⁄₁₆″ filler metal

Performance

The welder will demonstrate the correct procedure for welding a butt joint in overhead position.

Criteria

Welding technique, weld appearance, and weld strength as evaluated by the instructor.

Procedure

1. Tack weld the two pieces to form a butt joint. Position the joint so the workpiece is in overhead position.
2. Reduce the current 5% to 10% from the amount used for flat position.
3. Use the same procedure as for flat position. Refer to Exercise 4-3.
4. Repeat the exercise using the lay-wire technique.
5. Repeat each exercise using pulsed current.

exercise **4-11**

Welding a Lap Joint on Mild Steel in Overhead Position Using GTAW

Conditions

Overhead position
Refer to Exercise 4-1
Two pieces of 16-gauge, 1½″ × 6″
¹⁄₁₆″ filler metal

Performance

The welder will demonstrate the correct procedure for welding a lap joint in overhead position.

Criteria

Welding technique, weld appearance, and weld strength as evaluated by the instructor.

Procedure

1. Tack weld the two pieces to form a lap joint. Position the workpiece so the joint is in overhead position.
2. Reduce the current 5% to 10% from the amount used for flat position.
3. Refer to Exercise 4-5.

exercise 4-12

Welding a T-Joint on Mild Steel in Overhead Position Using GTAW

Conditions

Overhead position
Refer to Exercise 4-1
Two pieces of 16-gauge, 1½″ × 6″
¹⁄₁₆″ filler metal

Performance

The welder will demonstrate the correct procedure for welding a T-joint in overhead position.

Criteria

Welding technique, weld appearance, and weld strength as evaluated by the instructor.

Procedure

1. Tack weld the two pieces to form a T-joint. Position the workpiece so the joint is in overhead position.
2. Reduce the current 5% to 10% from the amount used for flat position.
3. Refer to Exercise 4-6 for procedure.

exercise 4-13

Depositing Beads on Aluminum in Flat Position Using GTAW

Conditions

Flat position
HF should be set to continuous
130 A to 150 A
⅛″ EWP or ³⁄₃₂″ EWZr (hemispherical tip), ⅛″ to ³⁄₁₆″ electrode stickout
Argon shielding gas 15 cfh to 20 cfm with 15-second postflow
⅛″ × 4″ × 6″ aluminum
⅛″ recommended filler metal

Performance

The welder will demonstrate the correct procedure for depositing beads on aluminum using GTAW.

Criteria

Welding technique, weld appearance, and weld strength as evaluated by the instructor.

Procedure

1. Clean the aluminum workpieces with a stainless steel wire brush.
2. To form the hemispherical tip on the electrode, switch the welding machine to DCEP and start the arc on a piece of copper while holding the welding torch at a 90° angle.
3. Start the arc by positioning the electrode ⅛″ away from the workpiece and activating the current flow with the foot- or hand-operated remote control. Do not touch the electrode to the workpiece.
4. Hold the welding torch at a 90° work angle and a 20° push angle, with the filler raised 20° above the workpiece.
5. Melt the base metal and add filler metal to form a bead ¼″ wide. Add filler metal to the leading edge of the weld pool using an in-and-out motion. Weave the torch slightly to distribute the heat to form the bead.
6. To fill the crater at the end of the weld, reduce the current and continue to add filler metal.
7. Deposit a series of beads on the workpiece approximately ⅜″ apart. Hold the torch in position at the end of each weld until the postflow times out.

exercise 4-14

Welding a Butt Joint on Aluminum in Flat Position Using GTAW

Conditions

Flat position
Refer to Exercise 4-13
Two pieces of ⅛″ × 1½″ × 6″ aluminum

Performance

The welder will demonstrate the correct procedure for welding a butt joint on aluminum in flat position.

Criteria

Welding technique, weld appearance, and weld strength as evaluated by the instructor.

Procedure

1. Refer to Exercise 4-13 for equipment setup and adjustment.
2. Use the procedures detailed for welding mild steel in Exercise 4-3. Brush the joint with a stainless steel wire brush immediately before.

exercise 4-15

Welding an Open Corner Joint on Aluminum in Flat Position Using GTAW

Conditions

Vertical position
Refer to Exercise 4-13
100 A to 110 A

Performance

The welder will demonstrate the correct procedure for depositing a fillet weld in an outside corner joint with and without filler metal.

Criteria

The weld with filler metal should be about 1 to 1½ electrode diameters wide. The autogenous weld should be about one electrode diameter wide. Both welds should have complete penetration through to the root side of the joint.

Procedure

1. Tack weld the two pieces to form an open corner joint and place it in the flat position.
2. Brush the joint with a stainless steel wire brush immediately before welding.
3. The work angle is 90° with a 15° to 20° push angle. The filler metal is raised 20° above the joint.
4. Activate the foot/hand remote control to start the arc.
5. Deposit the weld using the dip technique. Dip the filler metal into the leading edge of the weld pool and withdraw it while moving the torch with a steady push. Keep the end of the filler metal in the flow of shielding gas.
6. At the end of the joint, fill the crater, and hold the torch in position until the postflow times out.
7. Repeat the exercise without filler metal. Once the weld pool forms, move the torch with a steady push.

exercise 4-16

Welding a Lap Joint on Aluminum in Horizontal Position Using GTAW

Conditions

Horizontal position
Refer to Exercise 4-13
Two pieces of ⅛″ × 1½″ × 6″ aluminum

Performance

The welder will demonstrate the correct procedure for welding a lap joint on aluminum in horizontal position.

Criteria

Welding technique, weld appearance, and weld strength as evaluated by the instructor.

Procedure

1. Refer to Exercise 4-12 for equipment setup and adjustment.
2. Use the procedures listed for welding mild steel in Exercise 4-4.

exercise **4-17**

Welding a T-Joint on Aluminum in Horizontal Position Using GTAW

Conditions

Horizontal position
Refer to Exercise 4-13
Two pieces of ⅛″ × 1½″ × 6″ aluminum

Performance

The welder will demonstrate the correct procedure for welding a T-joint on aluminum in horizontal position.

Criteria

Welding technique, weld appearance, and weld strength as evaluated by the instructor.

Procedure

1. Refer to Exercise 4-12 for equipment setup and adjustment.
2. Use the procedures detailed for welding mild steel in Exercise 4-5.

advanced exercise **4-18**

Making a Multiple-Pass Fillet Weld, T-Joint, 2F Position on Carbon Steel Plate, Using GTAW

Conditions

Position: 2F
Materials: 2 pieces ¼″ × 2″ × 4″ mild steel
Electrodes: EWCe-2 or EWLa-X
Filler Metal: 3⁄32″ or ⅛″ ER70S-6
Shielding: 100% Ar
Polarity: DCEN
Amperage: 125 to 200 for 3⁄32″, 150 to 230 for ⅛″
Voltage: 17 to 18 for 3⁄32″, 18 to 19 for ⅛″

Performance

The welder will demonstrate the correct procedure and technique for depositing a weld in a T-joint.

Criteria

The weld will pass visual inspection criteria.

Procedure

1. Tack weld two pieces to form a T-joint. Position workpiece so joint is in 2F position.
2. Use a 45° work angle and a 5° to 10° push travel angle. Use a steady push travel to deposit root pass. Brush welds.
3. Deposit second pass on bottom toe of root pass with a 50° to 55° work angle and a 5° to 10° push travel angle. Use a steady push travel to cover root by one-half to two-thirds. Fuse into base metal evenly along toe. Brush weld.
4. Deposit third pass on top toe of root pass with a 30° to 35° work angle and a 5° to 10° push travel angle. Cover second pass by one-third to one-half. Fuse into base metal evenly along toe. Brush weld.
5. Deposit welds on other side of joint.

Additional Practice

Tack up new workpiece. Practice with Z-weave motion with nozzle in contact with joint pausing at each weld toe to deposit all of the passes. This is referred to as "walking the cup."

advanced exercise **4-19**

Making a Multiple-Pass Fillet Weld, T-Joint, 3F Position on Carbon Steel Plate, Using GTAW

Conditions

Position: 3F
Materials: 2 pieces ¼″ × 2″ × 4″ mild steel
Electrodes: EWCe-2 or EWLa-X
Filler Metal: 3⁄32″ or ⅛″ ER70S-6
Shielding: 100% Ar
Polarity: DCEN
Amperage: 125 to 200 for 3⁄32″, 150 to 230 for ⅛″
Voltage: 17 to 18 for 3⁄32″, 18 to 19 for ⅛″

Performance

The weld will pass visual inspection criteria.

Criteria

The weld will pass visual inspection criteria.

Procedure

1. Tack weld two pieces to form a T-joint. Position workpiece so joint is in 3F position.
2. Use a 45° work angle and a 5° to 10° push travel angle to deposit root pass. Brush weld.
3. Deposit second pass on either toe of root pass with a 50° to 55° work angle and a 5° to 10° push travel angle. Cover root pass by one-half to two-thirds. Fuse into base metal evenly along toe. Brush weld.
4. Deposit third pass on opposite toe of root pass with a 30° to 35° work angle and a 5° to 10° push travel angle. Cover second pass by one-third to one-half. Fuse into base metal evenly along toe. Brush weld.
5. Deposit welds on other side of joint.

Additional Practice

Tack up new workpiece. Practice with Z-weave motion with nozzle in contact with joint pausing at each weld toe to deposit all of the passes. This is referred to as "walking the cup."

advanced exercise **4-20**

Making a Multiple-Pass Fillet Weld, T-Joint, 4F Position on Carbon Steel Plate, Using GTAW

Conditions

Position: 4F
Materials: 2 pieces ¼″ × 2″ × 4″ mild steel
Electrodes: EWCe-2 or EWLa-X
Filler Metal: 3⁄32″ or ⅛″ ER70S-6
Shielding: 100% Ar
Polarity: DCEN
Amperage: 125 to 200 for 3⁄32″, 150 to 230 for ⅛″
Voltage: 17 to 18 for 3⁄32″, 18 to 19 for ⅛″

Performance

The weld will pass visual inspection criteria.

Criteria

The weld will pass visual inspection criteria.

Procedure

1. Tack weld two pieces to form a T-joint. Position workpiece so joint is in 4F position.
2. Use a 45° work angle and a 5° to 10° push travel angle to deposit root pass on both sides of joint. Brush weld.

3. Deposit second pass on bottom toe of root pass with a 50° to 55° work angle with a 5° to 10° push travel angle. Cover root by one-half to two-thirds. Fuse into base metal evenly along toe. Brush weld.

4. Deposit third pass on top toe of root pass with a 30° to 35° work angle with a 5° to 10° push travel angle. Cover second pass by one-third to one-half. Fuse into base metal evenly along toe. Brush weld.

5. Deposit welds on other side of joint.

Additional Practice

Tack up new workpiece. Practice with Z-weave motion with nozzle in contact with joint pausing at each weld toe to deposit all of the passes. This is referred to as "walking the cup."

advanced exercise **4-21**

Making Multiple-Pass Groove Welds, 3G Position on Carbon Steel Plate with and without Backing, Using GTAW

Conditions

Position: 3G
Materials: 2 pieces ⅜″ × 6″ × 7″ mild steel
Backing: ⅛″ × 1½″ × 9″ mild steel
Electrodes: EWCe-2 or EWLa-X
Filler Metal: 3⁄32″ or ⅛″ ER70S-6
Shielding: 100% Ar
Polarity: DCEN
Amperage: 125 to 200 for 3⁄32″, 150 to 230 for ⅛″
Voltage: 17 to 18 for 3⁄32″, 18 to 19 for ⅛″

Performance

The welder will demonstrate the correct procedure and technique for depositing welds in single-V groove joints in the vertical 3G position.

Criteria

The welds will pass visual inspection and AWS guided bend tests.

Procedure with Backing

1. Prepare a single-V groove joint with a 60° groove angle and backing strip. Tack assembly together with and a ¼″ root opening. Position workpiece so joint is in 3G position.

2. Use a 90° work angle and a 5° to 10° push travel angle. Move up joint with a Z-weave motion with nozzle in contact with joint ("walking the cup") to deposit root pass. Pause briefly at each toe to fuse into sides of joint. Brush weld.

3. Deposit second pass using the same electrode angles and weaving motion. Pause briefly at each toe to fuse into edges of joint. Brush weld.

4. Deposit remaining passes with the same electrode angles and weaving motion. Brush weld after each pass.
5. Face reinforcement should be 1⁄16″ maximum.
6. Root reinforcement should be 1⁄16″ maximum.
7. Prepare best weld for a guided bend test.

Procedure without Backing

1. Prepare a single-V groove open root joint with a 60° groove angle and a 3⁄32″ to 1⁄8″ root face. Tack assembly together with a 3⁄32″ to 1⁄8″ root opening. Position workpiece so joint is in 3G position.
2. Use a 90° work angle and a 5° to 10° push travel angle. Move up joint with a Z-weave motion with nozzle in contact with joint to deposit root pass. Pause briefly at each toe to fuse into sides of joint. Brush weld.
3. Use the same weaving motion to deposit the remaining passes. Pause briefly at each toe to ensure penetration into sides of joint. Brush after each pass.
4. Face reinforcement should be 1⁄16″ maximum.
5. Root reinforcement should be 1⁄16″ maximum.
6. Prepare best weld for a guided bend test.

advanced exercise **4-22**

Making Multiple-Pass Groove Welds, 4G Position on Carbon Steel Plate with and without Backing, Using GTAW

Conditions

Position: 4G
Materials: 2 pieces 3⁄8″ × 6″ × 7″ mild steel
Backing: 1⁄8″ × 1½″ × 9″ mild steel
Electrodes: EWCe-2 or EWLa-X
Filler Metal: 3⁄32″ or 1⁄8″ ER70S-6
Shielding: 100% Ar
Polarity: DCEN
Amperage: 125 to 200 for 3⁄32″, 150 to 230 for 1⁄8″
Voltage: 17 to 18 for 3⁄32″, 18 to 19 for 1⁄8″

Performance

The welder will demonstrate the correct procedure and technique for depositing welds in single-V groove joints in the overhead 4G position.

Criteria

The welds will pass visual inspection and AWS guided bend tests.

Procedure with Backing

1. Prepare a single-V groove joint with a 60° groove angle and backing strip. Tack assembly together with a ¼″ root opening. Position workpiece so joint is in 4G position.
2. Use a 90° work angle and a 5° to 10° push travel angle. Move along joint with a Z-weave motion with nozzle in contact with joint to deposit root pass. Pause briefly at each toe to fuse into sides of joint. Brush weld.
3. Deposit second pass using the same electrode angles and weaving motion. Pause briefly at each toe to fuse into edges of joint. Brush weld.
4. Deposit remaining passes with the same electrode angles and weaving motion. Brush weld after each pass.
5. Face reinforcement should be 1⁄16″ maximum.
6. Root reinforcement should be 1⁄16″ maximum.
7. Prepare best weld for a guided bend test.

Procedure without Backing

1. Prepare a single-V groove open root joint with a 60° groove angle and a 3⁄32″ to ⅛″ root face. Tack assembly together with a 3⁄32″ to ⅛″ root opening. Position workpiece so joint is in 4G position.
2. Use a 90° work angle and a 5° to 10° push travel angle. Move along joint with a Z-weave motion with nozzle in contact with joint to deposit root pass. Pause briefly at each toe to fuse into sides of joint. Brush weld.
3. Use a weaving motion to deposit the remaining passes. Pause briefly at each toe to ensure penetration in sides of joint. Brush after each pass.
4. Face reinforcement should be 1⁄16″ maximum.
5. Root reinforcement should be 1⁄16″ maximum.
6. Prepare best weld for a guided bend test.

advanced exercise **4-23**

Making a Multiple-Pass Fillet Weld, T-Joint, 2F Position on Carbon Steel with Stainless Steel Filler Metal, Using GTAW

Conditions

Position: 2F
Materials: 2 pieces ¼″ × 2″ × 4″ mild steel
Electrodes: EWCe-2 or EWLa-X
Filler Metal: 3⁄32″ or ⅛″ ER309
Shielding: 100% Ar
Polarity: DCEN
Amperage: 125 to 200 for 3⁄32″, 150 to 230 for ⅛″
Voltage: 17 to 18 for 3⁄32″, 18 to 19 for ⅛″

Performance

The welder will demonstrate the correct procedure and technique for depositing a weld in a T-joint.

Criteria

The weld will pass visual inspection criteria.

Procedure

1. Tack weld two pieces to form a T-joint. Position workpiece so joint is in 2F position.
2. Use a 45° work angle and a 5° to 10° push travel angle for root pass. Use a steady push to deposit root pass on both sides of joint. Brush welds.
3. Deposit second pass on bottom toe of root pass with a 50° to 55° work angle and a 5° to 10° push travel angle. Use a steady push travel. Cover root by one-half to two-thirds. Fuse into base metal evenly along toe. Brush weld.
4. Deposit third pass on top toe of root pass with a 30° to 35° work angle and a 5° to 10° push travel angle. Cover second pass by one-third to one-half. Fuse into base metal evenly along toe. Brush weld.
5. Deposit welds on other side of joint.

Additional Practice

Tack up new workpiece. Practice with Z-weave motion with nozzle in contact with joint pausing at each weld toe to deposit all of the passes. This is referred to as "walking the cup."

advanced exercise **4-24**

Making a Multiple-Pass Fillet Weld, T-Joint, 3F Position on Carbon Steel with Stainless Steel Filler Metal, Using GTAW

Conditions

Position: 3F
Materials: 2 pieces ¼″ × 2″ × 4″ mild steel
Electrodes: EWCe-2 or EWLa-X
Filler Metal: 3⁄32″ or ⅛″ ER309
Shielding: 100% Ar
Polarity: DCEN
Amperage: 125 to 200 for 3⁄32″, 150 to 230 for ⅛″
Voltage: 17 to 18 for 3⁄32″, 18 to 19 for ⅛″

Performance

Refer to Exercise 4-23.

Criteria

Refer to Exercise 4-23.

Procedure

1. Tack weld two pieces to form a T-joint. Position workpiece so joint is in 3F position.
2. Use a 45° work angle and a 5° to 10° push travel angle to deposit root pass on both sides of joint. Brush weld.
3. Deposit second pass on either toe of root pass with a 50° to 55° work angle and a 5° to 10° push travel angle. Cover root by one-half to two-thirds. Fuse into base metal evenly along toe. Brush weld.
4. Deposit third pass on opposite toe of root pass with a 30° to 35° work angle and a 5° to 10° push travel angle. Cover second pass by one-third to one-half. Fuse into base metal evenly along toe. Brush weld.
5. Deposit welds on other side of joint.

Additional Practice

Tack up new workpiece. Practice with Z-weave motion with nozzle in contact with joint pausing at each weld toe to deposit all of the passes. This is referred to as "walking the cup."

advanced exercise **4-25**

Making a Multiple-Pass Fillet Weld, T-Joint, 4F Position on Carbon Steel with Stainless Steel Filler Metal, Using GTAW

Conditions

Position: 4F
Materials: 2 pieces ¼″ × 2″ × 4″ mild steel
Electrodes: EWCe-2 or EWLa-X
Filler Metal: 3⁄32″ or ⅛″ ER309
Shielding: 100% Ar
Polarity: DCEN
Amperage: 125 to 200 for 3⁄32″, 150 to 230 for ⅛″
Voltage: 17 to 18 for 3⁄32″, 18 to 19 for ⅛″

Performance

Refer to Exercise 4-23.

Criteria

Refer to Exercise 4-23.

Procedure

1. Tack weld two pieces to form a T-joint. Position workpiece so joint is in 4F position.
2. Use a 45° work angle and a 5° to 10° push travel angle to deposit root pass on both sides of joint. Brush weld.

3. Deposit second pass on bottom toe of root pass with a 50° to 55° work angle and a 5° to 10° push travel angle. Cover root by one-half to two-thirds. Fuse into base metal evenly along toe. Brush weld.
4. Deposit third pass on top toe of root pass with a 30° to 35° work angle and a 5° to 10° push travel angle. Cover second pass by one-third to one-half. Fuse into base metal evenly along toe. Brush weld.
5. Deposit welds on other side of joint.

Additional Practice

Tack up new workpiece. Practice with Z-weave motion with nozzle in contact with joint pausing at each weld toe to deposit all of the passes. This is referred to as "walking the cup."

advanced exercise **4-26**

Making Multiple-Pass Groove Welds, 3G Position on Carbon Steel with Stainless Steel Filler Metal, Using GTAW with and without Backing

Conditions

Position: 3G
Materials: 2 pieces ⅜″ × 6″ × 7″ mild steel
Backing: ⅛″ × 1½″ × 9″ mild steel
Electrodes: EWCe-2 or EWLa-X
Filler Metal: 3⁄32″ or ⅛″ ER309
Shielding: 100% Ar
Polarity: DCEN
Amperage: 125 to 200 for 3⁄32″, 150 to 230 for ⅛″
Voltage: 17 to 18 for 3⁄32″, 18 to 19 for ⅛″

Performance

The welder will demonstrate the correct procedure and technique for depositing welds in single-V groove joints in the vertical 3G position.

Criteria

The welds will pass visual inspection and AWS guided bend tests.

Procedure with Backing

1. Prepare a single-V groove joint with a 60° groove angle and backing strip. Tack assembly together with a ¼″ root opening. Position workpiece so joint is in 3G position.
2. Use a 90° work angle and a 5° to 10° push travel angle. Move up joint with a Z-weave motion with nozzle in contact with joint to deposit root pass. Pause briefly at each toe to fuse into sides of joint. Brush weld.
3. Deposit second pass using the same electrode angles and weaving motion. Pause briefly at each toe to fuse into edges of joint. Brush weld.
4. Deposit remaining passes with the same electrode angles and weaving motion. Brush weld after each pass.

5. Face reinforcement should be 1/16″ maximum.
6. Root reinforcement should be 1/16″ maximum.
7. Prepare best weld for a guided bend test.

Procedure without Backing

1. Prepare a single-V groove open root joint with a 60° groove angle and a 3/32″ to 1/8″ root face. Tack assembly together with a 3/32″ to 1/8″ root opening. Position workpiece so joint is in 3G position.
2. Use a 90° work angle and a 5° to 10° push travel angle. Move up joint with a Z-weave motion with nozzle in contact with joint to deposit root pass. Pause briefly at each toe to fuse into sides of joint. Brush weld.
3. Use the same electrode angles and weaving motion to deposit the remaining passes. Pause briefly at each toe to ensure penetration into sides of joint. Brush after each weld.
4. Face reinforcement should be 1/16″ maximum.
5. Root reinforcement should be 1/16″ maximum.
6. Prepare best weld for a guided bend test.

advanced exercise 4-27

Making Multiple-Pass Groove Welds, 4G Position on Carbon Steel with Stainless Steel Filler Metal, Using GTAW with and without Backing

Conditions

Position: 4G
Materials: 2 pieces 3/8″ × 6″ × 7″ mild steel
Backing: 1/8″ × 1½″ × 9″ mild steel
Electrodes: EWCe-2 or EWLa-X
Filler Metal: 3/32″ or 1/8″ ER309
Shielding: 100% Ar
Polarity: DCEN
Amperage: 125 to 200 for 3/32″, 150 to 230 for 1/8″
Voltage: 17 to 18 for 3/32″, 18 to 19 for 1/8″

Performance

The welder will demonstrate the correct procedure and technique for depositing welds in single-V groove joints in the overhead 4G position.

Criteria

The welds will pass visual inspection and AWS guided bend tests.

Procedure with Backing

1. Prepare a single-V groove joint with a 60° groove angle and backing strip. Tack assembly together with a ¼″ root opening. Position workpiece so joint is in 4G position.
2. Use a 90° work angle and a 5° to 10° push travel angle. Move along joint with a Z-weave motion with nozzle in contact with joint to deposit root pass. Pause briefly at each toe to fuse into sides of joint. Brush weld.
3. Deposit second pass using the same electrode angles and weaving motion. Pause briefly at each toe to fuse into edges of joint. Brush weld.
4. Deposit remaining passes with the same electrode angles and weaving motion. Brush weld after each pass.
5. Face reinforcement should be 1⁄16″ maximum.
6. Root reinforcement should be 1⁄16″ maximum.
7. Prepare best weld for a guided bend test.

Procedure without Backing

1. Prepare a single-V groove open root joint with a 60° groove angle and a 3⁄32″ to ⅛″ root face. Tack assembly together with a 3⁄32″ to ⅛″ root opening. Position workpiece so joint is in 4G position.
2. Use a 90° work angle and a 5° to 10° push travel angle. Move up joint with a Z-weave motion with nozzle in contact with joint to deposit root pass. Pause briefly at each toe to fuse into sides of joint. Brush weld.
3. Deposit remaining passes with the same electrode angles and weaving motion. Brush weld after each pass.
4. Face reinforcement should be 1⁄16″ maximum.
5. Root reinforcement should be 1⁄16″ maximum.
6. Prepare best weld for a guided bend test.

advanced exercise **4-28**

Making a Multiple-Pass Fillet Weld, T-Joint, 2F Position on Aluminum, Using GTAW

Conditions

Position: 2F
Materials: 2 pieces ¼″ × 2″ × 4″ aluminum
Electrodes: EWZr-1
Filler Metal: 3⁄32″ or ⅛″ ER4XXX or ER5XXX
Shielding: 100% Ar
Polarity: AC
Amperage: 270 to 320
Voltage: 26 to 28

Performance

The welder will demonstrate the correct procedure and technique for depositing a weld in a T-joint.

Criteria

The weld will pass visual inspection criteria.

Procedure

1. Tack weld two pieces to form a T-joint. Position workpiece so joint is in 2F position.
2. Use a 45° work angle and a 5° to 10° push travel angle for root pass. Use a steady push to deposit root pass on both sides of joint. Brush welds with appropriate brush.
3. Deposit second pass on bottom toe of root pass with a 50° to 55° work angle and a 5° to 10° push travel angle. Use a steady push travel. Cover root by one-half to two-thirds. Fuse into base metal evenly along toe. Brush weld with appropriate brush.
4. Deposit third pass on top toe of root pass with a 30° to 35° work angle and a 5° to 10° push travel angle. Cover second pass by one-third to one-half. Fuse into base metal evenly along toe. Brush weld with appropriate brush.
5. Deposit welds on other side of joint.

Additional Practice

Tack up new workpiece. Practice with Z-weave motion with nozzle in contact with joint pausing at each weld toe to deposit all of the passes. This is referred to as "walking the cup."

advanced exercise **4-29**

Making a Multiple-Pass Fillet Weld, T-Joint, 3F Position on Aluminum, Using GTAW

Conditions

Position: 3F
Materials: 2 pieces ¼″ × 2″ × 4″ aluminum
Electrodes: EWZr-1
Filler Metal: ³⁄₃₂″ or ⅛″ ER4XXX or ER5XXX
Shielding: 100% Ar
Polarity: AC
Amperage: 270 to 320
Voltage: 26 to 28

Performance

Refer to Exercise 5-20.

Criteria

Refer to Exercise 5-20.

Procedure

1. Tack weld two pieces to form a T-joint. Position workpiece so joint is in 3F position.
2. Use a 45° work angle and a 5° to 10° push travel angle to deposit root pass on both sides of joint. Brush weld with appropriate brush.
3. Deposit second pass on either toe of root pass with a 50° to 55° work angle and a 5° to 10° push travel angle. Cover root by one-half to two-thirds. Fuse into base metal evenly along toe. Brush weld with appropriate brush.
4. Deposit third pass on opposite toe of root pass with a 30° to 35° work angle and a 5° to 10° push travel angle. Cover second pass by one-third to one-half. Fuse into base metal evenly along toe. Brush weld with appropriate brush.
5. Deposit welds on other side of joint.

Additional Practice

Tack up new workpiece. Practice with Z-weave motion with nozzle in contact with joint pausing at each weld toe to deposit all of the passes. This is referred to as "walking the cup."

advanced exercise **4-30**

Making a Multiple-Pass Fillet Weld, T-Joint, 4F Position on Aluminum, Using GTAW

Conditions

Position: 4F
Materials: 2 pieces ¼″ × 2″ × 4″ aluminum
Electrodes: EWZr-1
Filler Metal: 3⁄32″ or ⅛″ ER4XXX or ER5XXX
Shielding: 100% Ar
Polarity: AC
Amperage: 270 to 320
Voltage: 26 to 28

Performance

Refer to Exercise 5-20.

Criteria

Refer to Exercise 5-20.

Procedure

1. Tack weld two pieces to form a T-joint. Position workpiece so joint is in 4F position.
2. Use a 45° work angle and a 5° to 10° push travel angle to deposit root pass on both sides of joint. Brush weld with appropriate brush.

3. Deposit second pass on bottom toe of root pass with a 50° to 55° work angle and a 5° to 10° push travel angle. Cover root by one-half to two-thirds. Fuse into base metal evenly along toe. Brush weld with appropriate brush.
4. Deposit third pass on top toe of root pass with a 30° to 35° work angle and a 5° to 10° push travel angle. Cover second pass by one-third to one-half. Fuse into base metal evenly along toe. Brush weld with appropriate brush.
5. Deposit welds on other side of joint.

Additional Practice

Tack up new workpiece. Practice with Z-weave motion with nozzle in contact with joint pausing at each weld toe to deposit all of the passes. This is referred to as "walking the cup."

advanced exercise 4-31

Making Multiple-Pass Groove Welds, 3G Position on Aluminum with and without Backing, Using GTAW

Conditions

Position: 3G
Materials: 2 pieces 3/8″ × 6″ × 7″ aluminum
Backing: 1/4″ × 1½″ × 9″ aluminum
Electrodes: EWZr-1
Filler Metal: 3/32″ or 1/8″ ER4XXX or ER5XXX
Shielding: 100% Ar
Polarity: AC
Amperage: 235 to 380
Voltage: 29 to 30

Performance

The welder will demonstrate the correct procedure and technique for depositing welds in single-V groove joints in the vertical 3G position.

Criteria

The welds will pass visual inspection and AWS guided bend tests.

Procedure with Backing

1. Prepare a single-V groove joint with a 60° groove angle and backing strip. Tack assembly together with a 1/4″ root opening. Position workpiece so joint is in 3G position.
2. Use a 90° work angle and a 5° to 10° push travel angle. Move up joint with a Z-weave motion with nozzle in contact with joint to deposit root pass. Pause briefly at each toe to fuse into sides of joint. Brush weld with appropriate brush.

3. Deposit second pass using the same electrode angles and weaving motion. Pause briefly at each toe to fuse into edges of joint. Brush weld with appropriate brush.
4. Deposit remaining passes with the same electrode angles and weaving motion. Brush weld after each pass with appropriate brush.
5. Face reinforcement should be 1⁄16″ maximum.
6. Root reinforcement should be 1⁄16″ maximum.
7. Prepare best weld for a guided bend test.

Procedure without Backing

1. Prepare a single-V groove open root joint with a 60° groove angle and a 3⁄32″ to 1⁄8″ root face. Tack assembly together with a 3⁄32″ to 1⁄8″ root opening. Position workpiece so joint is in 3G position.
2. Use a 90° work angle and a 5° to 10° push travel angle. Move up joint with a Z-weave motion with nozzle in contact with joint to deposit root pass. Pause briefly at each toe to fuse into sides of joint. Brush weld with appropriate brush.
3. Deposit second pass using the same electrode angles and weaving motion. Pause briefly at each toe to fuse into edges of joint. Brush weld with appropriate brush.
4. Deposit remaining passes with the same electrode angles and weaving motion. Brush weld after each pass with appropriate brush.
5. Face reinforcement should be 1⁄16″ maximum.
6. Root reinforcement should be 1⁄16″ maximum.
7. Prepare best weld for a guided bend test.

advanced exercise **4-32**

Making Multiple-Pass Groove Welds, 4G Position on Aluminum with and without Backing, Using GTAW

Conditions

Position: 4G
Materials: 2 pieces 3⁄8″ × 6″ × 7″ aluminum
Backing: 1⁄4″ × 1½″ × 9″ aluminum
Electrodes: EWZr-1
Filler Metal: 3⁄32″ or 1⁄8″ ER4XXX or ER5XXX
Shielding: 100% Ar
Polarity: AC
Amperage: 325 to 380
Voltage: 29 to 30

Performance

The welder will demonstrate the correct procedure and technique for depositing welds in single-V groove joints in the overhead 4G position.

Criteria

The welds will pass visual inspection and AWS guided bend tests.

Procedure with Backing

1. Prepare a single-V groove joint with a 60° groove angle and backing strip. Tack assembly together with a ¼″ root opening. Position workpiece so joint is in 4G position.
2. Use a 90° work angle and a 5° to 10° push travel angle. Move up joint with a Z-weave motion with nozzle in contact with joint to deposit root pass. Pause briefly at each toe to fuse into sides of joint. Brush weld with appropriate brush.
3. Deposit second pass using the same electrode angles and weaving motion. Pause briefly at each toe to fuse into the edges of the joint. Brush weld with appropriate brush.
4. Deposit remaining passes with the same electrode angles and weaving motion. Brush weld after each pass with appropriate brush.
5. Face reinforcement should be 1/16″ maximum.
6. Root reinforcement should be 1/16″ maximum.
7. Prepare best weld for a guided bend test

Procedure without Backing

1. Prepare a single-V groove open root joint with a 60° groove angle and a 3/32″ to ⅛″ root face. Tack assembly together with a 3/32″ to ⅛″ root opening. Position workpiece so joint is in 4G position.
2. Use a 90° work angle and a 5° to 10° push travel angle. Move up joint with a Z-weave motion with nozzle in contact with joint to deposit root pass. Pause briefly at each toe to fuse into sides of joint. Brush weld with appropriate brush.
3. Deposit second pass using the same electrode angles and weaving motion. Pause briefly at each toe to fuse into edges of joint. Brush weld with appropriate brush.
4. Deposit remaining passes with the same electrode angles and weaving motion. Brush weld after each pass with appropriate brush.
5. Face reinforcement should be 1/16″ maximum.
6. Root reinforcement should be 1/16″ maximum.
7. Prepare best weld for a guided bend test.

advanced exercise **4-33**

Making a Multiple-Pass Fillet Weld, 2F Position, Carbon Steel Pipe to Plate, Using GTAW

Conditions

Position: 2F
Materials: 4″ to 6″ schedule 40 to 80 carbon steel pipe
¼″ carbon steel plate to fit
Electrodes: EWCe-2 or EWLa-X
Filler Metal: 3⁄32″ or ⅛″ ER70S-6
Shielding: 100% Ar
Polarity: DCEN
Amperage: 125 to 200 for 3⁄32″, 150 to 230 for ⅛″
Voltage: 17 to 18 for 3⁄32″, 18 to 19 for ⅛″

Performance

The welder will demonstrate the correct procedure and technique to weld pipe to plate in the 2F position.

Criteria

The weld will pass visual inspection criteria.

Procedure

1. Tack weld pipe to plate. Position workpiece so joint is in 2F position.
2. Use a 45° work angle and a 5° to 10° push travel angle to deposit root pass on both sides of joint. Brush weld.
3. Deposit second pass on bottom toe of root pass with a 50° to 55° work angle and a 5° to 10° push travel angle. Cover root by one-half to two-thirds. Fuse into base metal evenly along toe. Brush weld.
4. Deposit third pass on top toe of root pass with a 30° to 35° work angle and a 5° to 10° push travel angle. Cover second pass by one-third to one-half. Fuse into base metal evenly along toe. Brush weld.

Additional Practice

Tack up new workpiece. Practice with Z-weave motion with nozzle in contact with joint pausing at each weld toe to deposit all of the passes. This is referred to as "walking the cup."

advanced exercise **4-34**

Making a Multiple-Pass Fillet Weld, 4F Position, Carbon Steel Pipe to Plate, Using GTAW

Conditions

Position: 4F
Materials: 4″ to 6″ schedule 40 to 80 carbon steel pipe
¼″ carbon steel plate to fit

Electrodes: EWCe-2 or EWLa-X
Filler Metal: 3⁄32″ or 1⁄8″ ER70S-6
Shielding: 100% Ar
Polarity: DCEN
Amperage: 125 to 200 for 3⁄32″, 150 to 230 for 1⁄8″
Voltage: 17 to 18 for 3⁄32″, 18 to 19 for 1⁄8″

Performance

The welder will demonstrate the correct procedure and technique to weld pipe to plate in the 4F position.

Criteria

The weld will pass visual inspection criteria.

Procedures

1. Tack weld pipe to plate. Position workpiece so joint is in 4F position.
2. Use a 45° work angle and a 5° to 10° push travel angle to deposit root pass on both sides of joint. Brush weld.
3. Deposit second pass on bottom toe of root pass with a 50° to 55° work angle and a 5° to 10° push travel angle. Cover root by one-half to two-thirds. Fuse into base metal evenly along toe. Brush weld.
4. Deposit third pass on top toe of root pass with a 30° to 35° work angle and a 5° to 10° push travel angle. Cover second pass by one-third to one-half. Fuse into base metal evenly along toe. Brush weld.

Additional Practice

Tack up new workpiece. Practice with Z-weave motion with nozzle in contact with the joint pausing at each weld toe to deposit all of the passes. This is referred to as "walking the cup."

advanced exercise **4-35**

Making a Multiple-Pass Fillet Weld, 5F Position, Carbon Steel Pipe to Plate, Using GTAW

Conditions

Position: 5F
Materials: 4″ to 6″ schedule 40 to 80 carbon steel pipe
1⁄4″ carbon steel plate to fit
Electrodes: EWCe-2 or EWLa-X
Filler Metal: 3⁄32″ or 1⁄8″ ER70S-6
Shielding: 100% Ar
Polarity: DCEN
Amperage: 125 to 200 for 3⁄32″, 150 to 230 for 1⁄8″
Voltage: 17 to 18 for 3⁄32″, 18 to 19 for 1⁄8″

Performance

The welder will demonstrate the correct procedure and technique to weld pipe to plate in the 5F position.

Criteria

The weld will pass visual inspection criteria.

Procedure

1. Tack weld pipe to plate. Position workpiece so joint is in 5F position.
2. Use a 45° work angle and a 5° to 10° push travel angle to deposit root pass on both sides of joint. Brush weld.
3. Deposit second pass on either toe of root pass with a 50° to 55° work angle and a 5° to 10° push travel angle. Cover root by one-half to two-thirds. Fuse into base metal evenly along toe. Brush weld.
4. Deposit third pass on opposite toe of root pass with a 30° to 35° work angle and a 5° to 10° push travel angle. Cover second pass by one-third to one-half. Fuse into base metal evenly along toe. Brush weld.

Additional Practice

Tack up new workpiece. Practice with Z-weave motion with nozzle in contact with joint pausing at each weld toe to deposit all of the passes. This is referred to as "walking the cup."

advanced exercise **4-36**

Making Multiple-Pass Groove Welds, 2G Position, on Carbon Steel Pipe with and without Backing, Using GTAW

Conditions

Position: 2G
Materials: 4″ to 6″ schedule 40 to 80 carbon steel pipe
Backing ring to fit
Electrodes: EWCe-2 or EWLa-X
Filler Metal: 3⁄32″ or 1⁄8″ ER70S-6
Shielding: 100% Ar
Polarity: DCEN
Amperage: 125 to 200 for 3⁄32″, 150 to 230 for 1⁄8″
Voltage: 17 to 18 for 3⁄32″, 18 to 19 for 1⁄8″

Performance

The welder will demonstrate the correct procedure and technique for welding pipe in the 2G position.

Criteria

The welds will pass visual inspection criteria and AWS guided bend tests.

Procedure with Backing

1. Prepare a single-V groove joint with a 60° groove angle and backing strip. Tack assembly together with a ¼″ root opening. Position workpiece so joint is in 2G position.
2. Use a 90° work angle and a 5° to 10° push travel angle. Move around joint with a Z-weave motion with nozzle in contact with joint to deposit root pass. Pause briefly at each toe to fuse into sides of joint. Brush weld with appropriate brush.
3. Deposit second pass using the same electrode angles and weaving motion. Pause briefly at each toe to fuse into edges of joint. Brush weld with appropriate brush.
4. Deposit remaining passes with the same electrode angles and weaving motion. Brush weld after each pass with appropriate brush.
5. Face reinforcement should be 1⁄16″ maximum.
6. Root reinforcement should be 1⁄16″ maximum.
7. Prepare best weld for a guided bend test.

Procedure without Backing

1. Prepare a single-V groove open root joint with a 60° groove angle and a 3⁄32″ to ⅛″ root face. Tack assembly together with and 3⁄32″ to ⅛″ root opening. Position workpiece so joint is in 2G position.
2. Use a 90° work angle and a 5° to 10° push travel angle. Move up joint with a Z-weave motion with nozzle in contact with joint to deposit root pass. Pause briefly at each toe to fuse into sides of joint. Brush weld with appropriate brush.
3. Deposit second pass using the same electrode angles and weaving motion. Pause briefly at each toe to fuse into the edges of joint. Brush weld with appropriate brush.
4. Deposit remaining passes with the same electrode angles and weaving motion. Brush weld after each pass with appropriate brush.
5. Face reinforcement should be 1⁄16″ maximum.
6. Root reinforcement should be 1⁄16″ maximum.
7. Prepare best weld for a guided bend test.

advanced exercise **4-37**

Making Multiple-Pass Groove Welds, 5G Position, on Carbon Steel Pipe with and without Backing, Using GTAW

Conditions

Position: 5G
Materials: 4″ to 6″ schedule 40 to 80 carbon steel pipe
Backing ring to fit

Electrodes: EWCe-2 or EWLa-X
Filler Metal: 3⁄32″ or 1⁄8″ ER70S-6
Shielding: 100% Ar
Polarity: DCEN
Amperage: 125 to 200 for 3⁄32″, 150 to 230 for 1⁄8″
Voltage: 17 to 18 for 3⁄32″, 18 to 19 for 1⁄8″

Performance

The welder will demonstrate the correct procedure and technique for welding pipe in the 5G position.

Criteria

The welds will pass visual inspection criteria and AWS guided bend tests.

Procedure with Backing

1. Bevel two pieces of pipe with a 37.5° groove face and no root face. Insert backing ring and tack weld to form a single-V groove joint. Position workpiece so joint is in 5G position.
2. Use a 90° work angle and a 5° to 10° push travel angle. Move up joint with a Z-weave motion with nozzle in contact with joint to deposit root pass. Pause briefly at each toe to fuse into sides of joint. Brush weld.
3. Deposit second pass using the same electrode angles and weaving motion. Pause briefly at each toe to fuse into edges of joint. Brush weld.
4. Deposit remaining passes with the same electrode angles and weaving motion. Brush weld after each pass.
5. Face reinforcement should be 1⁄16″ maximum.
6. Root reinforcement should be 1⁄16″ maximum.
7. Prepare best weld for a guided bend test.

Procedure without Backing

1. Prepare a single-V groove open root joint with a 60° groove angle and a 3⁄32″ to 1⁄8″ root face. Tack assembly together with and 3⁄32″ to 1⁄8″ root opening. Position workpiece so joint is in 5G position.
2. Use a 90° work angle and a 5° to 10° push travel angle. Move up joint with a Z-weave motion with nozzle in contact with joint to deposit root pass. Pause briefly at each toe to fuse into sides of joint. Brush weld with appropriate brush.
3. Deposit second pass using the same electrode angles and weaving motion. Pause briefly at each toe to fuse into edges of joint. Brush weld with appropriate brush.
4. Use the same electrode angles and weaving motion to deposit the remaining passes. Pause briefly at each toe to ensure penetration into sides of joint. Brush after each weld with appropriate brush.
5. Face reinforcement should be 1⁄16″ maximum.

6. Root reinforcement should be 1⁄16″ maximum.

7. Prepare best weld for a guided bend test.

advanced exercise **4-38**

Making Multiple-Pass Groove Welds, 6G Position, on Carbon Steel Pipe with and without Backing, Using GTAW

Conditions

Position: 6G
Materials: 4″ to 6″ schedule 40 to 80 carbon steel pipe
Backing ring to fit
Electrodes: EWCe-2 or EWLa-X
Filler Metal: 3⁄32″ or 1⁄8″ ER70S-6
Shielding: 100% Ar
Polarity: DCEN
Amperage: 125 to 200 for 3⁄32″, 150 to 230 for 1⁄8″
Voltage: 17 to 18 for 3⁄32″, 18 to 19 for 1⁄8″

Performance

The welder will demonstrate the correct procedure and technique to weld pipe in the 6G position.

Criteria

The welds will pass visual inspection criteria and AWS guided bend tests.

Procedure with Backing

1. Bevel two pieces of pipe with a 37.5° groove face and no root face. Insert backing ring and tack weld to form a single-V groove joint. Position workpiece so joint is in 6G position.

2. Use a 90° work angle and a 5° to 10° push travel angle. Move up joint with a Z-weave motion with nozzle in contact with joint to deposit root pass. Pause briefly at each toe to fuse into sides of joint. Brush weld.

3. Deposit second pass using the same electrode angles and weaving motion. Pause briefly at each toe to fuse into edges of joint. Brush weld.

4. Deposit remaining passes with the same electrode angles and weaving motion. Brush weld after each pass.

5. Face reinforcement should be 1⁄16″ maximum.

6. Root reinforcement should be 1⁄16″ maximum.

7. Prepare best weld for a guided bend test.

Procedure without Backing

1. Prepare a new workpiece for open root welding. Use spacers and tack weld to establish a 3⁄32″ to 1⁄8″ root face and 3⁄32″ to 1⁄8″ root opening. Position workpiece so joint is in 6G position.
2. Use a 90° work angle and a 5° to 10° push travel angle. Move up joint with a Z-weave motion with nozzle in contact with joint to deposit root pass. Pause briefly at each toe to fuse into sides of joint. Brush weld.
3. Deposit second pass using the same electrode angles and weaving motion. Pause briefly at each toe to fuse into edges of joint. Brush weld.
4. Deposit remaining passes with the same electrode angles and weaving motion. Brush weld after each pass.
5. Face reinforcement should be 1⁄16″ maximum.
6. Root reinforcement should be 1⁄16″ maximum.
7. Prepare best weld for a guided bend test.

advanced exercise **4-39**

Making a Multiple-Pass Fillet Weld, 2F Position, on Carbon Steel Pipe to Plate, Using GTAW with Stainless Steel Filler

Conditions

Position: 2F
Materials: 4″ to 6″ schedule 40 to 80 carbon steel pipe
1⁄4″ carbon steel plate to fit
Electrodes: EWCe-2 or EWLa-X
Filler Metal: 3⁄32″ or 1⁄8″ ER309
Shielding: 100% Ar
Polarity: DCEN
Amperage: 125 to 200 for 3⁄32″, 150 to 230 for 1⁄8″
Voltage: 17 to 18 for 3⁄32″, 18 to 19 for 1⁄8″

Performance

The welder will demonstrate the correct procedure and technique to weld pipe to plate in the 2F position.

Criteria

The weld will pass visual inspection criteria.

Procedure

1. Tack weld pipe to plate. Position workpiece so joint is in 2F position.
2. Use a 45° work angle and a 5° to 10° push travel angle to deposit root pass on both sides of joint. Brush weld.

3. Deposit second pass on bottom toe of root pass with a 50° to 55° work angle and a 5° to 10° push travel angle. Cover root by one-half to two-thirds. Fuse into base metal evenly along toe. Brush weld.

4. Deposit third pass on top toe of root pass with a 30° to 35° work angle and a 5° to 10° push travel angle. Cover second pass by one-third to one-half. Fuse into base metal evenly along toe. Brush weld.

Additional Practice

Tack up new workpiece. Practice with Z-weave motion with nozzle in contact with joint pausing at each weld toe to deposit all of the passes. This is referred to as "walking the cup."

advanced exercise **4-40**

Making a Multiple-Pass Fillet Weld, 4F Position, on Carbon Steel Pipe to Plate, Using GTAW with Stainless Steel Filler

Conditions

Position: 4F
Materials: 4″ to 6″ schedule 40 to 80 carbon steel pipe
¼″ carbon steel plate to fit
Electrodes: EWCe-2 or EWLa-X
Filler Metal: 3⁄32″ or ⅛″ ER309
Shielding: 100% Ar
Polarity: DCEN
Amperage: 125 to 200 for 3⁄32″, 150 to 230 for ⅛″
Voltage: 17 to 18 for 3⁄32″, 18 to 19 for ⅛″

Performance

The welder will demonstrate the correct procedure and technique to weld pipe to plate in the 4F position.

Criteria

The weld will pass visual inspection criteria.

Procedure

1. Tack weld pipe to plate. Position workpiece so joint is in 4F position.

2. Use a 45° work angle and a 5° to 10° push travel angle to deposit root pass on both sides of joint. Brush weld.

3. Deposit second pass on bottom toe of root pass with a 50° to 55° work angle and a 5° to 10° push travel angle. Cover root by one-half to two-thirds. Fuse into base metal evenly along toe. Brush weld.

4. Deposit third pass on top toe of root pass with a 30° to 35° work angle and a 5° to 10° push travel angle. Cover second pass by one-third to one-half. Fuse into base metal evenly along toe. Brush weld.

Additional Practice

Tack up new workpiece. Practice with Z-weave motion with nozzle in contact with joint pausing at each weld toe to deposit all of the passes. This is referred to as "walking the cup."

advanced exercise **4-41**

Making a Multiple-Pass Fillet Weld, 5F Position, Carbon Steel Pipe to Plate, Using GTAW with Stainless Steel Filler

Conditions

Position: 5F
Materials: 4″ to 6″ schedule 40 to 80 carbon steel pipe
1/4″ carbon steel plate to fit
Electrodes: EWCe-2 or EWLa-X
Filler Metal: 3/32″ or 1/8″ ER309
Shielding: 100% Ar
Polarity: DCEN
Amperage: 125 to 200 for 3/32″, 150 to 230 for 1/8″
Voltage: 17 to 18 for 3/32″, 18 to 19 for 1/8″

Performance

The welder will demonstrate the correct procedure and technique to weld pipe to plate in the 5F position.

Criteria

The weld will pass visual inspection criteria.

Procedure

1. Tack weld pipe to plate. Position workpiece so joint is in 5F position.
2. Use a 45° work angle and a 5° to 10° push travel angle to deposit root pass on both sides of joint. Brush weld.
3. Deposit second pass on either toe of root pass with a 50° to 55° work angle and a 5° to 10° push travel angle. Cover root by one-half to two-thirds. Fuse into base metal evenly along toe. Brush weld.
4. Deposit third pass on opposite toe of root pass with a 30° to 35° work angle and a 5° to 10° push travel angle. Cover second pass by one-third to one-half. Fuse into base metal evenly along toe. Brush weld.

Additional Practice

Tack up new workpiece. Practice with Z-weave motion with nozzle in contact with joint pausing at each weld toe to deposit all of the passes. This is referred to as "walking the cup."

advanced exercise 4-42

Making Multiple-Pass Groove Welds, 2G Position, on Carbon Steel Pipe with and without Backing, Using GTAW with Stainless Steel Filler

Conditions

Position: 2G
Materials: 4″ to 6″ schedule 40 to 80 carbon steel pipe
Backing ring to fit
Electrodes: EWCe-2 or EWLa-X
Filler Metal: 3⁄32″ or 1⁄8″ ER309
Shielding: 100% Ar
Polarity: DCEN
Amperage: 125 to 200 for 3⁄32″, 150 to 230 for 1⁄8″
Voltage: 17 to 18 for 3⁄32″, 18 to 19 for 1⁄8″

Performance

The welder will demonstrate the correct procedure and technique for welding pipe in the 2G position.

Criteria

The welds will pass visual inspection criteria and AWS guided bend tests.

Procedure with Backing

1. Prepare a single-V groove joint with a 60° groove angle and backing strip. Tack assembly together with a 1⁄4″ root opening. Position workpiece so joint is in 2G position.
2. Use a 90° work angle and a 5° to 10° push travel angle. Move around joint with a Z-weave motion with nozzle in contact with joint to deposit root pass. Pause briefly at each toe to fuse into sides of joint. Brush weld with appropriate brush.
3. Deposit second pass using the same electrode angles and weaving motion. Pause briefly at each toe to fuse into the edges of the joint. Brush weld with appropriate brush.
4. Deposit remaining passes with the same electrode angles and weaving motion. Brush weld after each pass with appropriate brush.
5. Face reinforcement should be 1⁄16″ maximum.
6. Root reinforcement should be 1⁄16″ maximum.
7. Prepare best weld for a guided bend test.

Procedure without Backing

1. Prepare a new workpiece for open root welding. Use spacers and tack weld to establish a 3⁄32″ to 1⁄8″ root face and 3⁄32″ to 1⁄8″ root opening. Remove slag. Position workpiece so joint is in 2G position.

2. Use a 90° work angle and a 5° to 10° push travel angle. Move around joint with a Z-weave motion with nozzle in contact with joint to deposit root pass. Pause briefly at each toe to fuse into sides of joint. Brush weld with appropriate brush.

3. Deposit second pass using the same electrode angles and weaving motion. Pause briefly at each toe to fuse into edges of joint. Brush weld with appropriate brush.

4. Deposit remaining passes with the same electrode angles and weaving motion. Brush weld after each pass with appropriate brush.

5. Face reinforcement should be 1⁄16″ maximum.

6. Root reinforcement should be 1⁄16″ maximum.

7. Prepare best weld for a guided bend test.

advanced exercise **4-43**

Making Multiple-Pass Groove Welds, 5G Position, on Carbon Steel Pipe with and without Backing, Using GTAW with Stainless Steel Filler

Conditions

Position:	5G
Materials:	4″ to 6″ schedule 40 to 80 carbon steel pipe Backing ring to fit
Electrodes:	EWCe-2 or EWLa-X
Filler Metal:	3⁄32″ or 1⁄8″ ER309
Shielding:	100% Ar
Polarity:	DCEN
Amperage:	125 to 200 for 3⁄32″, 150 to 230 for 1⁄8″
Voltage:	17 to 18 for 3⁄32″, 18 to 19 for 1⁄8″

Performance

The welder will demonstrate the correct procedure and technique for welding pipe in the 5G position.

Criteria

The welds will pass visual inspection criteria and AWS guided bend tests.

Procedure with Backing

1. Bevel two pieces of pipe with a 37.5° groove face and no root face. Insert backing ring and tack weld to form a single-V groove joint. Position workpiece so joint is in 5G position.

2. Use a 90° work angle and a 5° to 10° push travel angle. Move up joint with a Z-weave motion with nozzle in contact with joint to deposit root pass. Pause briefly at each toe to fuse into sides of joint. Brush weld.

3. Deposit second pass using the same electrode angles and weaving motion. Pause briefly at each toe to fuse into edges of joint. Brush weld.
4. Deposit remaining passes with the same electrode angles and weaving motion. Brush weld after each pass.
5. Face reinforcement should be 1⁄16″ maximum.
6. Root reinforcement should be 1⁄16″ maximum.
7. Prepare best weld for a guided bend test.

Procedure without Backing

1. Prepare a new workpiece for open root welding. Use spacers and tack weld to establish a 3⁄32″ to 1⁄8″ root face and 3⁄32″ to 1⁄8″ root opening. Position workpiece so joint is in 5G position.
2. Use a 90° work angle and a 5° to 10° push travel angle. Move up joint with a Z-weave motion with nozzle in contact with joint to deposit root pass. Pause briefly at each toe to fuse into sides of joint. Brush weld.
3. Deposit second pass using the same electrode angles and weaving motion. Pause briefly at each toe to fuse into edges of joint. Brush weld.
4. Deposit remaining passes with the same electrode angles and weaving motion. Brush weld after each pass.
5. Face reinforcement should be 1⁄16″ maximum.
6. Root reinforcement should be 1⁄16″ maximum.
7. Prepare best weld for a guided bend test.

advanced exercise **4-44**

Making Multiple-Pass Groove Welds, 6G Position, on Carbon Steel Pipe with and without Backing, Using GTAW with Stainless Steel Filler

Conditions

Position: 6G
Materials: 4″ to 6″ schedule 40 to 80 carbon steel pipe
Backing ring to fit
Electrodes: EWCe-2 or EWLa-X
Filler Metal: 3⁄32″ or 1⁄8″ ER309
Shielding: 100% Ar
Polarity: DCEN
Amperage: 125 to 200 for 3⁄32″, 150 to 230 for 1⁄8″
Voltage: 17 to 18 for 3⁄32″, 18 to 19 for 1⁄8″

Performance

The welder will demonstrate the correct procedure and technique to weld pipe in the 6G position.

Criteria

The welds will pass visual inspection criteria and AWS guided bend tests.

Procedure with Backing

1. Bevel two pieces of pipe with a 37.5° groove face and no root face. Insert backing ring and tack weld to form a single-V groove joint. Position workpiece so joint is in 6G position.
2. Use a 90° work angle and a 5° to 10° push travel angle. Move up joint with a Z-weave motion with nozzle in contact with joint to deposit root pass. Pause briefly at each toe to fuse into sides of joint. Brush weld.
3. Deposit second pass using the same electrode angles and weaving motion. Pause briefly at each toe to fuse into edges of joint. Brush weld.
4. Deposit remaining passes with the same electrode angles and weaving motion. Brush weld after each pass.
5. Face reinforcement should be 1⁄16″ maximum.
6. Root reinforcement should be 1⁄16″ maximum.
7. Prepare best weld for a guided bend test.

Procedure without Backing

1. Prepare a new workpiece for open root welding. Use spacers and tack weld to establish a 3⁄32″ to 1⁄8″ root face and 3⁄32″ to 1⁄8″ root opening. Position workpiece so joint is in 6G position.
2. Use a 90° work angle and a 5° to 10° push travel angle. Move up joint with a Z-weave motion with nozzle in contact with joint to deposit root pass. Pause briefly at each toe to fuse into sides of joint. Brush weld.
3. Deposit second pass using the same electrode angles and weaving motion. Pause briefly at each toe to fuse into edges of joint. Brush weld.
4. Deposit remaining passes with the same electrode angles and weaving motion. Brush weld after each pass.
5. Face reinforcement should be 1⁄16″ maximum.
6. Root reinforcement should be 1⁄16″ maximum.
7. Prepare best weld for a guided bend test.

advanced exercise **4-45**

Making a Multiple-Pass Groove Weld, 6G Position, on Thick-Wall Carbon Steel Pipe with GTAW Root and SMAW Intermediate and Cover Passes

Conditions

Position: 6G
Materials: 2.75″ schedule 180 carbon steel pipe

Electrodes: EWCe-2 or EWLa-X
3⁄32″ or 1⁄8″ E7018
Filler Metal: 3⁄32″ or 1⁄8″ ER70S-6
Shielding: 100% Ar
Polarity: DCEN for GTAW (DCEP for SMAW)
Amperage: 125 to 200 for 3⁄32″, 150 to 230 for 1⁄8″ for GTAW
70 to 90 for 3⁄32″, 90 to 130 for 1⁄8″ for SMAW
Voltage: 17 to 18 for 3⁄32″, 18 to 19 for 1⁄8″ for GTAW
17 to 19 for 3⁄32″, 19 to 20 for 1⁄8″ for SMAW

Performance

The welder will demonstrate the correct procedure and technique to weld 2.75″ schedule 180 carbon steel pipe in the 6G position with GTAW root and SMAW intermediate and cover passes.

Criteria

The weld will pass visual inspection criteria and an AWS side bend test.

Procedure

1. Bevel two pieces of pipe with a 37.5° groove face and prepare a 3⁄32″ root face. Use a 3⁄32″ spacer to form a 3⁄32″ root opening and tack weld with GTAW to form a single-V groove joint with a 75° groove angle. Position workpiece so joint is in 6G position.

2. Use a 90° work angle and a 5° to 10° push travel angle. Move up joint with a Z-weave motion with nozzle in contact with joint to deposit root pass. Pause briefly at each toe to fuse into sides of joint. Brush weld.

3. Deposit an optional second pass with GTAW. Use a 90° work angle and a 5° to 10° push travel angle. Move up joint with same electrode angles to deposit second pass. Brush weld. Or switch to SMAW. Deposit second pass on bottom toe of root pass using a 3⁄32″ E7018 electrode with a 50° to 55° drag travel angle. Cover root by one-half to two-thirds. Fuse into base metal evenly along toe. Remove slag.

4. Deposit third pass on top toe of root pass with a 30° to 35° work angle and a 5° to 10° drag travel angle. Cover second pass by one-third to one-half. Fuse into base metal evenly along toe. Remove slag.

5. Switch to 1⁄8″ E7018 electrodes. Increase amperage and voltage as shown above to deposit remainder of intermediate passes and cover pass. Remove slag after each pass.

6. Face reinforcement should be 1⁄8″ maximum.

7. Root reinforcement should be 1⁄16″ maximum.

8. Prepare best weld for a guided bend test.

Name ______________________________ **Date** ____________________

exercise 5-1

Depositing Beads on Mild Steel Using GMAW

Conditions

Flat position
DCEP
90 A to 120 A
17 V to 20 V
0.035″ diameter wire, E70S-3
100% CO_2 or 75% Ar / 25% CO_2 shielding gas, 15 cfh to 25 cfh
Electrode extension, ¼″ to ⅜″
3⁄16″ to ¼″ × 4″ × 6″ mild steel

Performance

Deposit beads on a mild steel plate using GMAW.

Criteria

The beads should be approximately 5⁄16″ wide and ⅛″ high and consistent in width and straightness.

Procedure

1. Set the wire feeder for the correct current setting as detailed in the text.
2. Adjust the voltage for the correct setting.
3. Use a 90° work angle and a 10° to 15° drag angle.
4. Deposit a series of straight, consistent beads approximately ⅜″ apart.

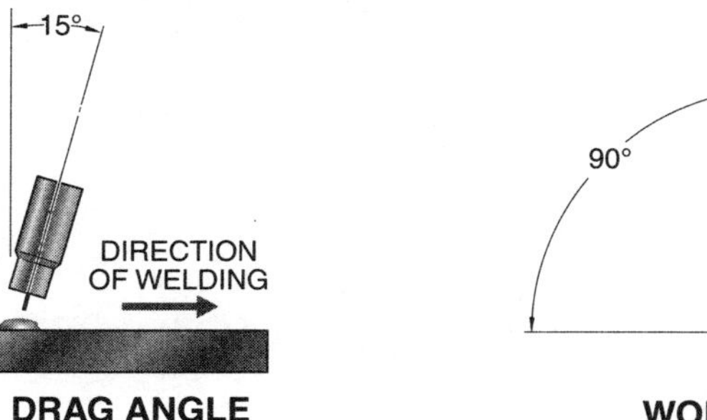

exercise **5-2**

Surfacing Using GMAW

Conditions

Flat position
Refer to Exercise 5-1

Performance

The welder will deposit successive overlapping beads to increase the thickness of the workpiece.

Criteria

Each bead should overlap the previous bead in half and completely penetrate the previous bead and the base metal without voids in the weld metal applied.

Procedure

1. Refer to Exercise 5-1 for equipment setup and adjustment.
2. Deposit a bead ¼″ from the edge of the workpiece and parallel to the length of the workpiece.
3. Use an 80° to 90° work angle and a 10° to 15° drag angle to overlap the first bead in half.
4. Deposit consistent, overlapping beads until the workpiece is covered.

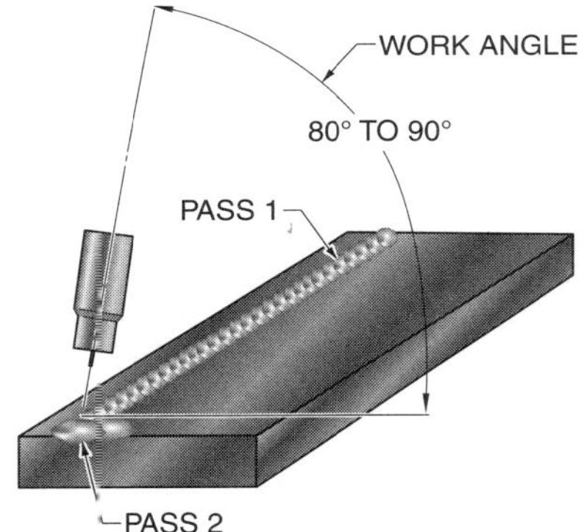

exercise **5-3**

Welding a Butt Joint on Mild Steel in Flat Position Using GMAW

Conditions

Flat position
Refer to Exercise 5-1
Two pieces of ¼″ × 1½″ × 6″ mild steel

Performance

The welder will demonstrate the correct procedure for welding a butt joint in flat position.

Criteria

Welding technique, weld appearance, and weld strength as evaluated by the instructor.

Procedure

1. Refer to Exercise 5-1 for equipment setup and adjustment.
2. Tack weld the two workpieces to form a butt joint. Allow a ³⁄₃₂″ root opening.
3. Use a 90° work angle and a 10° drag angle. A slight weaving motion may be used to control the weld pool.

exercise **5-4**

Welding a Lap Joint on Mild Steel in Flat Position Using GMAW

Conditions

Flat position
Refer to Exercise 5-3

Performance

The welder will demonstrate the correct procedure for welding a lap joint in flat position.

Criteria

Welding technique, weld appearance, and weld strength as evaluated by the instructor.

Procedure

1. Refer to Exercise 5-1 for equipment setup and adjustment.
2. Tack weld the two pieces to form a lap joint.
3. Secure the tack welded workpiece in a positioner, or rest it against a firebrick to obtain flat position.
4. Use a 45° work angle and a 10° to 15° drag angle with a slight weaving motion to deposit the bead.
5. The bead face should be flat to slightly convex.

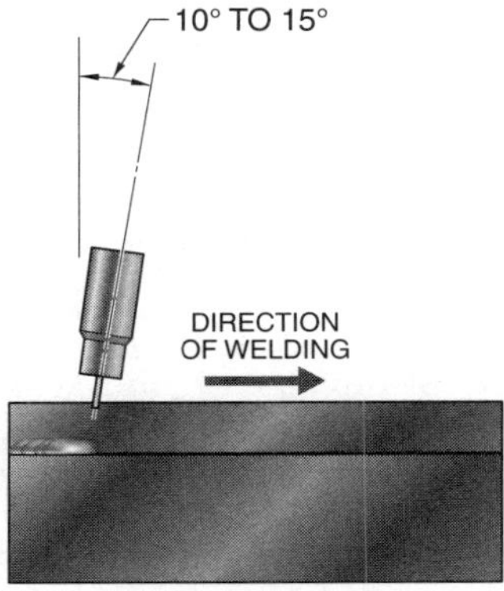

DRAG ANGLE

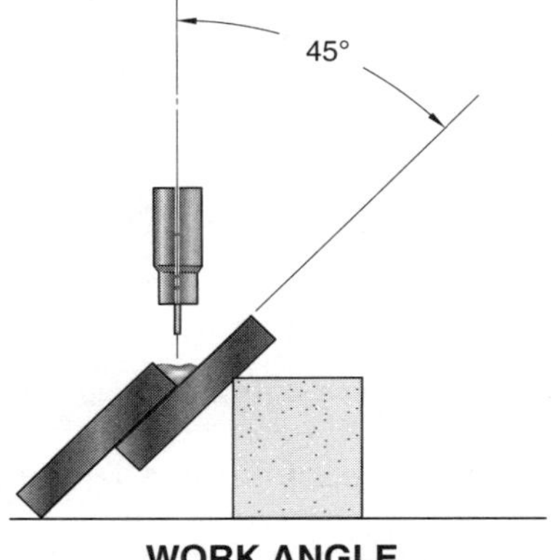

WORK ANGLE

exercise 5-5

Depositing Beads on Mild Steel in Horizontal Position Using GMAW

Conditions

Horizontal position
Refer to Exercise 5-1

Performance

The welder will demonstrate the correct procedure for depositing beads in horizontal position.

Criteria

The beads should be approximately 5⁄16″ wide and 1⁄8″ high. The beads should be parallel to the length of the workpiece and free from undercut or overlap.

Procedure

1. Refer to Exercise 5-1 for equipment setup and adjustment.
2. Position the workpiece so the beads can be deposited in horizontal position.
3. Use an 80° work angle and a 10° to 15° drag angle.
4. Deposit a series of straight, consistent beads approximately 3⁄8″ apart.

exercise 5-6

Welding a Multiple-Pass T-Joint in Horizontal Position Using GMAW

Conditions

Horizontal position
Refer to Exercise 5-1
Two pieces of 1⁄4″ × 2″ × 4″ mild steel

Performance

The welder will demonstrate the correct procedure for welding a multiple-pass T-joint in horizontal position.

Criteria

Welding technique, weld appearance, and weld strength as evaluated by the instructor.

Procedure

1. Refer to Exercise 5-1 for equipment setup and adjustment.
2. Tack weld the two pieces to form a T-joint.
3. Position the welding gun at a 45° work angle and a 10° to 15° drag angle.

4. Deposit the first pass on both sides of the T-joint.
5. Position the welding gun at a 55° work angle and a 10° to 15° drag angle.
6. Deposit the second pass, penetrating half of the first pass and the bottom workpiece on both sides of the joint.
7. Position the welding gun at a 35° work angle and a 10° to 15° drag angle.
8. Deposit the third pass, penetrating half of the first pass and the top workpiece on both sides of the joint.

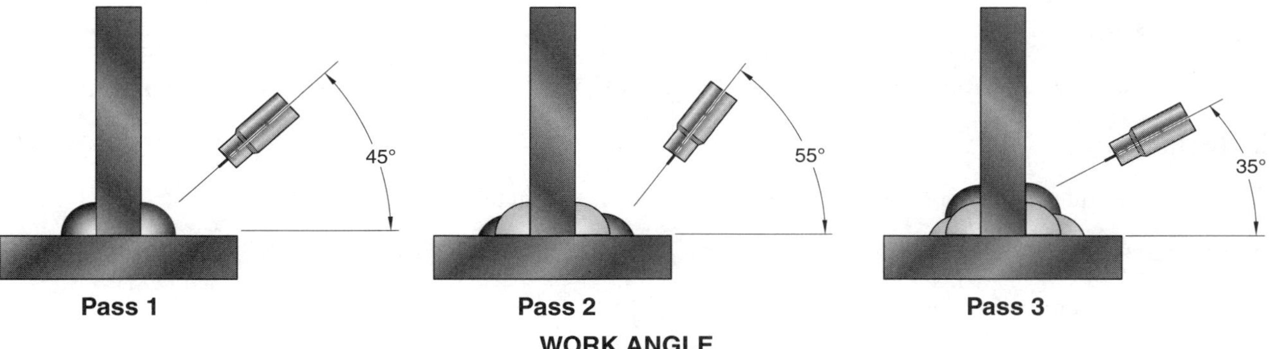

WORK ANGLE

exercise 5-7

Welding a Lap Joint on Mild Steel in Horizontal Position Using GMAW

Conditions

Horizontal position
Refer to Exercise 5-3

Performance

The welder will demonstrate the correct procedure for welding a lap joint in horizontal position.

Criteria

Welding technique, weld appearance, and weld strength as evaluated by the instructor.

Procedure

1. Refer to Exercise 5-3 for equipment setup and adjustment.
2. Tack weld the two pieces to form a lap joint.
3. Position the welding gun at a 45° work angle and a 10° to 15° drag angle.
4. Deposit the weld with a slight weaving motion. Pause at the toes of the weld to prevent undercutting.
5. Repeat the weld on the other side of the joint.

exercise 5-8

Welding a Butt Joint on Mild Steel in Horizontal Position Using GMAW

Conditions

Horizontal position
Refer to Exercise 5-3

Performance

The welder will demonstrate the correct procedure for welding a butt joint in horizontal position.

Criteria

Welding technique, weld appearance, and weld strength as evaluated by the instructor.

Procedure

1. Refer to Exercise 5-1 for equipment setup and adjustment.
2. Tack weld the two pieces to form a butt joint. Allow a 3⁄32″ root opening.
3. Position the workpiece so the joint is in horizontal position.
4. Use an 80° work angle and a 10° to 15° drag angle.
5. Deposit a bead approximately 5⁄16″ wide across the workpiece.

exercise 5-9

Depositing Beads on Mild Steel in Vertical Position Using GMAW

Conditions

Vertical position (downhill)
Refer to Exercise 5-6

Performance

The welder will demonstrate the correct procedure for depositing beads in both uphill and downhill positions.

Criteria

The beads should be approximately 1⁄4″ wide, 1⁄8″ high, and parallel to the length of the workpiece.

Procedure

1. Refer to Exercise 5-1 for equipment setup and adjustment.
2. Position the workpiece so the joint is in vertical position.
3. Using the downhill technique, position the welding gun with a 90° work angle and a 10° to 15° drag angle.

4. Deposit a series of consistent beads approximately ⅜″ apart.
5. On the other side of the workpiece, repeat the operation using the uphill technique.

exercises 5-10 and 5-11

Welding a Butt Joint on Mild Steel in Vertical Position Using GMAW

Conditions

Vertical position
Refer to Exercise 5-6
Exercise 5-10. Downhill
Exercise 5-11. Uphill

Performance

The welder will demonstrate the correct procedure for welding a butt joint in vertical position.

Criteria

Welding technique, weld appearance, and weld strength as evaluated by the instructor.

Procedure

1. Refer to Exercise 5-1 for equipment setup and adjustment.
2. Tack weld the two pieces to form a butt joint. Allow a 3⁄32″ root opening.
3. Position the workpiece so the joint is in vertical position.
4. Start the weld at the top (Exercise 5-10), or at the bottom (Exercise 5-11) of the joint.
5. Use a 90° work angle and a 15° drag angle with a slight weaving motion.

exercise 5-12

Welding a T-Joint on Mild Steel in Vertical Position (Downhill) Using GMAW

Conditions

Vertical position (downhill)
Refer to Exercise 5-6

Performance

The welder will demonstrate the correct procedure for welding a T-joint in vertical position.

Criteria

Welding technique, weld appearance, and weld strength as evaluated by the instructor.

Procedure

1. Refer to Exercise 5-1 for equipment setup and adjustment.
2. Tack weld the two pieces to form a T-joint. Position the workpiece so the joint is in vertical position.
3. Start at the top of the joint and use a 45° work angle and a 10° to 20° drag angle to deposit the bead.
4. Use a slight weaving motion, pausing at the toes to prevent undercutting.

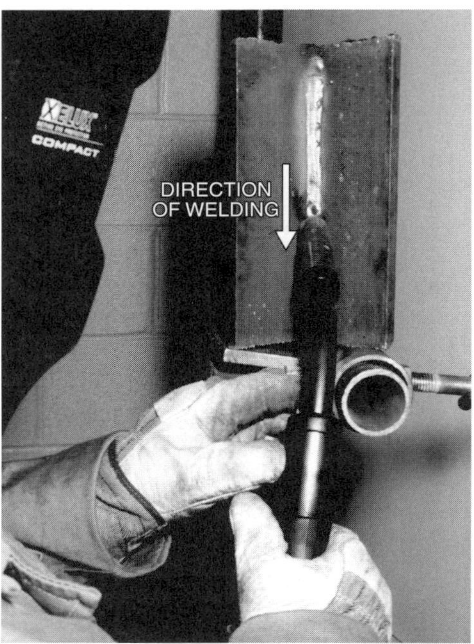

exercise **5-13**

Welding a Single-V-Groove Butt Joint on Mild Steel in Vertical Position Using GMAW

Conditions

Vertical position (downhill)
Refer to Exercise 5-6

Performance

The welder will demonstrate the correct procedure for welding a single-V butt joint in vertical position.

Criteria

Welding technique, weld appearance, and weld strength as evaluated by the instructor.

Procedure

1. Refer to Exercise 5-1 for equipment setup and adjustment.
2. Bevel one side of each plate 30° along its length with a 3⁄32″ root face.

3. Tack weld the two pieces to form a single-V-groove joint with a 60° groove angle. Allow a 3⁄32″ root opening.
4. Position the workpiece so the joint is in vertical position.
5. Start the weld at the top of the joint using a 90° work angle and a 15° drag angle.
6. Use a weaving motion to deposit additional passes required to cover the groove faces.

exercise **5-14**

Welding a Single-V-Groove Butt Joint on Mild Steel in Overhead Position Using GMAW

Conditions

Overhead position
Refer to Exercise 5-6

Performance

The welder will demonstrate the correct procedure for welding a single-V butt joint in overhead position.

Criteria

Welding technique, weld appearance, and weld strength as evaluated by the instructor.

Procedure

1. Refer to Exercise 5-13, Steps 1–3.
2. Position the workpiece so the joint is in overhead position.
3. Use a 90° work angle and a 10° to 15° drag angle to deposit the root pass.
4. Use a weaving motion to deposit additional passes required to cover the groove faces.

exercise **5-15**

Welding a Multiple-Pass T-Joint on Mild Steel in Overhead Position Using GMAW

Conditions

Overhead position
Refer to Exercise 5-6

Performance

The welder will demonstrate the correct procedure for welding a multiple-pass T-joint in overhead position.

Criteria

Welding technique, weld appearance, and weld strength as evaluated by the instructor.

Procedure

1. Refer to Exercise 5-1 for equipment setup and adjustment.
2. Tack weld the two pieces to form a T-joint.
3. Position the workpiece so the joint is in overhead position.
4. Deposit the first pass using a 45° work angle and a 5° to 10° drag angle on both sides of the joint.
5. Deposit the second pass using a 50° work angle and a 5° to 10° drag angle on both sides of the joint with a slight weaving motion.
6. Deposit the third pass using a 40° work angle and a 5° to 10° drag angle on both sides of the joint with a slight weaving motion.

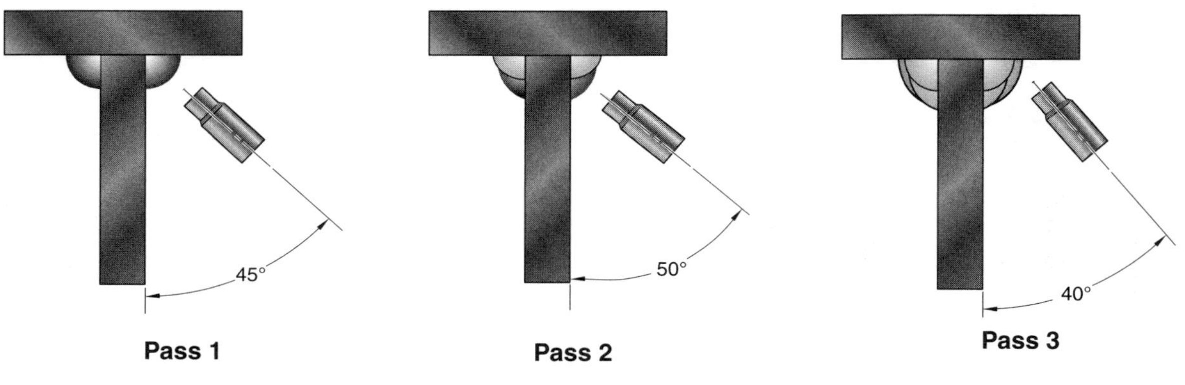

WORK ANGLE

exercise 5-16

Welding a Lap Joint on Mild Steel in Flat Position Using GMAW Spray Transfer

Conditions

Flat position
DCEP
235 A to 245 A
29 V to 30 V
0.045″ diameter wire, E70S-3
98% argon, 2% oxygen, 35 to 40 CFH
Electrode extension, ¾″ to 1″
Two pieces ¼″ × 4″ × 6″ mild steel

Performance

The welder will demonstrate the correct procedure for welding a lap joint in flat position with GMAW spray transfer.

Criteria

Welding technique will be observed, and welds will be visually examined by the instructor.

Procedure

1. Tack weld the two pieces to form a lap joint.
2. Position the workpiece so the joint is in flat position.
3. Position the electrode at a 45° work angle and a 5° to 10° push angle.
4. Use a steady push to maintain a consistent bead for the length of the plate. The finished weld should have ¼″ equal legs and a flat bead contour.
5. Deposit a fillet weld on the other side of the joint.

exercise 5-17

Welding a Lap Joint on Mild Steel in Horizontal Position Using GMAW Spray Transfer

Conditions

Horizontal position
Refer to Exercise 5-16

Performance

The welder will demonstrate the correct procedure for welding a lap joint in horizontal position with GMAW spray transfer.

Criteria

Welding technique will be observed, and welds will be visually examined by the instructor.

Procedure

1. Refer to Exercise 5-16 for equipment setup and adjustment.
2. Tack weld the two pieces to form a lap joint.
3. Position the workpiece so the joint is in horizontal position.
4. Position the electrode at a 45° work angle and a 10° to 15° push angle.
5. Use a steady push to maintain a consistent bead for the length of the plate. The finished weld should have ¼″ equal legs and a flat bead contour.
6. Deposit a fillet weld on the other side of the joint.

exercise 5-18

Welding a Single-V-Groove Joint with Backing on Mild Steel in Flat Position Using GMAW Spray Transfer

Conditions

Flat position
Refer to Exercise 5-17
Two pieces ⅜″ × 4″ × 6″ mild steel
On piece ¼″ × ½″ × 8″ mild steel for backing

Performance

The welder will demonstrate the correct procedure for welding a single-V-groove joint in flat position with GMAW spray transfer.

Criteria

Welding technique will be observed, and welds will be visually examined by the instructor.

Procedure

1. Refer to Exercise 5-16 for equipment setup and adjustment.
2. Bevel one side of each plate 35° along its length.
3. Center two pieces on a ¼″ backing strip and tack weld to form a single-V-groove joint with a 70° groove angle and a ¼″ root opening. The backing strip should form a 1″ starting and runoff tab.
4. Position the workpiece so the joint is in flat position.
5. Position the electrode at a 90° work angle and a 10° to 15° push angle and deposit the first pass with a steady push.
6. For the second pass, position the electrode on either toe of the root pass. The work angle is 85° with a 10° to 15° push angle. Cover the root pass by about one-half to two-thirds.
7. Deposit the third pass on the other toe using the same electrode angles.
8. Deposit a three-bead cover layer. Use a 90° work angle for each pass. Melt into the edges of the joint by about 1⁄16″ to ensure complete fusion along the toes.
9. The finished weld should be flat to slightly convex.

advanced exercise **5-19**

Making a Multiple-Pass Groove Weld, 1G Position with Backing on Carbon Steel, Using GMAW Spray Transfer

Conditions

Position: 1G
Materials: 2 pieces 1″ × 6″ × 7″ mild steel
Backing: ¼″ × 1½″ × 9″ mild steel
Electrodes: 0.035″ or 0.045″ ER70S-6
Shielding: 90% Ar/10% CO_2 or 98% Ar/2% O_2
Polarity: DCEP
WFS*: 500 to 520 for 0.035″, 375 to 390 for 0.045″
Amperage: 230 to 240 for 0.035″, 335 to 345 for 0.045″
Voltage: 28 to 29 for 0.035″, 30 to 31 for 0.045″
*wire feed speed

Performance

The welder will demonstrate the correct procedure and technique for depositing a weld in a single-V groove joint in the flat 1G position.

Criteria

The weld will pass visual inspection and an AWS side bend test.

Procedure

1. Prepare a single-V groove joint with a 60° groove angle and backing strip. Tack assembly together with a ¼″ root opening. Position workpiece so joint is in 1G position.
2. Use a 90° work angle and a 10° to 15° push travel angle. Move along joint with a steady push travel to deposit root pass. Brush weld.
3. Position electrode on either toe of root pass with 85° work angle and 10° to 15° push travel angle. Deposit first intermediate pass. Cover root pass by two-thirds. Brush weld.
4. Position electrode on opposite toe. Deposit weld with same electrode angles. Cover first intermediate pass by one-half to two-thirds. Allow workpiece to cool as needed before depositing second layer. Brush weld.
5. Number of passes per layer will increase as joint widens. Use same travel angle and adjust work angle as necessary. Allow workpiece to cool as needed after each layer. Brush weld
6. Last layer before cover should be about 1⁄16″ below joint edges.
7. Face reinforcement of finished weld should be ⅛″ maximum.
8. Root reinforcement of finished weld should be 1⁄16″ maximum.
9. Prepare best weld for a side bend test.

advanced exercise 5-20

Making a Multiple-Pass Fillet Weld, T-Joint, 2F Position, Using GMAW-P

Conditions

Position: 2F
Materials: 2 pieces ¼″ × 2″ × 4″ mild steel
Electrodes: 0.035″ or 0.045″ ER70S-6
Shielding: 90% Ar/10% CO_2 or 98% Ar/2% O_2
Polarity: DCEP
WFS*: 230 to 250 for 0.035″, 250 to 270 for 0.045″
Amperage: 130 to 140 for 0.035″, 140 to 150 for 0.045″
Voltage: 16 to 18 for 0.035″, 20 to 23 for 0.045″
*wire feed speed

Performance

The welder will demonstrate the correct procedure and technique for depositing a weld in a T-joint.

Criteria

The weld will pass visual inspection criteria.

Procedure

1. Tack weld two pieces to form a T-joint. Position workpiece so joint is in 2F position.
2. Use a 45° work angle and a 10° to 20° push travel angle for root pass. Use a steady push travel to deposit root pass on both sides of joint. Brush welds.
3. Deposit second pass on bottom toe of root pass with a 50° to 55° work angle. Use a steady push travel. Cover root by one-half to two-thirds. Fuse into base metal evenly along toe. Brush weld.
4. Deposit third pass on top toe of root pass with a 30° to 35° work angle. Cover second pass by one-third to one-half. Fuse into base metal evenly along toe. Brush weld.
5. Deposit welds on other side of joint.

Additional Practice

Repeat exercise using a Z-weave motion for second and third passes. Pause at each toe to ensure complete fusion into sides of joint.

advanced exercise 5-21

Making a Multiple-Pass Fillet Weld, T-Joint, 3F Position, Using GMAW-P

Conditions

Position: 3F
Materials: 2 pieces ¼″ × 2″ × 4″ mild steel
Electrodes: 0.035″ or 0.045″ ER70S-6
Shielding: 90% Ar/10% CO_2 or 98% Ar/2% O_2
Polarity: DCEP
WFS*: 230 to 250 for 0.035″, 250 to 270 for 0.045″
Amperage: 130 to 140 for 0.035″, 140 to 150 for 0.045″
Voltage: 16 to 18 for 0.035″, 20 to 23 for 0.045″
*wire feed speed

Performance

Refer to Exercise 5-20.

Criteria

Refer to Exercise 5-20.

Procedure

1. Tack weld two pieces to form a T-joint. Position workpiece so joint is in 3F position.
2. Use a 45° work angle and a 10° to 20° push travel angle to deposit root pass on both sides of joint. Brush weld.
3. Deposit second pass on either toe of root pass with a 50° to 55° work angle and a 10° to 20° push travel angle. Cover root by one-half to two-thirds. Fuse into base metal evenly along toe. Brush weld.
4. Deposit third pass on opposite toe of root pass with a 30° to 35° work angle and a 10° to 20° push travel angle. Cover second pass by one-third to one-half. Fuse into base metal evenly along toe. Brush weld.
5. Deposit welds on other side of joint.

Additional Practice

Repeat exercise using a Z-weave motion for second and third passes. Pause at each toe to ensure complete fusion into sides of joint.

advanced exercise **5-22**

Making a Multiple-Pass Fillet Weld, T-Joint, 4F Position, Using GMAW-P

Conditions

Position: 4F
Materials: 2 pieces ¼″ × 2″ × 4″ mild steel
Electrodes: 0.035″ or 0.045″ ER70S-6
Shielding: 90% Ar/10% CO_2 or 98% Ar/2% O_2
Polarity: DCEP
WFS*: 230 to 250 for 0.035″, 250 to 270 for 0.045″
Amperage: 130 to 140 for 0.035″, 140 to 150 for 0.045″
Voltage: 16 to 18 for 0.035″, 20 to 23 for 0.045″
*wire feed speed

Performance

Refer to Exercise 5-20.

Criteria

Refer to Exercise 5-20.

Procedure

1. Tack weld two pieces to form a T-joint. Position workpiece so joint is in 4F position.
2. Use a 45° work angle and a 5° to 10° slight push travel angle to deposit root pass on both sides of joint. Brush weld.
3. Deposit second pass on bottom toe of root pass with a 50° to 55° work angle and a 5° to 10° slight push travel angle. Cover root by one-half to two-thirds. Fuse into base metal evenly along toe. Brush weld.
4. Deposit third pass on top toe of root pass with a 30° to 35° work angle and a 5° to 10° slight push travel angle. Cover second pass by one-third to one-half. Fuse into base metal evenly along toe. Brush weld.
5. Deposit welds on other side of joint.

Additional Practice

Repeat exercise using a Z-weave motion for second and third passes. Pause at each toe to ensure complete fusion into sides of joint.

advanced exercise **5-23**

Making Multiple-Pass Groove Welds, 3G Position, with and without Backing, Using GMAW-P

Conditions

Position: 3G
Materials: 2 pieces 1″ × 6″ × 7″ mild steel
Backing: ¼″ × 1½″ × 9″ mild steel
Electrodes: 0.035″ or 0.045″ ER70S-6
Shielding: 90% Ar/10% CO_2 or 98% Ar/2% O_2
Polarity: DCEP
WFS*: 230 to 250 for 0.035″, 250 to 270 for 0.045″
Amperage: 130 to 140 for 0.035″, 140 to 150 for 0.045″
Voltage: 16 to 18 for 0.035″, 20 to 23 for 0.045″
*wire feed speed

Performance

The welder will demonstrate the correct procedure and technique for depositing welds in single-V groove joints in the vertical 3G position.

Criteria

The welds will pass visual inspection and AWS side bend tests.

Procedure with Backing

1. Prepare a single-V groove joint with a 60° groove angle and backing strip. Tack assembly together with a ¼″ root opening. Position workpiece so joint is in 3G position.
2. Use a 90° work angle and a 10° to 20° push travel angle. Move up joint with a Z-weave motion to deposit root pass. Pause briefly at each toe to fuse into sides of joint. Brush weld.
3. Deposit second pass using same electrode angles and weaving motion. Pause briefly at each toe to fuse into edges of joint. Brush weld.
4. Deposit remaining passes with same electrode angles and weaving motion. Brush weld after each pass.
5. Face reinforcement should be ⅛″ maximum.
6. Root reinforcement should be ¹⁄₁₆″ maximum.
7. Prepare best weld for a side bend test.

Procedure without Backing

1. Prepare a single-V groove open root joint with a 60° groove angle and a ³⁄₃₂″ to ⅛″ root face. Tack assembly together with a ³⁄₃₂″ to ⅛″ root opening. Position workpiece so joint is in 3G position.

2. Use a 90° work angle and a 10° to 20° push travel angle. Move up joint with a steady push angle. Control weld pool to ensure complete joint penetration. Brush weld.
3. Use a weaving motion to deposit remaining passes. Pause briefly at each toe to ensure penetration in sides of joint. Brush after each weld.
4. Face reinforcement should be ⅛″ maximum.
5. Root reinforcement should be 1⁄16″ maximum.
6. Prepare best weld for a side bend test.

advanced exercise **5-24**

Making Multiple-Pass Groove Welds, 4G Position, with and without Backing, Using GMAW-P

Conditions

Position: 4G
Materials: 2 pieces 1″ × 6″ × 7″ mild steel
Backing: ¼″ × 1½″ × 9″ mild steel
Electrodes: 0.035″ or 0.045″ ER70S-6
Shielding: 90% Ar/10% CO_2 or 98% Ar/2% O_2
Polarity: DCEP
WFS*: 230 to 250 for 0.035″, 250 to 270 for 0.045″
Amperage: 130 to 140 for 0.035″, 140 to 150 for 0.045″
Voltage: 16 to 18 for 0.035″, 20 to 23 for 0.045″
*wire feed speed

Performance

The welder will demonstrate the correct procedure and technique for depositing welds in single-V groove joints in the overhead 4G position.

Criteria

The welds will pass visual inspection and AWS side bend tests.

Procedure with Backing

1. Prepare a single-V groove joint with a 60° groove angle and backing strip. Tack assembly together with a ¼″ root opening. Position workpiece so joint is in 4G position.
2. Use a 90° work angle and a 5° to 10° slight push travel angle. Move up joint with a steady push to deposit root pass. A slight oscillation may help to fuse into sides of joint. Brush weld.
3. Deposit second and third pass on either toe using an 85° work angle and a 5° to 10° slight push travel angle. Cover root by one-half to two-thirds. Fuse into base metal evenly along toe. Brush weld.
4. Deposit third pass on opposite toe with same work and electrode angles. Cover second pass by one-third to one-half. Fuse into base metal evenly along toe. Brush weld.
5. Face reinforcement should be ⅛″ maximum.
6. Root reinforcement should be ¹⁄₁₆″ maximum.
7. Prepare best weld for a side bend test.

Procedure without Backing

1. Prepare a single-V groove open root joint with a 60° groove angle and a 3⁄32″ to 1⁄8″ root face. Tack assembly together with a 3⁄32″ to 1⁄8″ root opening. Position workpiece so joint is in 4G position.
2. Use a 90° work angle and a 5° to 10° slight push travel angle. Move up joint with a steady push. Control weld pool to ensure complete joint penetration. Brush weld.
3. Use a weaving motion to deposit remaining passes. Pause briefly at each toe to ensure penetration into sides of joint. Brush after each weld.
4. Face reinforcement should be 1⁄8″ maximum.
5. Root reinforcement should be 1⁄16″ maximum.
6. Prepare best weld for a side bend test.

advanced exercise **5-25**

Making a Multiple-Pass Fillet Weld, T-Joint, 2F Position, with Stainless Steel Electrode Wire, Using GMAW-P

Conditions

Position: 2F
Materials: 2 pieces ¼″ × 2″ × 4″ mild steel
Electrodes: 0.035″ or 0.045″ ER309
Shielding: 90% He/8% Ar/2% O_2
Polarity: DCEP
WFS*: 156 to 225 for 0.035″, 140 to 190 for 0.045″
Amperage: 60 to 100 for 0.035″, 100 to 175 for 0.045″
Voltage: 15 to 17 for 0.035″, 17 to 19 for 0.045″
*wire feed speed

Performance

The welder will demonstrate the correct procedure and technique for depositing a weld in a T-joint.

Criteria

The weld will pass visual inspection criteria.

Procedure

1. Tack weld two pieces to form a T-joint. Position workpiece so joint is in 2F position.
2. Use a 45° work angle and a 10° to 20° push travel angle for root pass. Use a steady push to deposit root pass on both sides of joint. Brush welds.
3. Deposit second pass on bottom toe of root pass with a 50° to 55° work angle and a 10° to 20° push travel angle. Cover root by one-half to two-thirds. Fuse into base metal evenly along toe. Brush weld.
4. Deposit third pass on top toe of root pass with a 30° to 35° work angle and a 10° to 20° push angle. Cover second pass by one-third to one-half. Fuse into base metal evenly along toe. Brush weld.
5. Deposit welds on other side of joint using same work and travel angles.

Additional Practice

Repeat exercise using a Z-weave motion for second and third passes. Pause at each toe to ensure complete fusion into sides of joint.

advanced exercise **5-26**

Making a Multiple-Pass Fillet Weld, T-Joint, 3F Position, with Stainless Steel Electrode Wire, Using GMAW-P

Conditions

Position: 3F
Materials: 2 pieces ¼″ × 2″ × 4″ mild steel
Electrodes: 0.035″ or 0.045″ ER309
Shielding: 90% He/8% Ar/2% O_2
Polarity: DCEP
WFS*: 156 to 225 for 0.035″, 140 to 190 for 0.045″
Amperage: 60 to 100 for 0.035″, 100 to 175 for 0.045″
Voltage: 15 to 17 for 0.035″, 17 to 19 for 0.045″
*wire feed speed

Performance

Refer to Exercise 5-25.

Criteria

Refer to Exercise 5-25.

Procedure

1. Tack weld two pieces to form a T-joint. Position workpiece so joint is in 3F position.
2. Use a 45° work angle and a 10° to 20° push travel angle to deposit root pass on both sides of joint. Brush weld.
3. Deposit second pass on either toe of root pass with a 50° to 55° work angle and a 10° to 20° push travel angle. Cover root by one-half to two-thirds. Fuse into base metal evenly along toe. Brush weld.
4. Deposit third pass on opposite toe of root pass with a 30° to 35° work angle and a 10° to 20° push travel angle. Cover second pass by one-third to one-half. Fuse into base metal evenly along toe. Brush weld.
5. Deposit welds on other side of joint.

Additional Practice

Repeat exercise using a Z-weave motion for second and third passes. Pause at each toe to ensure complete fusion into sides of joint.

advanced exercise 5-27

Making a Multiple-Pass Fillet Weld, T-Joint, 4F Position, with Stainless Steel Electrode Wire, Using GMAW-P

Conditions

Position: 4F
Materials: 2 pieces ¼″ × 2″ × 4″ mild steel
Electrodes: 0.035″ or 0.045″ ER309
Shielding: 90% He/8% Ar/2% O2
Polarity: DCEP
WFS*: 156 to 225 for 0.035″, 140 to 190 for 0.045″
Amperage: 60 to 100 for 0.035″, 100 to 175 for 0.045″
Voltage: 15 to 17 for 0.035″, 17 to 19 for 0.045″
*wire feed speed

Performance

Refer to Exercise 5-25.

Criteria

Refer to Exercise 5-25.

Procedure

1. Tack weld two pieces to form a T-joint. Position workpiece so joint is in 4F position.
2. Use a 45° work angle and a 5° to 10° push travel angle to deposit root pass on both sides of joint. Brush weld.
3. Deposit second pass on bottom toe of root pass with a 50° to 55° work angle and a 5° to 10° push travel angle. Cover root by one-half to two-thirds. Fuse into base metal evenly along toe. Brush weld.
4. Deposit third pass on top toe of root pass with a 30° to 35° work angle and a 5° to 10° push travel angle. Cover second pass by one-third to one-half. Fuse into base metal evenly along toe. Brush weld.
5. Deposit welds on other side of joint.

Additional Practice

Repeat exercise using a Z-weave motion for second and third passes. Pause at each toe to ensure complete fusion into sides of joint.

advanced exercise **5-28**

Making Multiple-Pass Groove Welds, 3G Position with and without Backing, Using GMAW-P with Stainless Steel Electrode Wire

Conditions

Position:	3G
Materials:	2 pieces 1″ × 6″ × 7″ mild steel
Backing:	¼″ × 1½″ × 9″ mild steel
Electrodes:	0.035″ or 0.045″ ER309
Shielding:	90% He/8% Ar/2% O_2
Polarity:	DCEP
WFS*:	225 to 312 for 0.035″, 190 to 280 for 0.045″
Amperage:	100 to 140 for 0.035″, 175 to 210 for 0.045″
Voltage:	17 to 21 for 0.035″, 20 to 22 for 0.045″
	*wire feed speed

Performance

The welder will demonstrate the correct procedure and technique for depositing welds in single-V groove joints in the vertical 3G position.

Criteria

The welds will pass visual inspection and AWS side bend tests.

Procedure with Backing

1. Prepare a single-V groove joint with a 60° groove angle and backing strip. Tack assembly together with a ¼″ root opening. Position workpiece so joint is in 3G position.
2. Use a 90° work angle and a 10° to 20° push travel angle. Move up joint with a steady push to deposit root pass. Brush weld.
3. Deposit second pass using same electrode angles and a Z-weave motion. Pause briefly at each toe to fuse into edges of joint. Brush weld.
4. Deposit remaining passes with same electrode angles and weaving motion. Brush weld after each pass.
5. Face reinforcement should be ⅛″ maximum.
6. Root reinforcement should be 1⁄16″ maximum.
7. Prepare best weld for a side bend test.

Procedure without Backing

1. Prepare a single-V groove open root joint with a 60° groove angle, a 3⁄32″ to ⅛″ root face, and a 3⁄32″ to ⅛″ root opening. Position workpiece so joint is in 3G position.

2. Use a 90° work angle and a 10° to 20° push travel angle. Move up joint with a steady push. Control weld pool to ensure complete joint penetration. Brush weld.

3. Use a weaving motion to deposit remaining passes. Pause briefly at each toe to ensure complete fusion into sides of joint. Brush after each weld.

4. Face reinforcement should be ⅛″ maximum.

5. Root reinforcement should be $\frac{1}{16}$″ maximum.

6. Prepare best weld for a side bend test.

advanced exercise **5-30**

Making a Multiple-Pass Fillet Weld, T-Joint, 1F Position on Aluminum Plate, Using GMAW-P

Conditions

Position:	1F
Materials:	2 pieces ¼″ × 2″ × 4″ aluminum
Electrodes:	³⁄₆₄″ or ¹⁄₁₆″ ER4XXX or ER5XXX
Shielding:	100% Ar
Polarity:	DCEP
WFS*:	350 to 375 for ³⁄₆₄″, 170 to 185 for ¹⁄₁₆″
Amperage:	180 to 210
Voltage:	24 to 25
	*wire feed speed

Performance

The welder will demonstrate the correct procedure and technique for depositing a weld in a T-joint.

Criteria

The weld will pass visual inspection criteria.

Procedure

1. Tack weld two pieces to form a T-joint. Position workpiece so joint is in the 1F position.
2. Use a 90° work angle and a 5° to 10° push travel angle for root pass. Use a steady push travel to deposit root pass on both sides of joint. Brush welds with appropriate wire brush.
3. Deposit second pass on either toe of root pass with an 80° to 85° work angle and a 5° to 10° push travel angle. Use a steady push travel. Cover root by one-half to two-thirds. Fuse into base metal evenly along toe. Brush weld with appropriate wire brush.
4. Deposit third pass on opposite toe of root pass with a 80° to 85° work angle and a 5° to 10° push travel angle. Cover second pass by one-third to one-half. Fuse into base metal evenly along toe. Brush weld with appropriate wire brush.
5. Deposit welds on other side of joint.

Additional Practice

Repeat exercise using a Z-weave motion for second and third passes. Pause at each toe to ensure complete fusion into sides of joint.

advanced exercise **5-31**

Making a Multiple-Pass Fillet Weld, T-Joint, 2F Position on Aluminum Plate, Using GMAW-P

Conditions

Position: 2F
Materials: 2 pieces ¼″ × 2″ × 4″ aluminum
Electrodes: ³⁄₆₄″ or ¹⁄₁₆″ ER4XXX or ER5XXX
Shielding: 100% Ar
Polarity: DCEP
WFS*: 350 to 375 for ³⁄₆₄″, 170 to 185 for ¹⁄₁₆″
Amperage: 180 to 210
Voltage: 24 to 25
*wire feed speed

Performance

The welder will demonstrate the correct procedure and technique for depositing a weld in a T-joint.

Criteria

The weld will pass visual inspection criteria.

Procedure

1. Tack weld two pieces to form a T-joint. Position workpiece so joint is in 2F position.
2. Use a 45° work angle and a 5° to 10° push travel angle for root pass. Use a steady push to deposit root pass on both sides of joint. Brush weld with appropriate wire brush.
3. Deposit second pass on bottom toe of root pass with a 50° to 55° work angle and a 5° to 10° push travel angle. Use a steady push travel. Cover root by one-half to two-thirds. Fuse into base metal evenly along toe. Brush weld with appropriate wire brush.
4. Deposit third pass on top toe of root pass with a 30° to 35° work angle and a 5° to 10° push travel angle. Cover second pass by one-third to one-half. Fuse into base metal evenly along toe. Brush weld with appropriate wire brush.
5. Deposit welds on other side of joint.

Additional Practice

Repeat exercise using a Z-weave motion for second and third passes. Pause at each toe to ensure complete fusion into sides of joint.

advanced exercise **5-32**

Making a Multiple-Pass Groove Weld, 1G Position on Aluminum with Backing, Using GMAW Spray Transfer

Conditions

Position: 1G
Materials: 2 pieces ⅜″ × 6″ × 7″ aluminum
Backing: ¼″ × 1½″ × 9″ aluminum
Electrodes: 3⁄64″ or 1⁄16″ ER4XXX or ER5XXX
Shielding: 100% Ar
Polarity: DCEP
WFS*: 450 to 480 for 3⁄64″, 220 to 230 for 1⁄16″
Amperage: 220 to 250
Voltage: 26 to 28
*wire feed speed

Performance

The welder will demonstrate the correct procedure and technique for depositing a weld in a single-V groove joint in the flat 1G position.

Criteria

The weld will pass visual inspection and an AWS side bend test.

Procedure

1. Prepare a single-V groove joint with a 60° groove angle and backing strip. Tack assembly together with a ¼″ root opening. Position workpiece so joint is in 1G position.
2. Use a 90° work angle and a 10° to 15° push travel angle. Move along joint with a steady push travel to deposit root pass. Brush weld with an appropriate wire brush.
3. Position electrode on either toe of root pass with 85° work angle and 10° to 15° push travel angle. Deposit first intermediate pass. Cover root pass by two-thirds. Brush weld with an appropriate wire brush.
4. Position electrode on opposite toe. Deposit weld with same electrode angles. Cover first intermediate pass by one-half to two-thirds. Allow workpiece to cool as needed before depositing second layer. Brush weld with appropriate wire brush.
5. Number of passes per lay will increase as joint widens. Use same travel angle and adjust work angle as necessary. Clean each weld with appropriate wire brush. Allow workpiece to cool as needed after each layer.
6. The last layer before cover should be about 1⁄16″ below joint edges.
7. The cover pass should be approximately ⅛″ wider than original weld groove.
8. Face reinforcement of finished weld should be ⅛″ maximum.
9. Root reinforcement should be 1⁄16″ maximum.
10. Prepare best weld for a side bend test.

advanced exercise 5-33

Making a Multiple-Pass Fillet Weld, 2F Position, Carbon Steel Pipe to Plate, Using GMAW-S

Conditions

Position: 2F
Materials: 4″ to 6″ schedule 40 to 80 carbon steel pipe
¼″ carbon steel plate to fit
Electrodes: 0.035″ or 0.045″ ER70S-6
Shielding: 75% Ar/25% CO_2 or 100% CO_2
Polarity: DCEP
WFS*: 180 to 230 for 0.035″, 125 to 200 for 0.045″
Amperage: 90 to 140 for 0.035″, 130 to 200 for 0.045″
Voltage: 15 to 17 for 0.035″, 18 to 20 for 0.045″
*wire feed speed

Performance

The welder will demonstrate the correct procedure and technique for depositing a weld to weld pipe to plate in the 2F position.

Criteria

The weld will pass visual inspection criteria.

Procedure

1. Tack weld pipe to plate. Position workpiece so joint is in 2F position.
2. Use a 45° work angle and a 5° to 10° push travel angle to deposit root pass on both sides of joint. Brush weld.
3. Deposit second pass on bottom toe of root pass with a 50° to 55° work angle and a 5° to 10° push travel angle. Cover root by one-half to two-thirds. Fuse into base metal evenly along toe. Brush weld.
4. Deposit third pass on top toe of root pass with a 30° to 35° work angle and a 5° to 10° push travel angle. Cover second pass by one-third to one-half. Fuse into base metal evenly along toe. Brush weld.

Additional Practice

Repeat exercise using a Z-weave motion for second and third passes. Pause at each toe to ensure complete fusion into sides of joint.

advanced exercise 5-34

Making a Multiple-Pass Fillet Weld, 4F Position, Carbon Steel Pipe to Plate, Using GMAW-S

Conditions

Position: 4F
Materials: 4″ to 6″ schedule 40 to 80 carbon steel pipe
¼″ carbon steel plate to fit
Electrodes: 0.035″ or 0.045″ ER70S-6
Shielding: 75% Ar/25% CO_2 or 100% CO_2
Polarity: DCEP
WFS*: 180 to 230 for 0.035″, 125 to 200 for 0.045″
Amperage: 90 to 140 for 0.035″, 130 to 200 for 0.045″
Voltage: 15 to 17 for 0.035″, 18 to 20 for 0.045″
*wire feed speed

Performance

The welder will demonstrate the correct procedure and technique for depositing a weld to weld pipe to plate in the 4F position.

Criteria

The weld will pass visual inspection criteria.

Procedure

1. Tack weld pipe to plate. Position workpiece so joint is in 4F position.
2. Use a 45° work angle and a 5° to 10° push travel angle to deposit root pass on both sides of joint. Brush weld.
3. Deposit second pass on bottom toe of root pass with a 50° to 55° work angle and a 5° to 10° push travel angle. Cover root by one-half to two-thirds. Fuse into base metal evenly along toe. Brush weld.
4. Deposit third pass on top toe of root pass with a 30° to 35° work angle and a 5° to 10° push travel angle. Cover second pass by one-third to one-half. Fuse into base metal evenly along toe. Brush weld.

Additional Practice

Repeat exercise using a Z-weave motion for second and third passes. Pause at each toe to ensure complete fusion into sides of joint.

advanced exercise **5-35**

Making a Multiple-Pass Fillet Weld, 5F Position, Carbon Steel Pipe to Plate, Using GMAW-S

Conditions

Position:	5F
Materials:	4″ to 6″ schedule 40 to 80 carbon steel pipe ¼″ carbon steel plate to fit
Electrodes:	0.035″ or 0.045″ ER70S-6
Shielding:	75% Ar/25% CO_2 or 100% CO_2
Polarity:	DCEP
WFS*:	180 to 230 for 0.035″, 125 to 200 for 0.045″
Amperage:	90 to 140 for 0.035″, 130 to 200 for 0.045″
Voltage:	15 to 17 for 0.035″, 18 to 20 for 0.045″ *wire feed speed

Performance

The welder will demonstrate the correct procedure and technique for depositing a weld to weld pipe to plate in the 5F position.

Criteria

The weld will pass visual inspection criteria.

Procedure

1. Tack weld pipe to plate. Position workpiece so joint is in 5F position.
2. Use a 45° work angle and a 5° to 10° push travel angle to deposit root pass on both sides of joint. Brush weld.
3. Deposit second pass on bottom toe of root pass with a 50° to 55° work angle and a 5° to 10° push travel angle. Cover root by one-half to two-thirds. Fuse into base metal evenly along toe. Brush weld.
4. Deposit third pass on top toe of root pass with a 30° to 35° work angle and a 5° to 10° push travel angle. Cover second pass by one-third to one-half. Fuse into base metal evenly along toe. Brush weld.

Additional Practice

Repeat exercise using a Z-weave motion for second and third passes. Pause at each toe to ensure complete fusion into sides of joint.

advanced exercise **5-36**

Making Multiple-Pass Groove Welds, 2G Position, on Carbon Steel Pipe with and without Backing, Using GMAW-S

Conditions

Position:	2G
Materials:	4″ to 6″ schedule 40 to 80 carbon steel pipe
	Backing ring to fit
Electrodes:	0.035″ or 0.045″ ER70S-6
Shielding:	75% Ar/25% CO_2 or 100% CO_2
Polarity:	DCEP
WFS*:	180 to 230 for 0.035″, 125 to 200 for 0.045″
Amperage:	90 to 140 for 0.035″, 130 to 200 for 0.045″
Voltage:	15 to 17 for 0.035″, 18 to 20 for 0.045″
	*wire feed speed

Performance

The welder will demonstrate the correct procedure and technique for depositing welds to weld pipe to plate in the 2G position.

Criteria

The welds will pass visual inspection and AWS side bend tests.

Procedure with Backing

1. Bevel two pieces of pipe with a 37.5° groove face and no root face. Insert backing ring and tack weld to form a single-V groove joint. Position workpiece so joint is in 2G position.
2. Use a 90° work angle and a 5° to 10° push travel angle to deposit root pass on both sides of joint. Grind high points and/or brush weld.
3. Deposit second pass on bottom toe of root pass with a 50° to 55° work angle and a 5° to 10° push travel angle. Cover root by one-half to two-thirds. Fuse into base metal evenly along toe. Grind high points and/or brush weld.
4. Deposit third pass on top toe of root pass with a 30° to 35° work angle and a 5° to 10° push travel angle. Cover second pass by one-third to one-half. Fuse into base metal evenly along toe. Grind high points and/or brush weld.
5. Face reinforcement should be ⅛″ maximum.
6. Root reinforcement should be 1⁄16″ maximum.
7. Prepare best weld for a side bend test.

Procedure without Backing

1. Prepare a new workpiece for open root welding. Use spacers and tack weld to establish a 3⁄32″ to 1⁄8″ root face and a 3⁄32″ to 1⁄8″ root opening. Position workpiece so joint is in 2G position.

2. Use a 90° work angle with a 5° to 10° push travel angle to deposit root pass on both sides of joint. Grind high points and/or brush weld.

3. Deposit second pass on bottom toe of root pass with a 50° to 55° work angle and a 5° to 10° push travel angle. Cover root by one-half to two-thirds. Fuse into base metal evenly along toe. Grind high points and/or brush weld.

4. Deposit third pass on top toe of root pass with a 30° to 35° work angle and a 5° to 10° push travel angle. Cover second pass by one-third to one-half. Fuse into base metal evenly along toe. Grind high points and/or brush weld.

5. Face reinforcement should be 1⁄8″ maximum.

6. Root reinforcement should be 1⁄16″ maximum.

7. Prepare best weld for a side bend test.

advanced exercise 5-37

Making Multiple-Pass Groove Welds, 5G Position, on Carbon Steel Pipe with and without Backing, Using GMAW-S

Conditions

Position: 5G
Materials: 4″ to 6″ schedule 40 to 80 carbon steel pipe
Backing ring to fit
Electrodes: 0.035″ or 0.045″ ER70S-6
Shielding: 75% Ar/25% CO_2 or 100% CO_2
Polarity: DCEP
WFS*: 180 to 230 for 0.035″, 125 to 200 for 0.045″
Amperage: 90 to 140 for 0.035″, 130 to 200 for 0.045″
Voltage: 15 to 17 for 0.035″, 18 to 20 for 0.045″
*wire feed speed

Performance

The welder will demonstrate the correct procedure and technique for depositing welds to weld pipe to plate in the 5G position.

Criteria

The welds will pass visual inspection and AWS side bend tests.

Procedure with Backing

1. Bevel two pieces of pipe with a 37.5° groove face and no root face. Insert backing ring and tack weld to form a single-V groove joint. Position workpiece so joint is in 5G position.

2. Use a 90° work angle and a 5° to 10° push travel angle to deposit root pass on both sides of joint. Grind high points and/or brush weld.

3. Deposit second pass on either toe of root pass with a 50° to 55° work angle and a 5° to 10° push travel angle. Cover root by one-half to two-thirds. Fuse into base metal evenly along toe. Grind high points and/or brush weld.

4. Deposit third pass on opposite toe of root pass with a 30° to 35° work angle and a 5° to 10° push travel angle. Cover second pass by one-third to one-half. Fuse into base metal evenly along toe. Grind high points and/or brush weld.

5. Face reinforcement should be ⅛″ maximum.

6. Root reinforcement should be 1⁄16″ maximum.

7. Face reinforcement should be ⅛″ maximum.

8. Prepare best weld for a bend test.

Procedure without Backing

1. Prepare a new workpiece for open root welding. Use spacers and tack weld to establish a 3⁄32″ to 1⁄16″ root face and a 3⁄32″ to 1⁄8″ root opening

2. Use a 90° work angle with a 5° to 10° push travel angle to deposit root pass on both sides of joint. Grind high points and/or brush weld.

3. Deposit second pass on either toe of root pass with a 50° to 55° work angle and a 5° to 10° push travel angle. Cover root by one-half to two-thirds. Fuse into base metal evenly along toe. Grind high points and/or brush weld.

4. Deposit third pass on opposite toe of root pass with a 30° to 35° work angle and a 5° to 10° push travel angle. Cover second pass by one-third to one-half. Fuse into base metal evenly along toe. Grind high points and/or brush weld.

5. Face reinforcement should be 1⁄8″ maximum.

6. Root reinforcement should be 1⁄16″ maximum.

7. Prepare best weld for a bend test.

advanced exercise **5-38**

Making Multiple-Pass Groove Welds, 6G Position, on Carbon Steel Pipe with and without Backing, Using GMAW-S

Conditions

Position: 6G
Materials: 4″ to 6″ schedule 40 to 80 carbon steel pipe
Backing ring to fit
Electrodes: 0.035″ or 0.045″ ER70S-6
Shielding: 75% Ar/25% CO_2 or 100% CO_2
Polarity: DCEP
WFS*: 180 to 230 for 0.035″, 125 to 200 for 0.045″
Amperage: 90 to 140 for 0.035″, 130 to 200 for 0.045″
Voltage: 15 to 17 for 0.035″, 18 to 20 for 0.045″
*wire feed speed

Performance

The welder will demonstrate the correct procedure and technique for depositing welds to weld pipe to plate in the 6G position.

Criteria

The welds will pass visual inspection and AWS guided bend tests.

Procedure with Backing

1. Bevel two pieces of pipe with a 37.5° groove face and no root face. Insert backing ring and tack weld to form a single-V groove joint. Position workpiece so joint is in 6G position.

2. Use a 90° work angle and a 5° to 10° push travel angle to deposit root pass on both sides of joint. Grind high points and/or brush weld.

3. Deposit second pass on bottom toe of root pass with a 50° to 55° work angle and a 5° to 10° push travel angle. Cover root by one-half to two-thirds. Fuse into base metal evenly along toe. Grind high points and/or brush weld.

4. Deposit third pass on top toe of root pass with a 30° to 35° work angle and a 5° to 10° push travel angle. Cover second pass by one-third to one-half. Fuse into base metal evenly along toe. Grind high points and/or brush weld.

5. Face reinforcement should be ⅛″ maximum.

6. Root reinforcement should be 1⁄16″ maximum.

7. Prepare best weld for a guided bend test.

Procedure without Backing

1. Prepare a new workpiece for open root welding. Use spacers and tack weld to establish a 3/32″ to 1/8″ root face and a 3/32″ to 1/8″ root opening. Position workpiece so joint is in 6G position.
2. Use a 90° work angle with a 5° to 10° travel angle to deposit root pass on both sides of joint. Grind high points and/or brush weld.
3. Deposit second pass on bottom toe of root pass with a 50° to 55° work angle and a 5° to 10° push travel angle. Cover root by one-half to two-thirds. Fuse into base metal evenly along toe. Grind high points and/or brush weld.
4. Deposit third pass on top toe of root pass with a 30° to 35° work angle and a 5° to 10° push travel angle. Cover second pass by one-third to one-half. Fuse into base metal evenly along toe. Grind high points and/or brush weld.
5. Face reinforcement should be 1/8″ maximum.
6. Root reinforcement should be 1/16″ maximum.
7. Prepare best weld for a guided bend test.

advanced exercise 5-39

Making a Multiple-Pass Fillet Weld, 2F Position, Carbon Steel Pipe to Plate, Using GMAW Spray Transfer

Conditions

Position: 2F
Materials: 4″ to 6″ schedule 40 to 80 carbon steel pipe
¼″ carbon steel plate to fit
Electrodes: 0.035″ or 0.045″ ER70S-6
Shielding: 90% Ar/10% CO_2 or 98% Ar/2% O_2
Polarity: DCEP
WFS*: 500 to 520 for 0.035″, 465 to 485 for 0.045″
Amperage: 230 to 240 for 0.035″, 335 to 345 for 0.045″
Voltage: 28 to 29 for 0.035″, 30 to 31 for 0.045″
*wire feed speed

Performance

The welder will demonstrate the correct procedure and technique for depositing a weld to weld pipe to plate in the 2F position.

Criteria

The weld will pass visual inspection criteria.

Procedure

1. Tack weld pipe to plate. Position workpiece so joint is in 2F position.
2. Use a 45° work angle and a 5° to 10° push travel angle to deposit root pass on both sides of joint. Brush weld.
3. Deposit second pass on bottom toe of root pass with a 50° to 55° work angle and a 5° to 10° push travel angle. Cover root by one-half to two-thirds. Fuse into base metal evenly along toe. Brush weld.
4. Deposit third pass on top toe of root pass with a 30° to 35° work angle and a 5° to 10° push travel angle. Cover second pass by one-third to one-half. Fuse into base metal evenly along toe. Brush weld.

advanced exercise **5-40**

Making a Multiple-Pass Groove Weld, 1G Position on Carbon Steel Pipe, Using GMAW-S and GMAW Spray Transfer

Conditions

Position: 1G
Materials: 4″ to 6″ schedule 40 to 80 carbon steel pipe
Backing: Backing ring to fit
Electrodes: 0.035″ or 0.045″ ER70S-6
Shielding: 75% Ar/25% CO_2 or 100% CO_2 for GMAW-S
90% Ar/10% CO_2 or 98% Ar/2% O_2 for GMAW Spray
Polarity: DCEP
WFS*: 180 to 230 for 0.035″ or 125-200 for 0.045″ GMAW-S
500 to 520 for 0.035″ or 465 to 485 for 0.045″ GMAW Spray
Amperage: 90 to 140 for 0.035″ or 130-200 for 0.045″ GMAW-S
230 to 240 for 0.035″ or 335 to 345 for 0.045″ GMAW Spray
Voltage: 15 to 17 for 0.035″ or 18 to 20 for 0.045″ GMAW-S
28 to 29 for 0.035″ or 30 to 31 for 0.045″ GMAW Spray
*wire feed speed

Performance

The welder will demonstrate the correct procedure and technique for depositing a weld in a single-V groove joint in the flat 1G position on carbon steel pipe with a GMAW-S root and the remainder with a GMAW Spray.

Criteria

The weld will pass visual inspection and an AWS side bend test.

Procedure

1. Bevel two pieces of pipe with a 37.5° groove face and a 3⁄32″ to 1⁄16″ root face. Insert backing ring and tack weld to form a single-V groove joint with a 3⁄32″ to 1⁄16″ root opening. Position workpiece so joint is in 1G position.
2. For root pass with GMAW-S, use a 90° work angle and a 10° to 15° push travel angle. As pipe is rotated use a steady push travel to deposit root pass. Grind high points and/or brush weld.
3. Switch to GMAW spray transfer and deposit first intermediate pass as pipe is rotated. Position electrode on either toe of root pass with an 85° work angle and a 10° to 15° push travel angle. Cover root pass by two-thirds. Brush weld.
4. Position electrode on opposite toe. As pipe is rotated, deposit weld with same electrode angles. Cover first intermediate pass by one-half to two-thirds. Allow workpiece to cool as needed before depositing second layer. Brush weld.
5. Last layer before cover should be about 1⁄16″ below joint edges.
6. Face reinforcement of finished weld should be 1⁄8″ maximum.
7. Root reinforcement of finished weld should be 1⁄16″ maximum.
8. Prepare best weld for a side bend test.

Name ______________________ **Date** ______________

1. Explain the following plastic welding techniques.

Hot Gas ______________________

Heated Tool ______________________

Induction ______________________

Friction ______________________

2. List and define the five major components of a robotic welding system.

3. Use arrows to show the direction of joint movement of the six-axis robot shown.

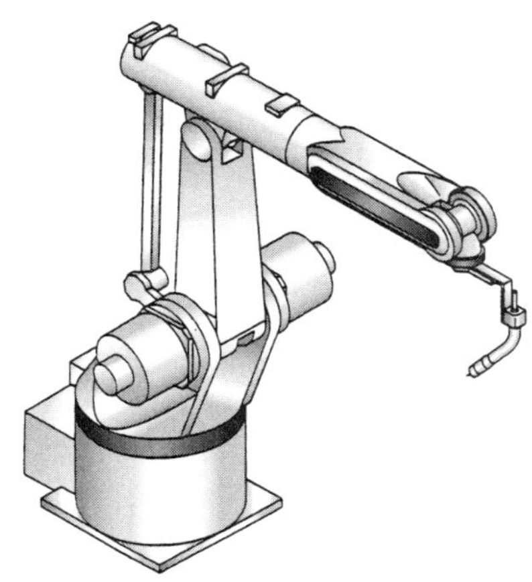

AXIS MOTION

exercises 6-1 and 6-2

Depositing Beads on Mild Steel in Flat Position Using FCAW

Conditions

Exercise 6-1
Gas-shielded (FCAW-G)
Flat position
Two pieces ¼″ to ⅜″ × 4″ × 6″ mild steel
0.045″ E71T-1 or E71T-1M
DCEP
165 A to 195 A
250 ipm to 300 ipm
24 V to 27 V (CO_2)
22 V to 25 V (mixed)
100% CO_2 or 75% Ar/25% CO_2, 35 to 40 cfh
Electrode extension: ½″ to ¾″

Exercise 6-2
Self-shielded (FCAW-S)
Flat position
Two pieces ¼″ to ⅜″ × 4″ × 6″ mild steel
0.045″ E71T-11
DCEN
140 A to 170 A
90 ipm to 130 ipm
16 V to 18 V
Electrode extension: ⅜″ to ¾″

Performance

The welder will demonstrate the correct procedure for depositing beads on mild steel plate using FCAW-G and FCAW-S.

Criteria

Each bead should be approximately ¼″ to ⅜″ wide, ⅛″ to 3/16″ high, and parallel to the edge of the workpiece.

Procedure

1. Place the workpiece on the work surface in flat position.
2. Position the fume extractor as close as possible to the fume source.
3. Adjust the wire feed speed to the correct current and set the voltage by running test beads on scrap material.
4. Position the electrode at a 90° work angle and a 10° to 20° drag angle.
5. Use a steady drag to deposit a series of beads approximately ⅜″ apart. Remove the slag after each bead. Allow the workpiece to cool periodically. A second workpiece can be welded while the first workpiece is cooling.
6. Repeat the exercise with E71T-11.

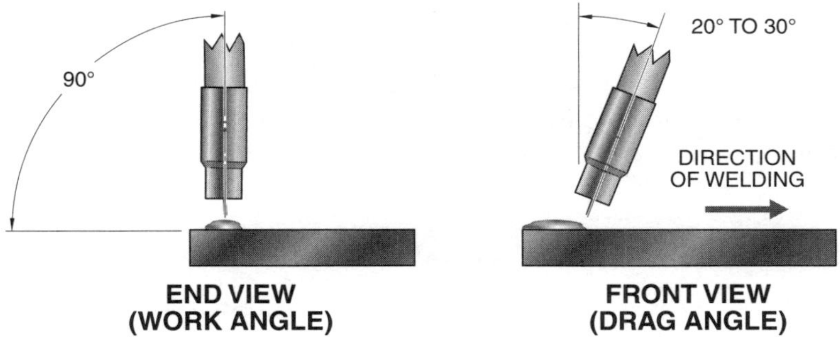

exercises 6-3 and 6-4

Surfacing using FCAW

Conditions

Flat position
Refer to Exercises 6-1 and 6-2

Performance

The welder will deposit successive overlapping beads to increase the thickness of the workpiece using FCAW-G and FCAW-S.

Criteria

Each bead should overlap the previous bead by about one-half and fuse evenly into the base metal along the toe without voids in the weld metal.

Procedure

1. Refer to Exercise 6-1 for equipment setup and adjustment.
2. Overlap each bead by about one-half. Remove the slag after each bead. Allow the workpiece to cool periodically. A second workpiece can be welded while the first workpiece is cooling.
3. Repeat the exercise with E71T-11. Refer to Exercise 6-2.

exercises 6-5 and 6-6

Welding a Lap Joint on Mild Steel in Flat Position Using FCAW

Conditions

Exercise 6-5
Gas-shielded (FCAW-G)
Flat position
Two pieces ¾″ to 1″ × 2½″ × 12″ mild steel
0.052″ E71T-1 or E71T-1M
DCEP
215 A to 245 A
200 ipm to 250 ipm
24 V to 27 V (CO_2)
22 V to 25 V (mixed)
100% CO_2 or 75% Ar/25% CO_2, 35 to 40 cfh
Electrode extension: ½″ to 1″

Exercise 6-6
Self-shielded (FCAW-S)
Flat position
Two pieces ¾″ to 1″ × 2½″ × 12″ mild steel
1⁄16″ E71T-11
DCEN
180 A to 260 A
80 ipm to 140 ipm
18 V to 20 V
Electrode extension: ½″ to 1″

Performance

The welder will demonstrate the correct procedure for depositing a multiple-pass fillet weld in a lap joint in flat position using FCAW-G and FCAW-S.

Criteria

The finished weld should be flat to slightly convex and melt into the edges of the joint with no edge weld undercut.

Procedure

1. Tack weld two pieces to form a lap. Position the workpiece so the joint is in horizontal position.
2. Position the fume extractor as close as possible to the fume source.
3. Position the electrode at a 90° work angle and a 5° to 10° drag angle. Deposit the root pass with a steady drag.
4. Deposit the root pass on the other side of the joint.
5. Use the same electrode angles to deposit a two-bead intermediate layer on both sides of the joint. For the first bead, center the electrode on either toe. It should cover the root pass by about one-half to two-thirds. For the second bead, center the electrode on the other toe. It should cover the first intermediate pass by about one-half to two-thirds. Remove the slag after each weld. Both beads should fuse evenly into the sides of the joint.
6. Deposit additional intermediate layers until the joint is filled to within about 1⁄16″ of the top of the joint edges. The number of beads per layer will increase as the joint widens. Allow the workpiece to cool briefly after each layer. A second workpiece can be welded while the first workpiece is cooling.

7. Deposit a cover layer. Melt into the edges of the joint by about 1⁄16″ to ensure complete fusion along the toes. The finished weld should be flat to slightly convex.
8. Deposit welds on the other side of the joint.
9. Repeat the exercise on a new workpiece with E71T-11.

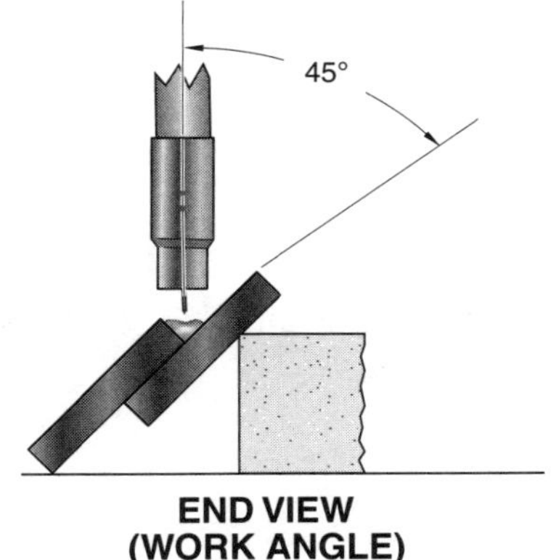

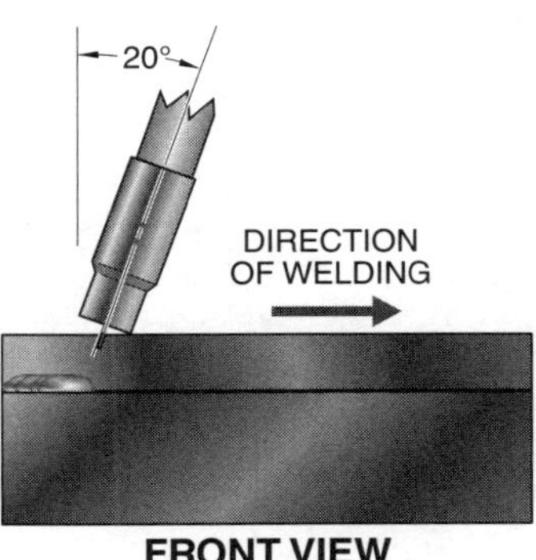

exercises 6-7 and 6-8

Welding a Lap Joint on Mild Steel in Flat Position Using FCAW

Conditions

Exercise 6-7
Gas-shielded (FCAW-G)
Flat position
Two pieces 3⁄4″ to 1″ × 2½″ × 12″ mild steel
1⁄16″ E71T-1 or E71T-1M
DCEP
235 A to 275 A
200 ipm to 250 ipm
23 V to 26 V (CO_2)
21 V to 24 V (mixed)
100% CO_2 or 75% Ar/25% CO_2, 35 to 40 cfh
Electrode extension: ½″ to 1″

Exercise 6-8
Self-shielded (FCAW-S)
Flat position
Two pieces 3⁄4″ to 1″ × 2½″ × 12″ mild steel
0.068″ E71T-11
DCEN
190 A to 270 A
75 ipm to 130 ipm
18 V to 20 V
Electrode extension: ½″ to 1″

Performance

The welder will demonstrate the correct procedure for depositing a multiple-pass fillet weld in a lap joint in the flat position using FCAW-G and FCAW-S.

Criteria

The finished weld should be flat to slightly convex, and melt into the edges of the joint with no edge weld undercut.

Procedure

1. Repeat Exercises 6-5 and 6-6 with 1⁄16″ E71T-1 and E71T-11. The number of beads per layer and the number of layers may vary due to the larger electrode diameter.

exercises 6-9 and 6-10

Welding a T-Joint on Mild Steel in Horizontal Position Using FCAW

Conditions

Horizontal position
Two pieces ¼″ × 2″ × 6″ mild steel
Refer to Exercises 6-1 and 6-2

Performance

The welder will demonstrate the correct procedure for depositing a multiple-pass fillet weld in a T-joint in horizontal position using FCAW-G and FCAW-S.

Criteria

The finished weld should be flat to slightly convex, with ½″ equal legs, complete fusion along the toes, and no edge weld undercut or overlap.

Procedure

1. Tack weld two pieces to form a T-joint. Position the workpiece so the joint is in horizontal position.
2. Position the fume extractor as close as possible to the fume source.
3. Position the electrode at a 45° work angle and a 10° to 20° drag angle. Deposit the root pass with a steady drag.
4. Deposit the root pass on the other side of the joint. Remove the slag from each bead.
5. For the second pass, center the electrode on the bottom toe of the root pass. Cover the root pass by one-half to two-thirds. Deposit the second pass on the other side of the joint. Remove the slag.
6. For the third pass, center the electrode on the top toe of the root pass. Cover the second pass by about one-half. Deposit the third pass on the other side of the joint. Remove the slag.
7. Repeat the exercise with E70T1-11.

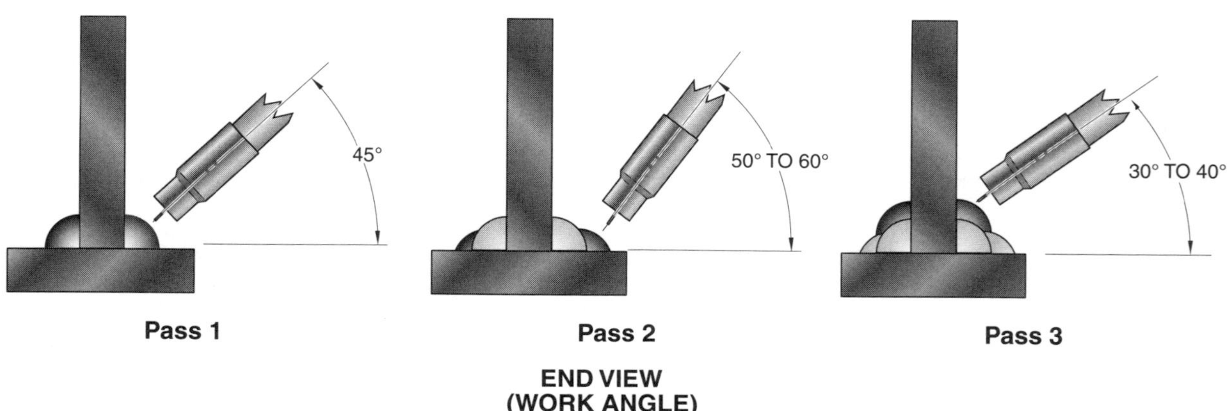

END VIEW (WORK ANGLE)

exercises 6-11 and 6-12

Depositing Beads on Mild Steel in Horizontal Position Using FCAW

Conditions

Horizontal position
Two pieces ¼″ to ⅜″ × 4″ × 6″ mild steel
Refer to Exercises 6-1 and 6-2

Performance

The welder will demonstrate the correct procedure for depositing beads on mild steel plate using FCAW-G and FCAW-S.

Criteria

Each bead should be approximately ¼″ to ⅜″ wide, ⅛″ to $^3\!/_{16}$″ high, and parallel to the edge of the workpiece.

Procedure

1. Position the workpiece so the joint is in horizontal position.
2. Position the fume extractor as close as possible to the fume source.
3. Use an 80° to 85° work angle and a 10° to 20° drag travel.
4. Use a steady drag to deposit a series of bead approximately ⅜″ apart. Remove the slag after each bead. Allow the workpiece to cool periodically. A second workpiece can be welded while the first workpiece is cooling.
5. Repeat the exercise with E71T-11.

exercises 6-13 and 6-14

Welding a T-Joint on Mild Steel in Vertical Position Using FCAW

Conditions

Vertical position
Two pieces ¼″ × 2″ × 6″ mild steel
Refer to Exercises 6-1 and 6-2

Performance

The welder will demonstrate the correct procedure for depositing a multiple-pass fillet weld in a T-joint in vertical position using FCAW-G and FCAW-S.

Criteria

The finished weld should be flat to slightly convex, with a consistent ripple pattern and ½″ equal legs. The weld should have complete fusion along the toes, and no edge weld undercut or overlap.

Procedure

1. Tack weld two pieces to form a T-joint. Position the workpiece so the joint is in vertical position.
2. Position the fume extractor as close as possible to the fume source.
3. Position the electrode at a 45° work angle and a 5° to 10° push angle. Deposit the root pass with a steady push.
4. Deposit the root pass on the other side of the joint. Remove the slag from each bead.
5. For the second pass (intermediate pass), center the electrode on either toe of the root pass. Travel up the joint weaving from toe to toe in ⅛″ increments. Pause briefly at each toe to fuse into the edges of the joint and prevent undercut. Remove the slag.
6. For the third pass (cover pass), center the electrode on either toe of the second pass. Use the same weave technique as the second pass. Remove the slag.
7. Repeat the exercise with E70T1-11.

exercises 6-15 and 6-16

Welding a T-Joint on Mild Steel in Overhead Position Using FCAW

Conditions

Overhead position
Two pieces ¼″ × 2″ × 6″ mild steel
Refer to Exercises 6-1 and 6-2

Performance

The welder will demonstrate the correct procedure for depositing a multiple-pass fillet weld in a T-joint in overhead position using FCAW-G and FCAW-S.

Criteria

The finished weld should be flat to slightly convex, with ½″ equal legs, complete fusion along the toes, and no edge weld undercut or overlap.

Procedure

1. Tack weld two pieces to form a T-joint. Position the workpiece so the joint is in vertical position.
2. Position the fume extractor as close as possible to the fume source.

3. Position the electrode at a 45° work angle and a 10° to 20° drag angle. Deposit the root pass with a steady drag. Deposit the root pass on the other side of the weld, and remove the slag from both welds.

4. For the second weld, center the electrode on the bottom toe of the root pass with the same electrode angles. Use a steady drag and cover the root pass by about one-half to two-thirds. Deposit the second pass on the other side of the joint. Remove the slag.

5. For the third pass, center the electrode on the top toe of the root pass with the same electrode angles. Use a steady drag, and cover the second pass by about one-half. Deposit the third pass on the other side of the joint. Remove the slag.

6. Repeat the exercise with E70T-11.

exercises 6-17 and 6-18

Welding a Single-V-Groove Joint with Backing on Mild Steel in Flat Position Using FCAW

Conditions

Flat Position
Refer to Exercises 6-1 and 6-2
3⁄8″ × 4″ × 6″ mild steel
3⁄16″ × 1½″ × 6″ mild steel backing

Performance

The welder will demonstrate the correct procedure for depositing a multiple-pass groove weld in a single-V-groove butt joint with backing in flat position using FCAW-G and FCAW-S.

Criteria

The finished weld should be flush to slightly convex, with complete fusion along the toes, and no edge weld undercut or overlap.

Procedure

1. Bevel one side of each plate 35° along its length. Place two pieces in the center of a 3⁄16″ backing strip and tack weld to form a single-V-groove joint with a 70° groove angle and a 3⁄16″ root opening. The backing strip will form 1″ starting and runoff tabs.

2. Position the workpiece so the joint is in flat position and position the fume extractor as close as possible to the fume source.

3. Position the electrode on the starting tab with a 90° work angle and a 5° to 15° drag angle. Use a steady drag to deposit the root pass. Remove the slag.

4. Deposit a two-bead intermediate layer. Position the electrode on the either toe of the root pass with the same electrode angles and deposit the first intermediate pass. Use the same electrode angles to deposit a weld on the other toe. Remove the slag from each weld.

5. Deposit a three-bead intermediate layer. Use the same electrode angles for each bead. Remove the slag after each weld. The layer should fill the joint within about 1⁄16″ of the surface. Add additional beads as necessary to raise the surface of the layer.

6. Deposit a four-bead cover layer. Use the same electrode angles for each bead. Remove the slag after each weld. The first and last bead of the layer should melt into the edges of the joint by about 1⁄16″ to ensure complete fusion.

7. Repeat the exercise on a new workpiece with E71T-11.

exercises 6-19 and 6-20

Welding a Single-V-Groove Joint with Backing on Mild Steel in Horizontal Position Using FCAW

Conditions

Horizontal Position
Refer to Exercises 6-1 and 6-2
3⁄8″ × 4″ × 6″ mild steel
3⁄16″ × 1½″ × 6″ mild steel backing

Performance

The welder will demonstrate the correct procedure for depositing a multiple-pass groove weld in a single-V-groove butt joint with backing in horizontal position using FCAW-G and FCAW-S.

Criteria

The finished weld should be flush to slightly convex, with complete fusion along the toes, and no edge weld undercut or overlap.

Procedure

1. Prepare a single-V-groove joint as in step 1 of Exercises 6-17 and 6-18.

2. Position the workpiece so the joint is in horizontal position and position the fume extractor as close as possible to the fume source.

3. Position the electrode on the starting tab with an 85° work angle and a 10° to 20° drag angle. Use a steady drag to deposit the root pass. Remove the slag.

4. Deposit a two-bead intermediate layer. Position the electrode on the either toe of the root pass with the same electrode angles and deposit the first intermediate pass. Use the same electrode angles to deposit a weld on the other toe. Remove the slag from each weld.

5. Deposit a three-bead intermediate layer. Use the same electrode angles for each bead. Remove the slag after each weld. The layer should fill the joint to within about 1⁄16″ of the surface. Add additional beads as necessary to raise the surface of the layer.

6. Deposit a four-bead cover layer. Use the same electrode angles for each bead. Remove the slag after each weld. The welds at the top and bottom of the layer should melt into the edges of joint by 1⁄16″ to ensure complete fusion.

7. Repeat the exercise on a new workpiece with E71T-11.

exercises 6-21 and 6-22

Welding a Single-V-Groove Joint with Backing on Mild Steel in Vertical Position Using FCAW

Conditions

Vertical Position
Refer to Exercises 6-1 and 6-2
3⁄8″ × 4″ × 6″ mild steel
3⁄16″ × 1½″ × 6″ mild steel backing

Performance

The welder will demonstrate the correct procedure for depositing a multiple-pass groove weld in a single-V-groove butt joint with backing in vertical position using FCAW-G and FCAW-S.

Criteria

The finished weld should be flush to slightly convex, with complete fusion along the toes, and no edge weld undercut or overlap.

Procedure

1. Prepare a single-V-groove joint as in step 1 of Exercises 6-17 and 6-18.

2. Position the workpiece so the joint is in vertical position, and position the fume extractor as close as possible to the fume source.

3. Position the electrode on the starting tab at the bottom of the joint with a 90° work angle and a 5° to 15° push angle. Use a steady push to deposit the root pass. Remove the slag.

4. For the second pass (intermediate pass), center the electrode on either toe of the root pass. Travel up the joint weaving from toe to toe in 1⁄8″ increments. Pause briefly at each toe to fuse into the edges of the joint and prevent undercut. Remove the slag.

5. For the third pass (intermediate pass), center the electrode on either toe of the second pass. Use the same weave technique as the second pass. Remove the slag.

6. Use the same technique for the cover pass. Melt into the edges of the joint by about 1⁄16″ to ensure complete fusion. Remove the slag.

7. Repeat the exercise with E70T1-11.

exercises 6-23 and 6-24

Welding a Single-V-Groove Joint with Backing on Mild Steel in Overhead Position Using FCAW

Conditions

Overhead Position
Refer to Exercises 6-1 and 6-2
⅜″ × 4″ × 6″ mild steel
³⁄₁₆″ × 1½″ × 6″ mild steel backing

Performance

The welder will demonstrate the correct procedure for depositing a multiple-pass groove weld in a single-V-groove butt joint with backing in overhead position using FCAW-G and FCAW-S.

Criteria

The finished weld should be flush to slightly convex, with complete fusion along the toes, and no edge weld undercut or overlap.

Procedure

1. Prepare a single-V-groove joint as in step 1 of Exercises 6-17 and 6-18.
2. Position the workpiece so the joint is in overhead position and position the fume extractor as close as possible to the fume source.
3. Position the electrode on the starting tab with a 90° work angle and a 5° to 15° drag angle. Use a steady push to deposit the root pass. Remove the slag.
4. For the second pass (intermediate pass), center the electrode on either toe of the root pass with the same angles. Travel along the joint weaving from toe to toe in ⅛″ increments. Pause briefly at each toe to fuse into the edges of the joint and prevent undercut. Remove the slag.
5. For the third pass (intermediate pass), center the electrode on either toe of the second pass. Use the same weave technique as the second pass. Remove the slag.
6. Use the same technique for the cover pass. Melt into the edges of the joint by about ¹⁄₁₆″ to ensure complete fusion. Remove the slag.
7. Repeat the exercise with E70T1-11.

advanced exercise 6-25

Making a Multiple-Pass Fillet Weld, T-Joint, 2F Position, Using FCAW-G

Conditions

Position: 2F
Materials: 2 pieces ¼″ × 2″ × 4″ mild steel

Electrodes: 0.045″, 0.052″, or 1⁄16″ E71T-1
Shielding: 100% CO_2 or 75% Ar/25% CO_2
Polarity: DCEP
WFS*: 250 to 325 for 0.045″, 150 to 225 for 0.052″, or 110 to 195 for 1⁄16″
Amperage: 110 to 140 for 0.045″, 130 to 160 for 0.052″, or 135 to 175 for 1⁄16″
Voltage: 22 to 23 for 0.045″, 24 to 25 for 0.052″, or 26 to 27 for 1⁄16″
*wire feed speed

Performance

The welder will demonstrate the correct procedure and technique for depositing a weld in a T-joint.

Criteria

The weld will pass visual inspection criteria.

Procedure

1. Tack weld two pieces to form a T-joint. Remove slag. Position workpiece so joint is in 2F position.
2. Use a 45° work angle and a 10° to 20° drag travel angle for root pass. Use a steady drag to deposit root pass on both sides of joint. Remove slag.
3. Deposit second pass on bottom toe of root pass with a 50° to 55° work angle and a 10° to 20° drag travel angle. Using a steady drag travel, cover root by one-half to two-thirds. Fuse into base metal evenly along toe. Remove slag.
4. Deposit third pass on top toe of root pass with a 30° to 35° work angle and a 10° to 20° drag travel angle. Cover second pass by one-third to one-half. Fuse into base metal evenly along toe. Remove slag.
5. Deposit welds on other side of joint.

Additional Practice

Tack up new workpiece. Use same technique for root pass. Practice with Z-weave motion pausing at each weld toe to deposit second and third passes.

advanced exercise **6-26**

Making a Multiple-Pass Fillet Weld, T-Joint, 3F Position, Using FCAW-G

Conditions

Position: 3F
Materials: 2 pieces 1⁄4″ × 2″ × 4″ mild steel
Electrodes: 0.045″, 0.052″, or 1⁄16″ E71T-1
Shielding: 100% CO_2 or 75% Ar/25% CO_2
WFS*: 250 to 325 for 0.045″, 150 to 225 for 0.052″, or 110 to 195 for 1⁄16″

Amperage: 110 to 140 for 0.045″, 130 to 160 for 0.052″, or 135 to 175 for 1⁄16″
Voltage: 22 to 23 for 0.045″, 24 to 25 for 0.052″, or 26 to 27 for 1⁄16″
*wire feed speed

Performance

Refer to Exercise 5-20.

Criteria

Refer to Exercise 5-20.

Procedure

1. Tack weld two pieces to form a T-joint. Remove slag. Position workpiece so joint is in 3F position.
2. Use a 45° work angle and a 10° to 20° push travel angle to deposit root pass on both sides of joint. Remove slag.
3. Deposit second pass on either toe of root pass with a 50° to 55° work angle and a 10° to 20° push travel angle. Cover root by one-half to two-thirds. Fuse into base metal evenly along toe. Remove slag.
4. Deposit third pass on opposite toe of root pass with a 30° to 35° work angle and a 10° to 20° push travel angle. Cover second pass by one-third to one-half. Fuse into base metal evenly along toe. Remove slag.
5. Deposit welds on other side of joint.

Additional Practice

Tack up new workpiece. Use same technique for root pass. Practice with Z-weave motion pausing at each weld toe to deposit second and third passes.

advanced exercise **6-27**

Making a Multiple-Pass Fillet Weld, T-Joint, 4F Position, Using FCAW-G

Conditions

Position: 4F
Materials: 2 pieces 1⁄4″ × 2″ × 4″ mild steel
Electrodes: 0.045″, 0.052″, or 1⁄16″ E71T-1
Shielding: 100% CO_2 or 75% Ar/25% CO_2
Polarity: DCEP
WFS*: 250 to 325 for 0.045″, 150 to 225 for 0.052″, or 110 to 195 for 1⁄16″
Amperage: 110 to 140 for 0.045″, 130 to 160 for 0.052″, or 135 to 175 for 1⁄16″
Voltage: 22 to 23 for 0.045″, 24 to 25 for 0.052″, or 26 to 27 for 1⁄16″
*wire feed speed

Performance

Refer to Exercise 5-20.

Criteria

Refer to Exercise 5-20.

Procedure

1. Tack weld two pieces to form a T-joint. Remove slag. Position workpiece so joint is in 4F position.
2. Use a 45° work angle and a 5° to 10° drag travel angle to deposit root pass on both sides of joint. Remove slag.
3. Deposit second pass on bottom toe of root pass with a 50° to 55° work angle and a 5° to 10° drag travel angle. Cover root by one-half to two-thirds. Fuse into base metal evenly along toe. Remove slag.
4. Deposit third pass on top toe of root pass with a 30° to 35° work angle and a 5° to 10° drag travel angle. Cover second pass by one-third to one-half. Fuse into base metal evenly along toe. Remove slag.
5. Deposit welds on other side of joint.

Additional Practice

Tack up new workpiece. Use same technique for root pass. Practice with Z-weave motion pausing at each weld toe to deposit second and third passes.

advanced exercise **6-28**

Making Multiple-Pass Groove Welds, 3G Position with and without Backing, Using FCAW-G

Conditions

Position: 3G
Materials: 2 pieces 1″ × 6″ × 7″ mild steel
Backing: ¼″ × 1½″ × 9″ mild steel
Electrodes: 0.045″, 0.052″, or 1⁄16″ E71T-1
Shielding: 100% CO_2 or 75% Ar/25% CO_2
Polarity: DCEP
WFS*: 250 to 325 for 0.045″, 150 to 225 for 0.052″, or 110 to 195 for 1⁄16″
Amperage: 110 to 140 for 0.045″, 130 to 160 for 0.052″, or 135 to 175 for 1⁄16″
Voltage: 22 to 23 for 0.045″, 24 to 25 for 0.052″, or 26 to 27 for 1⁄16″
*wire feed speed

Performance

The welder will demonstrate the correct procedure and technique for depositing welds in single-V groove joints in the vertical 3G position.

Criteria

The welds will pass visual inspection and AWS side bend tests.

Procedure with Backing

1. Prepare a single-V groove joint with a 60° groove angle and backing strip. Tack assembly together with a ¼″ root opening. Remove slag. Position workpiece so joint is in 3G position.
2. Use a 90° work angle and a 10° to 20° push travel angle. Move up joint with a Z-weave motion to deposit root pass. Pause briefly at each toe to fuse into sides of joint. Remove slag.
3. Deposit second pass using same electrode angles and weaving motion. Pause briefly at each toe to fuse into edges of joint. Remove slag.
4. Deposit remaining passes with same electrode angles and weaving motion. Remove slag after each pass.
5. Face reinforcement should be ⅛″ maximum.
6. Root reinforcement should be 1⁄16″ maximum.
7. Face reinforcement should be ⅛″ maximum.
8. Prepare best weld for a side bend test.

Procedure without Backing

1. Prepare a single-V groove open root joint with a 60° groove angle and a 3⁄32″ to ⅛″ root face. Tack assembly together with a 3⁄32″ to ⅛″ root opening. Remove slag. Position workpiece so joint is in 3G position.
2. Use a 90° work angle and a 10° to 20° push travel angle. Move up joint with a steady push. Control weld pool to ensure complete joint penetration. Remove slag.
3. Use a weaving motion to deposit remaining passes. Pause briefly at each toe to ensure penetration in sides of joint. Remove slag after each weld.
4. Face reinforcement should be ⅛″ maximum.
5. Root reinforcement should be 1⁄16″ maximum.
6. Prepare best weld for a side bend test.

advanced exercise **6-29**

Making Multiple-Pass Groove Welds, 4G Position with and without Backing, Using FCAW-G

Conditions

Position: 4G
Materials: 2 pieces 1″ × 6″ × 7″ mild steel
Backing: ¼″ × 1½″ × 9″ mild steel
Electrodes: 0.045″, 0.052″, or 1⁄16″ E71T-1
Shielding: 100% CO_2 or 75% Ar/25% CO_2
Polarity: DCEP

WFS*:	250 to 325 for 0.045″, 150 to 225 for 0.052″, or 110 to 195 for 1⁄16″
Amperage:	110 to 140 for 0.045″, 130 to 160 for 0.052″, or 135 to 175 for 1⁄16″
Voltage:	22 to 23 for 0.045″, 24 to 25 for 0.052″, or 26 to 27 for 1⁄16″
	*wire feed speed

Performance

The welder will demonstrate the correct procedure and technique for depositing welds in single-V groove joints in the overhead 4G position.

Criteria

The welds will pass visual inspection and AWS side bend tests.

Procedure with Backing

1. Prepare a single-V groove joint with a 60° groove angle and backing strip. Tack assembly together with a 1⁄4″ root opening. Remove slag. Position workpiece so joint is in 4G position.
2. Use a 90° work angle and a 5° to 10° drag travel angle. Move along joint with a steady drag to deposit root pass. A slight oscillation may help to fuse into sides of joint. Remove slag.
3. Deposit second pass on either toe using an 85° work angle and a 5° to 10°drag travel angle. Cover root by one-half to two-thirds. Fuse into base metal evenly along toe. Remove slag.
4. Deposit third pass on opposite toe with same electrode angles. Cover second pass by one-third to one-half. Fuse into base metal evenly along toe. Remove slag.
5. Face reinforcement should be 1⁄8″ maximum.
6. Root reinforcement should be 1⁄16″ maximum.
7. Prepare best weld for a side bend test.

Procedure without Backing

1. Prepare a single-V groove open root joint with a 60° groove angle and a 3⁄32″ to 1⁄8″ root face. Tack assembly together with a 3⁄32″ to 1⁄8″ root opening. Remove slag. Position workpiece so joint is in 3G position.
2. Use a 90° work angle and a 10° to 20° push travel angle. Move up joint with a steady push. Control weld pool to ensure complete joint penetration. Remove slag.
3. Use a weaving motion to deposit remaining passes. Pause briefly at each toe to ensure penetration into sides of joint. Remove slag after each weld.
4. Face reinforcement should be 1⁄8″ maximum.
5. Root reinforcement should be 1⁄16″ maximum.
6. Prepare best weld for a side bend test.

advanced exercise 6-30

Making a Multiple-Pass Fillet Weld, T-Joint, 2F Position, Using FCAW-S

Conditions

Position: 2F
Materials: 2 pieces ¼″ × 2″ × 4″ mild steel
Electrodes: 1⁄16″, 0.072″, or 5⁄64″ E71T-8
Polarity: DCEN
WFS*: 125 to 150 for 1⁄16″, 100 to 130 for 0.072″, or 90 to 110 for 5⁄64″
Amperage: 160 to 180 for 1⁄16″, 170 to 205 for 0.072″, or 175 to 215 for 5⁄64″
Voltage: 18 to 19 for 1⁄16″, 19 to 20 for 0.072″, or 20 to 21 for 5⁄64″
*wire feed speed

Performance

The welder will demonstrate the correct procedure and technique for depositing a weld in a T-joint.

Criteria

The weld will pass visual inspection criteria.

Procedure

1. Tack weld two pieces to form a T-joint. Remove slag. Position workpiece so joint is in 2F position.
2. Use a 45° work angle and a 10° to 20° drag travel angle for root pass. Use a steady drag to deposit root pass on both sides of joint. Remove slag.
3. Deposit second pass on bottom toe of root pass with a 50° to 55° work angle and a 10° to 20° drag travel angle. Use a steady drag travel. Cover root by one-half to two-thirds. Fuse into base metal evenly along toe. Remove slag.
4. Deposit third pass on top toe of root pass with a 30° to 35° work angle and a 10° to 20° drag travel angle. Cover second pass by one-third to one-half. Fuse into base metal evenly along toe. Remove slag.
5. Deposit welds on other side of joint.

Additional Practice

Tack up new workpiece. Use same technique for root pass. Practice with Z-weave motion pausing at each weld toe to deposit second and third passes.

advanced exercise 6-31

Making a Multiple-Pass Fillet Weld, T-Joint, 3F Position, Using FCAW-S

Conditions

Position: 3F
Materials: 2 pieces ¼″ × 2″ × 4″ mild steel

Electrodes: 1⁄16″, 0.072″, or 5⁄64″ E71T-8
Polarity: DCEN
WFS*: 125 to 150 for 1⁄16″, 100 to 130 for 0.072″, or 90 to 110 for 5⁄64″
Amperage: 160 to 180 for 1⁄16″, 170 to 205 for 0.072″, or 175 to 215 for 5⁄64″
Voltage: 18 to 19 for 1⁄16″, 19 to 20 for 0.072″, or 20 to 21 for 5⁄64″
*wire feed speed

Performance

Refer to Exercise 5-20.

Criteria

Refer to Exercise 5-20.

Procedure

1. Tack weld two pieces to form a T-joint. Remove slag. Position workpiece so joint is in 3F position.
2. Use a 45° work angle and a 10° to 20° push travel angle to deposit root pass on both sides of joint. Remove slag.
3. Deposit second pass on either toe of root pass with a 50° to 55° work angle and a 10° to 20° push travel angle. Cover root by one-half to two-thirds. Fuse into base metal evenly along toe. Remove slag.
4. Deposit third pass on opposite toe of root pass with a 30° to 35° work angle and a 10° to 20° push travel angle. Cover second pass by one-third to one-half. Fuse into base metal evenly along toe. Remove slag.
5. Deposit welds on other side of joint.

Additional Practice

Tack up new workpiece. Use same technique for root pass. Practice with Z-weave motion pausing at each weld toe to deposit second and third passes.

advanced exercise **6-32**

Making a Multiple-Pass Fillet Weld, T-Joint, 4F Position, Using FCAW-S

Conditions

Position: 4F
Materials: 2 pieces 1⁄4″ × 2″ × 4″ mild steel
Electrodes: 1⁄16″, 0.072″, or 5⁄64″ E71T-8
Polarity: DCEN
WFS*: 125 to 150 for 1⁄16″, 100 to 130 for 0.072″, or 90 to 110 for 5⁄64″
Amperage: 160 to 180 for 1⁄16″, 170 to 205 for 0.072″, or 175 to 215 for 5⁄64″
Voltage: 18 to 19 for 1⁄16″, 19 to 20 for 0.072″, or 20 to 21 for 5⁄64″
*wire feed speed

Performance

Refer to Exercise 5-20.

Criteria

Refer to Exercise 5-20.

Procedure

1. Tack weld two pieces to form a T-joint. Remove slag. Position workpiece so joint is in 4F position.
2. Use a 45° work angle and a 5° to 10° drag travel angle to deposit root pass on both sides of joint. Remove slag.
3. Deposit second pass on bottom toe of root pass with a 50° to 55° work angle and a 5° to 10° drag travel angle. Cover root by one-half to two-thirds. Fuse into base metal evenly along toe. Remove slag.
4. Deposit third pass on top toe of root pass with a 30° to 35° work angle and a 5° to 10° drag travel angle. Cover second pass by one-third to one-half. Fuse into base metal evenly along toe. Remove slag.
5. Deposit welds on other side of joint.

Additional Practice

Tack up new workpiece. Use same technique for root pass. Practice with Z-weave motion pausing at each weld toe to deposit second and third passes.

advanced exercise **6-33**

Making Multiple-Pass Groove Welds, 3G Position with and without Backing, Using FCAW-S

Conditions

Position: 3G
Materials: 2 pieces 1″ x 6″ x 7″ mild steel
Backing: ¼″ × 1½″ × 9″ mild steel
Electrodes: 1⁄16″, 0.072″, or 5⁄64″ E71T-8
Polarity: DCEN
WFS*: 125 to 150 for 1⁄16″, 100 to 130 for 0.072″, or 90 to 110 for 5⁄64″
Amperage: 160 to 180 for 1⁄16″, 170 to 205 for 0.072″, or 175 to 215 for 5⁄64″
Voltage: 18 to 19 for 1⁄16″, 19 to 20 for 0.072″, or 20 to 21 for 5⁄64″
*wire feed speed

Performance

The welder will demonstrate the correct procedure and technique for depositing welds in single-V groove joints in the vertical 3G position.

Criteria

The welds will pass visual inspection and AWS side bend tests.

Procedure with Backing

1. Prepare a single-V groove joint with a 60° groove angle and backing strip. Tack assembly together with a ¼″ root opening. Remove slag. Position workpiece so joint is in 3G position.
2. Use a 90° work angle and a 10° to 20° push travel angle. Move up joint with a Z-weave motion to deposit root pass. Pause briefly at each toe to fuse into sides of joint. Remove slag.
3. Deposit second pass using same electrode angles and weaving motion. Pause briefly at each toe to fuse into edges of joint. Remove slag.
4. Deposit remaining passes with same electrode angles and weaving motion. Remove slag weld after each pass.
5. Face reinforcement should be ⅛″ maximum.
6. Root reinforcement should be 1⁄16″ maximum
7. Prepare best weld for a side bend test.

Procedure without Backing

1. Prepare a single-V groove open root joint with a 60° groove angle and a 3⁄32″ to ⅛″ root face. Tack assembly together with a 3⁄32″ to ⅛″ root opening. Remove slag. Position workpiece so joint is in 3G position.
2. Use a 90° work angle and a 10° to 20° push travel angle. Move up joint with a steady push. Control weld pool to ensure complete joint penetration. Remove slag.
3. Use a weaving motion to deposit remaining passes. Pause briefly at each toe to ensure penetration in sides of joint. Remove slag after each weld.
4. Face reinforcement should be ⅛″ maximum.
5. Root reinforcement should be 1⁄16″ maximum.
6. Prepare best weld for a side bend test.

advanced exercise **6-34**

Making Multiple-Pass Groove Welds, 4G Position with and without Backing, Using FCAW-S

Conditions

Position: 4G
Materials: 2 pieces 1″ × 6″ × 7″ mild steel
Backing: ¼″ × 1½″ × 9″ mild steel
Electrodes: 1⁄16″, 0.072″, or 5⁄64″ E71T-8

Polarity: DCEN
WFS*: 125 to 150 for 1⁄16″, 100 to 130 for 0.072″, or 90 to 110 for 5⁄64″
Amperae: 160 to 180 for 1⁄16″, 170 to 205 for 0.072″, or 175 to 215 for 5⁄64″
Voltage: 18 to 19 for 1⁄16″, 19 to 20 for 0.072″, or 20 to 21 for 5⁄64″
*wire feed speed

Performance

The welder will demonstrate the correct procedure and technique for depositing welds in single-V groove joints in the overhead 4G position.

Criteria

The welds will pass visual inspection and AWS side bend tests.

Procedure with Backing

1. Prepare a single-V groove joint with a 60° groove angle and backing strip. Tack assembly together with a 1⁄4″ root opening. Remove slag. Position workpiece so joint is in 4G position.
2. Use a 90° work angle and a 5° to 10° drag travel angle. Move along joint with a steady drag to deposit root pass. A slight oscillation may help to fuse into sides of joint. Remove slag.
3. Deposit second pass on either toe using an 85° work angle and a 5° to 10° drag travel angle. Cover root by one-half to two-thirds. Fuse into base metal evenly along toe. Remove slag.
4. Deposit third pass on opposite toe using same electrode angles. Cover second pass by one-third to one-half. Fuse into base metal evenly along toe. Remove slag.
5. Face reinforcement should be 1⁄8″ maximum.
6. Root reinforcement should be 1⁄16″ maximum.
7. Prepare best weld for a side bend test.

Procedure without Backing

1. Prepare a single-V groove open root joint with a 60° groove angle and a 3⁄32″ to 1⁄8″ root face. Tack assembly together with a 3⁄32″ to 1⁄8″ root opening. Remove slag. Position workpiece so joint is in 3G position.
2. Use a 90° work angle and a 5° to 10° drag travel angle. Move up joint with a steady push. Control weld pool to ensure complete joint penetration. Remove slag.
3. Use a weaving motion to deposit remaining passes. Pause briefly at each toe to ensure penetration into sides of joint. Remove slag after each weld.
4. Face reinforcement should be 1⁄8″ maximum.
5. Root reinforcement should be 1⁄16″ maximum.
6. Prepare best weld for a side bend test.

advanced exercise **6-35**

Making a Multiple-Pass Fillet Weld, 2F Position, Carbon Steel Pipe to Plate, Using FCAW-G

Conditions

Position: 2F
Materials: 4″ to 6″ schedule 40 to 80 carbon steel pipe
¼″ carbon steel plate to fit
Electrodes: 0.045″, 0.052″, or 1⁄16″ E71T-1
Shielding: 100% CO_2 or 75% Ar/25% CO_2
Polarity: DCEP
WFS*: 250 to 325 for 0.045″, 150 to 225 for 0.052″, or 110 to 195 for 1⁄16″
Amperage: 110 to 140 for 0.045″, 130 to 160 for 0.052″, or 135 to 175 for 1⁄16″
Voltage: 22 to 23 for 0.045″, 24 to 25 for 0.052″, or 26 to 27 for 1⁄16″
*wire feed speed

Performance

The welder will demonstrate the correct procedure and technique for depositing a weld to weld pipe to plate in the 2F position.

Criteria

The weld will pass visual inspection criteria.

Procedure

1. Tack weld pipe to plate. Remove slag. Position workpiece so joint is in 2F position.
2. Use a 45° work angle and a 5° to 10° drag travel angle to deposit root pass on both sides of joint. Remove slag.
3. Deposit second pass on bottom toe of root pass with a 50° to 55° work angle and a 5° to 10° drag travel angle. Cover root by one-half to two-thirds. Fuse into base metal evenly along toe. Remove slag.
4. Deposit third pass on top toe of root pass with a 30° to 35° work angle and a 5° to 10°drag travel angle. Cover second pass by one-third to one-half. Fuse into base metal evenly along toe. Remove slag.

Additional Practice

Tack up new workpiece. Use same technique for root pass. Practice with Z-weave motion pausing at each weld toe to deposit second and third passes.

advanced exercise **6-36**

Making a Multiple-Pass Fillet Weld, 4F Position, Carbon Steel Pipe to Plate, Using FCAW-G

Conditions

Position: 4F
Materials: 4″ to 6″ schedule 40 to 80 carbon steel pipe
¼″ carbon steel plate to fit

Electrodes: 0.045″, 0.052″, or 1⁄16″ E71T-1
Shielding: 100% CO_2 or 75% Ar/25% CO_2
Polarity: DCEP
WFS*: 250 to 325 for 0.045″, 150 to 225 for 0.052″, or 110 to 195 for 1⁄16″
Amperage: 110 to 140 for 0.045″, 130 to 160 for 0.052″, or 135 to 175 for 1⁄16″
Voltage: 22 to 23 for 0.045″, 24 to 25 for 0.052″, or 26 to 27 for 1⁄16″
*wire feed speed

Performance

The welder will demonstrate the correct procedure and technique for depositing a weld to weld pipe to plate in the 4F position.

Criteria

The weld will pass visual inspection criteria.

Procedure

1. Tack weld pipe to plate. Remove slag. Position workpiece so joint is in 4F position.
2. Use a 45° work angle and a 5° to 10° drag travel angle to deposit root pass on both sides of joint. Remove slag.
3. Deposit second pass on bottom toe of root pass with a 50° to 55° work angle and a 5° to 10° drag travel angle. Cover root by one-half to two-thirds. Fuse into base metal evenly along toe. Remove slag.
4. Deposit third pass on top toe of root pass with a 30° to 35° work angle and a 5° to 10° drag travel angle. Cover second pass by one-third to one-half. Fuse into base metal evenly along toe. Remove slag.

Additional Practice

Tack up new workpiece. Use same technique for root pass. Practice with Z-weave motion pausing at each weld toe to deposit second and third passes.

advanced exercise **6-37**

Making a Multiple-Pass Fillet Weld, 5F Position, Carbon Steel Pipe to Plate, Using FCAW-G

Conditions

Position: 5F
Materials: 4″ to 6″ schedule 40 to 80 carbon steel pipe
1⁄4″ carbon steel plate to fit
Electrodes: 0.045″, 0.052″, or 1⁄16″ E71T-1
Shielding: 100% CO_2 or 75% Ar/25% CO_2
Polarity: DCEP
WFS*: 250 to 325 for 0.045″, 150 to 225 for 0.052″, or 110 to 195 for 1⁄16″

Amperage: 110 to 140 for 0.045″, 130 to 160 for 0.052″, or 135 to 175 for 1⁄16″
Voltage: 22 to 23 for 0.045″, 24 to 25 for 0.052″, or 26 to 27 for 1⁄16″
*wire feed speed

Performance

The welder will demonstrate the correct procedure and technique for depositing a weld to weld pipe to plate in the 5F position.

Criteria

The weld will pass visual inspection criteria.

Procedure

1. Tack weld pipe to plate. Remove slag. Position workpiece so joint is in 5F position.
2. Use a 45° work angle and a 5° to 10° push travel angle to deposit root pass on both sides of joint. Remove slag.
3. Deposit second pass on bottom toe of root pass with a 50° to 55° work angle and a 5° to 10° push travel angle. Cover root by one-half to two-thirds. Fuse into base metal evenly along toe. Remove slag.
4. Deposit third pass on top toe of root pass with a 30° to 35° work angle and a 5° to 10° push travel angle. Cover second pass by one-third to one-half. Fuse into base metal evenly along toe. Remove slag.

Additional Practice

Tack up new workpiece. Use same technique for root pass. Practice with Z-weave motion pausing at each weld toe to deposit second and third passes.

advanced exercise **6-38**

Making Multiple-Pass Groove Welds, 2G Position, on Carbon Steel Pipe with and without Backing, Using FCAW-G

Conditions

Position: 2G
Materials: 4″ to 6″ schedule 40 to 80 carbon steel pipe
Backing ring to fit
Electrodes: 0.045″, 0.052″, or 1⁄16″ E71T-1
Shielding: 100% CO_2 or 75% Ar/25% CO_2
Polarity: DCEP
WFS*: 250 to 325 for 0.045″, 150 to 225 for 0.052″, or 110 to 195 for 1⁄16″
Amperage: 110 to 140 for 0.045″, 130 to 160 for 0.052″, or 135 to 175 for 1⁄16″
Voltage: 22 to 23 for 0.045″, 24 to 25 for 0.052″, or 26 to 27 for 1⁄16″
*wire feed speed

Performance

The welder will demonstrate the correct procedure and technique for depositing welds to weld pipe to plate in the 2G position.

Criteria

The welds will pass visual inspection and AWS side bend tests.

Procedure with Backing

1. Bevel two pieces of pipe with a 37.5° groove face and no root face. Insert backing ring and tack weld to form a single-V groove joint. Remove slag. Position workpiece so joint is in 2G position.
2. Use a 90° work angle and a 5° to 10° drag travel angle to deposit root pass on both sides of joint. Remove slag.
3. Deposit second pass on bottom toe of root pass with a 50° to 55° work angle and a 5° to 10° drag travel angle. Cover root by one-half to two-thirds. Fuse into base metal evenly along toe. Remove slag.
4. Deposit third pass on top toe of root pass with a 30° to 35° work angle and a 5° to 10° drag travel angle. Cover second pass by one-third to one-half. Fuse into base metal evenly along toe. Remove slag.
5. Face reinforcement should be ⅛″ maximum.
6. Root reinforcement should be ¹⁄₁₆″ maximum.
7. Prepare best weld for a side bend test.

Procedure without Backing

1. Prepare a new workpiece for open root welding. Use spacers and tack weld to establish a ³⁄₃₂″ to ⅛″ root face and a ³⁄₃₂″ to ⅛″ root opening. Remove slag. Position workpiece so joint is in 2G position.
2. Use a 90° work angle with a 5° to 10° drag travel angle to deposit root pass.
3. Deposit second pass on bottom toe of root pass with a 50° to 55° work angle and a 5° to 10° drag travel angle. Cover root by one-half to two-thirds. Fuse into base metal evenly along toe.
4. Deposit third pass on top toe of root pass with a 30° to 35° work angle and a 5° to 10° drag travel angle. Cover second pass by one-third to one-half. Fuse into base metal evenly along toe.
5. Face reinforcement should be ⅛″ maximum.
6. Root reinforcement should be ¹⁄₁₆″ maximum.
7. Prepare best weld for a side bend test.

advanced exercise **6-39**

Making Multiple-Pass Groove Welds, 5G Position, on Carbon Steel Pipe with and without Backing, Using FCAW-G

Conditions

Position: 5G
Materials: 4″ to 6″ schedule 40 to 80 carbon steel pipe
Backing ring to fit
Electrodes: 0.045″, 0.052″, or 1⁄16″ E71T-1
Shielding: 100% CO_2 or 75% Ar/25% CO_2
Polarity: DCEP
WFS*: 250 to 325 for 0.045″, 150 to 225 for 0.052″, or 110 to 195 for 1⁄16″
Amperage: 110 to 140 for 0.045″, 130 to 160 for 0.052″, or 135 to 175 for 1⁄16″
Voltage: 22 to 23 for 0.045″, 24 to 25 for 0.052″, or 26 to 27 for 1⁄16″
*wire feed speed

Performance

The welder will demonstrate the correct procedure and technique for depositing welds to weld pipe to plate in the 5G position.

Criteria

The welds will pass visual inspection and AWS side bend tests.

Procedure with Backing

1. Bevel two pieces of pipe with a 37.5° groove face and no root face. Insert backing ring and tack weld to form a single-V groove joint. Remove slag. Position workpiece so joint is in the 5G position.
2. Use a 90° work angle and a 5° to 10° push travel angle to deposit root pass on both sides of joint. Remove slag.
3. Deposit second pass on either toe of root pass with a 50° to 55° work angle and a 5° to 10° push travel angle. Cover root by one-half to two-thirds. Fuse into base metal evenly along toe. Remove slag.
4. Deposit third pass on opposite toe of root pass with a 30° to 35° work angle and a 5° to 10° push travel angle. Cover second pass by one-third to one-half. Fuse into base metal evenly along toe. Remove slag.
5. Face reinforcement should be 1⁄8″ maximum.
6. Root reinforcement should be 1⁄16″ maximum.
7. Prepare best weld for a side bend test.

Procedure without Backing

1. Prepare a new workpiece for open root welding. Use spacers and tack weld to establish a 3⁄32″ to 1⁄8″ root face and a 3⁄32″ to 1⁄8″ root opening. Remove slag. Position workpiece so joint is in 5G position.

Criteria

Cutting technique, cut appearance, and cut accuracy will be evaluated by the instructor.

Procedure

1. Obtain pieces of mild steel plate. Use a soapstone to mark cutting lines.
2. Inspect the OFC equipment and report any safety issues.
3. Mount the correct cutting tip in the cutting torch and set up the equipment.
4. Adjust the gas pressures as recommended.
5. Light the torch and adjust the preheat flame to neutral.
6. With the oxygen lever ON, adjust the cutting flame so that the preheat cones are burning with a neutral flame.
7. Practice piercing holes and cutting straight square-edge cuts, square-edge curved and shape cuts, straight bevel-edge cuts, and round-bar steel cuts.
8. Repeat straight square-edge and bevel-edge cuts with a track burner (mechanized OFC).

advanced exercise 6-46 through 6-49

Advanced Manual and Mechanized Oxyfuel Gas Cutting

Conditions

Position: Horizontal and vertical positions
Materials: Oxyfuel gas cutting equipment
Pipe beveler
¼″ to ½″ steel plate (or larger)
Carbon steel structural shapes (angle, channel, square, or rectangular tubing)
4″ to 8″ schedule 40 to 80 carbon steel pipe

Performance

The welder will perform a safety inspection of OFC equipment, any necessary operator maintenance, and setup. The welder will then demonstrate the correct procedures using OFC to perform the following exercises:

Exercise 6-46. Straight Square-Edge Cuts

Exercise 6-47. Straight Bevel-Edge Cuts

Exercise 6-48. Straight Structural Shape Cuts (angle, channel, square, or rectangular tubing)

Exercise 6-49. Straight Square-Edge and Bevel-Edge Cuts on Pipe

Criteria

Cutting technique, cut appearance, and cut accuracy will be evaluated by the instructor.

Procedure

1. Follow steps 1 through 6 for exercises 6-41 through 6-45.
2. Practice making straight square-edge cuts, bevel-edge cuts, and structural shape cuts (angle, channel, square or rectangular tubing) in the horizontal and vertical positions.
3. Practice making straight square-edge and bevel-edge cuts on pipe, with the pipe axis in the horizontal and vertical positions.
4. Practice using a pipe beveler (mechanized OFC) to make square edge and bevel edge cuts on pipe in the flat position.

Caution: Take appropriate safety precautions to avoid falling dross when cutting in the horizontal and vertical positions.

exercises 6-50 through 6-51

Cutting Mild Steel, Stainless Steel, and Aluminum Using Plasma Arc Cutting (PAC)

Conditions

Position: Flat positon
Materials: Plasma arc cutting equipment
Compressed air
10- to 16-gauge (or thicker) mild steel, austenitic stainless steel, and aluminum plate

Performance

The welder will perform a safety inspection of Plasma Arc Cutting (PAC) equipment, any necessary operator maintenance, and setup. The welder will then demonstrate the correct procedures using PAC to perform the following exercises:

Exercise 6-50. Straight Square-Edge Cuts

Exercise 6-51. Shape Square-Edge Cuts

Criteria

Cutting technique, cut appearance, and cut accuracy will be evaluated by the instructor.

Procedure

1. Obtain pieces of mild steel, austenitic stainless steel, and aluminum. Mark pieces using a straight edge to serve as a guide for cutting.

2. Inspect the PAC equipment and report any safety issues before completing setup.
3. Use proper ventilation and personal protective equipment.
4. Practice making straight square-edge cuts on each type of metal in flat position.
5. Practice making shape square-edge cuts on each type of metal in flat position.

advanced exercises 6-52 through 6-56

Advanced Plasma Arc Cutting

Conditions

Positions: Horizontal and vertical positions
Materials: Plasma arc cutting equipment
Compressed air
10- to 16-gauge (or thicker) mild steel, austenitic stainless steel, and aluminum plate
Carbon steel, stainless steel, and aluminum structural shapes (angle, channel, square, or rectangular tubing)
4″ to 8″ schedule 40 to 80 carbon steel pipe (stainless steel and aluminum pipe, optional)

Performance

The welder will perform a safety inspection of Plasma Arc Cutting (PAC) equipment, any necessary operator maintenance, and setup. The welder will then demonstrate the correct procedures using PAC to perform the following exercises:

Exercise 6-52. Straight Square-Edge Cuts

Exercise 6-53. Shape Square-Edge Cuts

Exercise 6-54. Straight Structural Shape Cuts (angle, channel, square, or rectangular tubing)

Exercise 6-55. Straight Square-Edge Cuts on Pipe

Exercise 6-56. Straight Bevel-Edge Cuts on Pipe

Criteria

Cutting technique, cut appearance, and cut accuracy will be evaluated by the instructor.

Procedure

1. Follow steps 1 through 3 for exercises 6-50 through 6-51.
2. Practice using PAC to make straight square-edge cuts, bevel-edge cuts, and structural shape cuts (angle, channel, square, or rectangular tubing) on each type of metal in the horizontal and vertical positions.
3. Practice using PAC to make square-edge and bevel-edge cuts on pipe, with the pipe axis in the horizontal and vertical positions.

Caution: Take appropriate safety precautions to avoid falling dross when cutting in the horizontal and vertical positions.

exercises 6-57 through 6-59

Washing and Gouging Mild Steel Using Air Carbon Arc Cutting (CAC-A)

Conditions

Position: Flat and horizontal positions
Materials: Constant current power source equipped for air carbon arc (CAC-A) cutting
Compressed air
¼″ to 1″ mild steel plate (previously welded and unwelded)

Performance

The welder will perform a safety inspection of air carbon arc (CAC-A) equipment, any necessary operator maintenance, and setup. The welder will then demonstrate the correct procedures using CAC-A to perform the following exercises:

Exercise 6-57. Washing (Scarfing)

Exercise 6-58. Gouging Root Side of Groove Welds

Exercise 6-59. Gouging (Removing) Weld Metal

Criteria

Cutting technique, cut appearance, and cut accuracy will be evaluated by the instructor.

Procedure

1. Obtain pieces of mild steel from previous welding exercises (groove and fillet welds as well as new material).
2. Inspect the CAC-A equipment and report any safety issues before completing setup.
3. Adjust the power source to the correct polarity and current.
4. Set compressed air to the required pressure.
5. Adjust the electrode to extend a maximum of 6″ from the electrode holder.
6. Check to see that the air jet orifices are positioned under the electrode.
7. Use proper ventilation and personal protective equipment.
8. Practice washing (scarfing) backing strips from groove welds in the flat and horizontal positions. Adjust push travel angle as necessary.
9. Practice gouging the root side of groove welds in flat and horizontal position to prepare for back welds. Adjust push travel angle as necessary. (Practice depositing back welds if time permits.)
10. Practice gouging (removing) weld metal from groove and fillet welds so that the material can be salvaged for future use. Adjust the push travel angle as necessary.

Caution: Take appropriate safety precautions to avoid falling dross when scarfing or gouging in the horizontal position.

advanced exercises 6-60 through 6-63

Advanced Air Carbon Arc Cutting (CAC-A) Cutting, Gouging, and Beveling

Conditions

Position: Horizontal and vertical (downhill) positions
Materials: Constant current power source equipped for air carbon arc (CAC-A) cutting
Compressed air
¼″ to 1″ mild steel plate (previously welded and unwelded)
4″ to 8″ schedule 40 to 80 carbon steel pipe (previously welded and unwelded)

Performance

The welder will perform a safety inspection of air carbon arc (CAC-A) equipment, any necessary operator maintenance, and setup. The welder will then demonstrate the correct procedures using CAC-A to perform the following exercises:

Exercise 6-60. Straight Square-edge and Bevel-edge Cuts on Carbon Steel Plate

Exercise 6-61. Gouging (Removing) Weld Metal and Base Metal

Exercise 6-62. Straight Structural Shape Cuts (angle, channel, square, or rectangular tubing)

Exercise 6-63. Straight Square-Edge and Bevel-Edge Cuts on Pipe

Criteria

Cutting technique, cut appearance, and cut accuracy will be evaluated by the instructor.

Procedure

1. Follow steps 1 through 7 for exercises 6-57 through 6-59. Also obtain pieces of previously welded and unwelded pipe.

2. Practice straight square-edge and bevel-edge cuts in the flat, horizontal, and vertical down positions on carbon steel plate. Adjust push travel angle as necessary.

3. Practice gouging base metal and weld metal in the horizontal and vertical down positions. Adjust push travel angle as necessary.

4. Practice straight structural shape cutting and gouging on carbon steel (angle, channel, square, or rectangular tubing) in the horizontal and vertical down positions. Adjust the push travel angle as necessary.

5. Perform straight square edge and bevel cuts on carbon steel pipe with the pipe axis in the horizontal and vertical positions. Adjust push travel angle as necessary.

Caution: Take appropriate safety precautions to avoid falling dross when cutting or gouging in the horizontal and vertical positions.

exercises 6-64 through 6-66

Brazing Joints on Mild Steel Using Oxyacetylene

Conditions

Flat position
Oxyacetylene welding equipment
Brazing rod (prefluxed)
Two pieces of $1/16''$ × $1\frac{1}{2}''$ × $5''$ mild steel

Performance

The welder will demonstrate the correct procedure for brazing the following joints:

Exercise 6-64. Lap Joint

Exercise 6-65. Butt Joint with a Backing Bar

Exercise 6-66. Scarfed Butt Joint

Criteria

Brazing technique, brazed joint appearance, and brazed joint strength will be evaluated by the instructor.

Procedure

1. Refer to Exercise 2-1 for oxyacetylene equipment and setup.
2. Remove all dirt, grease, oil, and oxides from the workpieces to be brazed.
3. Align the workpieces in the correct position. Use jigs if necessary.
4. Adjust the oxyacetylene flame to neutral or slightly carburizing.
5. Preheat the workpieces uniformly.
6. Touch the brazing filler metal to the preheated piece. As the filler metal melts, apply it with a forehand technique. Use a circular motion with the torch to distribute the filler metal and to form the bead.
7. Remove all flux residue.

exercise 6-67

Braze Welding a Single-V Butt Joint on Mild Steel Using Oxyacetylene

Conditions

Flat position
Refer to Exercises 6-7 through 6-9
Two pieces of $3/16''$ × $2''$ × $4''$ mild steel

Performance

The welder will demonstrate the correct procedure for braze welding a single-V butt joint.

Criteria

Braze welding technique, braze weld joint appearance, and braze weld joint strength will be evaluated by the instructor.

Procedure

1. Bevel the pieces to a 60° groove angle with a 3⁄32″ root face.
2. Refer to Exercises 6-7 through 6-9, Steps 1 through 5.
3. Hold the torch at the start of the joint. The metal should be heated until it turns dull red.
4. Melt a small amount of brazing filler metal and spread it across the entire joint. This tinning operation prepares the joint to accept the filler metal necessary for joint strength.
5. Use a forehand technique and a circular torch motion to fill the groove of the joint.
6. The metal should be heated only enough to melt the filler metal. Do not overheat the base metal.

exercises 6-68 and 6-69

Soldering Joints on Mild Steel Using a Soldering Copper

Conditions

Flat position
Soldering copper
50/50 solder
Two pieces of 22-gauge, 2″ × 6″ mild steel

Performance

The welder will demonstrate the correct procedure for soldering the following joints:

Exercise 6-68. Seam Solder Butt Joint

Exercise 6-69. Sweat Solder Lap Joint

Criteria

Soldering technique, soldered joint appearance, and soldered joint strength will be evaluated by the instructor.

Procedure

1. Remove all dirt, grease, oil, and oxides from the workpieces to be soldered.
2. Align the workpieces in the correct position. Use jigs if necessary.

3. Apply flux to the areas to be tinned.
4. Prepare and tin the soldering copper.
5. Tack the fluxed workpieces together in several places (seam soldering).
6. Hold the soldering copper on the joint until the flux begins to sizzle. Begin applying solder behind the copper while carefully advancing the copper across the joint. The solder should melt from the heat of the metal, not the copper (seam soldering).
7. Apply a uniform coating of solder on the surfaces to be joined (sweat soldering).
8. Place the surfaces together with the soldered sides in contact (sweat soldering).
9. Place the flat side of the copper on the seam. As the solder begins to melt and flow out of the joint, draw the copper slowly across the joint. Use a punch to hold the workpieces in contact until sufficient cooling has taken place and the solder has solidified (sweat soldering).

exercise **6-70**

Surfacing Mild Steel Using SMAW

Conditions

Flat position
Refer to Exercise 3-25
Surfacing electrode
Worn part or steel plate

Performance

The welder will demonstrate the correct procedure for surfacing the mild steel provided by the instructor.

Criteria

Surfacing technique, surfacing appearance, and surfacing strength will be evaluated by the instructor.

Procedure

1. Obtain a workpiece to be surfaced.
2. Clean the surface by removing rust, scale, or other foreign matter.
3. Use only enough current to maintain the arc necessary to heat the workpiece.
4. Arrange the workpiece in flat position.
5. Maintain a long arc length. Avoid touching the electrode to the area being surfaced.
6. A weaving motion may be used in areas where a thin deposit is required.
7. Remove all slag from the surface before depositing additional layers.

exercises 6-71 through 6-76

Welding Pipe with SMAW

Exercise 6-71. See Exercise 3-107 2F position. (See page 227.)

Exercise 6-72. See Exercise 3-108 4F position. (See page 228.)

Exercise 6-73. See Exercise 3-109 5F position. (See page 229.)

Exercise 6-74. See Exercise 3-110 2G position. (See page 230.)

Exercise 6-75. See Exercise 3-111 5G position. (See page 231.)

Exercise 6-76. See Exercise 3-112 6G position. (See page 232.)

exercises 6-77 through 6-89

Welding Pipe with GTAW

Exercise 6-77. See Exercise 4-33 2F position Carbon Steel. (See page 263.)

Exercise 6-78. See Exercise 4-34 4F position Carbon Steel. (See page 263.)

Exercise 6-79. See Exercise 4-35 5F position Carbon Steel. (See page 264.)

Exercise 6-80. See Exercise 4-36 2G position Carbon Steel. (See page 265.)

Exercise 6-81. See Exercise 4-37 5G position Carbon Steel. (See page 266.)

Exercise 6-82. See Exercise 4-38 6G position Carbon Steel. (See page 268.)

Exercise 6-83. See Exercise 4-39 2F position Stainless Steel. (See page 269.)

Exercise 6-84. See Exercise 4-40 4F position Stainless Steel. (See page 270.)

Exercise 6-85. See Exercise 4-41 5F position Stainless Steel. (See page 271.)

Exercise 6-86. See Exercise 4-42 2G position Stainless Steel. (See page 272.)

Exercise 6-87. See Exercise 4-43 5G position Stainless Steel. (See page 273.)

Exercise 6-88. See Exercise 4-44 6G position Stainless Steel. (See page 274.)

Exercise 6-89. See Exercise 4-45 6G position Thick Wall, GTAW, and SMAW. (See page 275.)

exercises 6-90 through 6-97

Welding Pipe with GMAW

Exercise 6-90. See Exercise 5-33 2F position GMAW-S. (See page 307.)

Exercise 6-91. See Exercise 5-34 4F position GMAW-S. (See page 308.)

Exercise 6-92. See Exercise 5-35 5F position GMAW-S. (See page 309.)

Exercise 6-93. See Exercise 5-36 2G position GMAW-S. (See page 310.)

Exercise 6-94. See Exercise 5-37 5G position GMAW-S. (See page 312.)

Exercise 6-95. See Exercise 5-38 6G position GMAW-S. (See page 314.)

Exercise 6-96. See Exercise 5-39 2F position GMAW spray transfer. (See page 316.)

Exercise 6-97. See Exercise 5-40 1G position GMAW spray transfer. (See page 317.)

exercises 6-98 through 6-103

Welding Pipe with FCAW-G

Exercise 6-98. See Exercise 6-35 2F position. (See page 341.)

Exercise 6-99. See Exercise 6-36 4F position. (See page 341.)

Exercise 6-100. See Exercise 6-37 5F position. (See page 342.)

Exercise 6-101. See Exercise 6-38 2G position. (See page 343.)

Exercise 6-102. See Exercise 6-39 5G position. (See page 345.)

Exercise 6-103. See Exercise 6-40 6G position. (See page 346.)

Section 7 Activities

Weld Evaluation and Testing

Name ______________________________ Date ____________________

Identify the characteristic discontinuity/defect in the welds.

______________ **1.** Slag inclusion

______________ **2.** Incomplete fusion

______________ **3.** Undercutting

______________ **4.** Porosity

______________ **5.** Cracking

______________ **6.** Incomplete penetration

______________ **7.** Underfill

______________ **8.** Melt-through

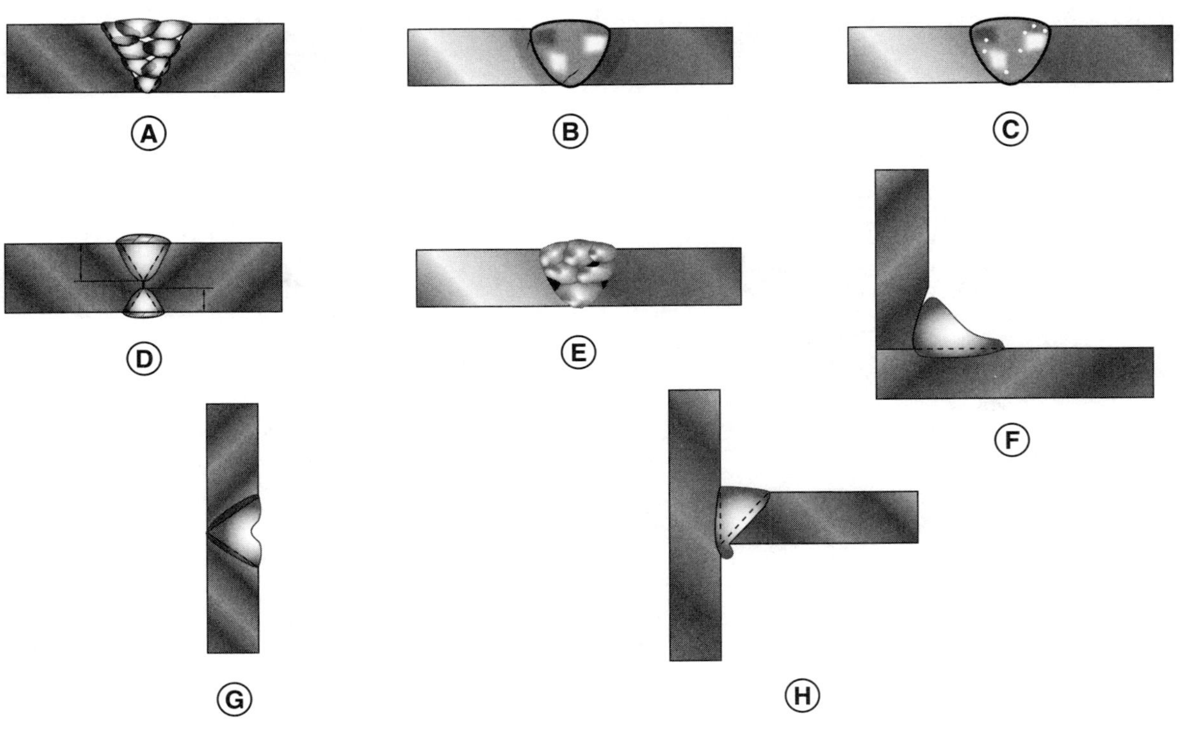

Refer to the welding procedure specification (WPS) form.

____________________ **9.** The welding position required is ___.

____________________ **10.** The welding process specified is GMAW, which stands for ___.

____________________ **11.** The electrode specified is ___.

____________________ **12.** The material specified is ASTM ___.

____________________ **13.** Filler metal specified is AWS filler metal F No. ___.

____________________ **14.** The weld type specified is a(n) ___ weld.

____________________ **15.** The shielding gas flow is ___ cubic feet per hour.

____________________ **16.** The shielding gas specified is 95% ___.

____________________ **17.** A 400°F ___ temperature is specified.

WELDING PROCEDURE SPECIFICATION (WPS)

Identification 302
Date 2/4 Revision NA
Company name National Fabricators, Inc., Atlanta, GA
Supporting PQR no.(s) 1 Type - Manual ( ) Semi-Automatic (X)
Welding process(es) GMAW Mechanized ( ) Automatic ( )
Backing: Yes ( ) No (X)
Backing material (type) NA
Material number M-1 Group M-1 To material number M-1 Group M-1
Material spec. type and grade ASTM A36 To material spec. type and grade ASTM A36
Base metal thickness range: Groove 1/2 inch Fillet NA
Deposited weld metal thickness range As requlred
Filler metal F no. 8 A no. 1
Spec. no. (AWS) A5.18 and A5.28 Flux tradename NA
Electrode-flux (Class) E70S-5 Type NA
Consumable insert: Yes ( ) No (X) Classifications NA
Shape NA
Position(s) of joint Flat 1G Size NA
Welding progression: Up (—) Down (—) Ferrite number (when reqd.) NA

PREHEAT:
Preheat temp., min 400°F
Interpass temp., max NA
(continuous or special heating, where applicable, should be recorded)

GAS:
Shielding gas(es) Argon/Oxygen
Percent composition 95/5
Flow rate 40-50 CFH

Refer to the welding procedure specification (WPS) form.

__________________ **18.** The diameter of the electrode is ___.

__________________ **19.** The wire feed speed is ___ inches per minute.

__________________ **20.** The electrode polarity is ___.

__________________ **21.** The metal transfer mode is ___ transfer.

__________________ **22.** The travel speed is ___ inches per minute.

__________________ **23.** A(n) ___° push angle is specified.

Welding progression: Up (—) Down (—)
Ferrite number (when reqd.) NA

PREHEAT:
Preheat temp., min 400°F
Interpass temp., max NA
(continuous or special heating, where applicable, should be recorded)

GAS:
Shielding gas(es) Argon/Oxygen
Percent composition 95/5
Flow rate 40-50 CFH
Root shielding gas NA
Trailing gas composition NA
Trailing gas flow rate NA

POSTWELD HEAT TREATMENT:
Temperature range NA
Time range NA

Tungsten electrode, type and size NA
Mode of metal transfer for GMAW: Short-circuiting ( ) Globular ( ) Spray (X)
Electrode wire feed speed range: 105-110 IPM
Stringer bead ( ) Weave bead ( ) Peening: Yes ( ) No (X)
Oscillation As required
Standoff distance 5/8—7/8
Multiple ( ) or single electrode (X)
Other

Weld layer(s)	Filler metal: Process	Class	Dia.	Current: Type & polarity	Amp range	Volt range	Travel speed range	
ALL	GMAW	E70S-5	3/32"	DCEP	400-425	26-27	30 IPM	10° PUSH ANGLE

Approved for Production by __________
Employer

Note: Those items that are not applicable should be marked N.A.

24. Obtain a piece of ferrous metal. Perform a spark test on the metal and identify the type of metal using the Spark Chart in the Appendix.

Indicate the meaning of the following welding symbols. See Appendix.

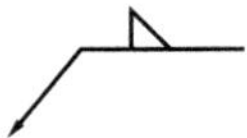

14. ____________ **15.** ____________ **16.** ____________

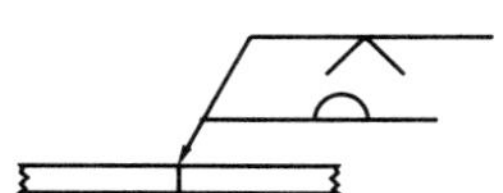

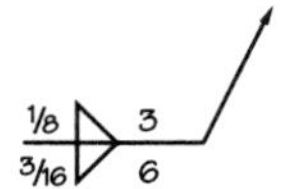

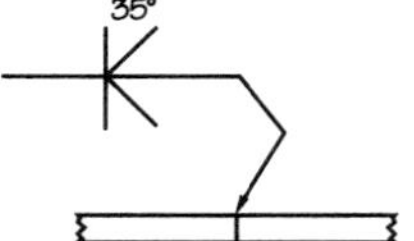

17. ____________ **18.** ____________ **19.** ____________

Draw the weld symbol that describes the weld shown. See Appendix.

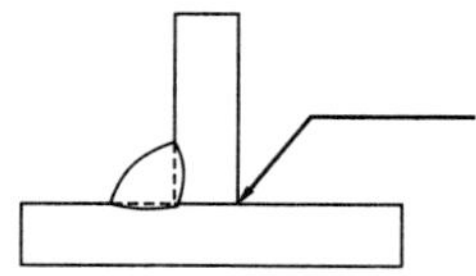

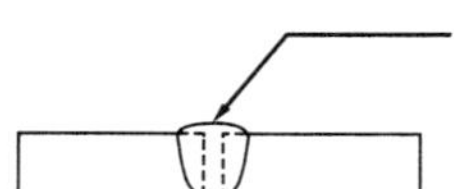

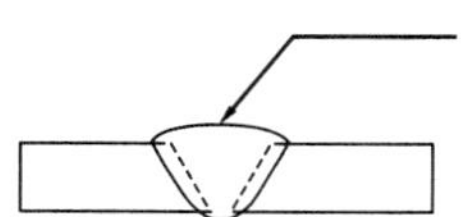

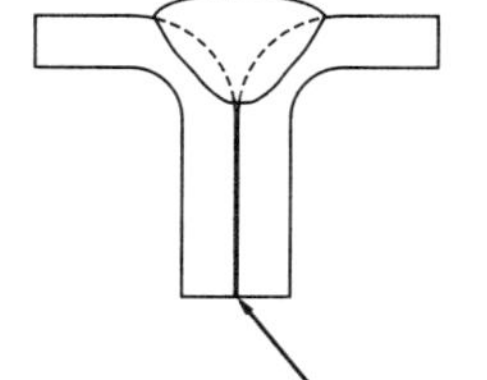

20. ____________ **21.** ____________ **22.** ____________ **23.** ____________

Identify the contour and finishing (mechanical) method required by the welding symbols shown.

C M G H

24. ____________ **25.** ____________ **26.** ____________ **27.** ____________

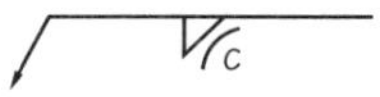

28. ______________ **29.** ______________ **30.** ______________ **31.** ______________

Indicate the meaning of the following nondestructive examination symbols. See Appendix.

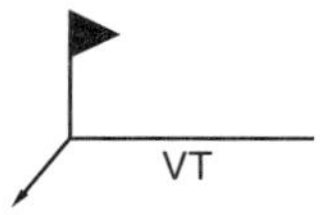

VT

RT

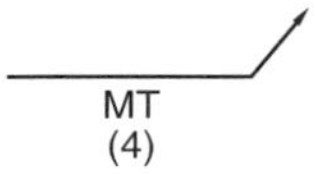

MT
(4)

32. ______________ **33.** ______________ **34.** ______________

ET

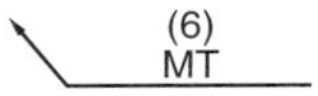

(6)
MT

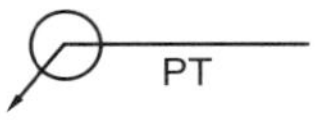

PT

35. ______________ **36.** ______________ **37.** ______________

Identify the properties of metal with a P (physical property), M (mechanical property), or C (chemical property).

______________ **38.** Chemical inhomogeneity

______________ **39.** Hardness

______________ **40.** Strength

______________ **41.** Corrosion resistance

______________ **42.** Solubility

______________ **43.** Electrical conductivity

______________ **44.** Coefficient of linear expansion

Match each family of stainless steels with its description.

______________ **45.** Martensitic

______________ **46.** Austenitic

______________ **47.** Precipitation hardening

______________ **48.** Duplex

______________ **49.** Cast

A. microstructure of ferrite and austenite

B. most widely used stainless steel

C. heat treated to highest strengths

D. C series and H series

E. least corrosion-resistant

Identify the components of the phase diagram shown.

______________ **50.** Completely molten

______________ **51.** Liquidus

______________ **52.** Liquid first appears

______________ **53.** Solidus

______________ **54.** Cooling

______________ **55.** Heating

______________ **56.** Solids first appear

______________ **57.** Completely solid

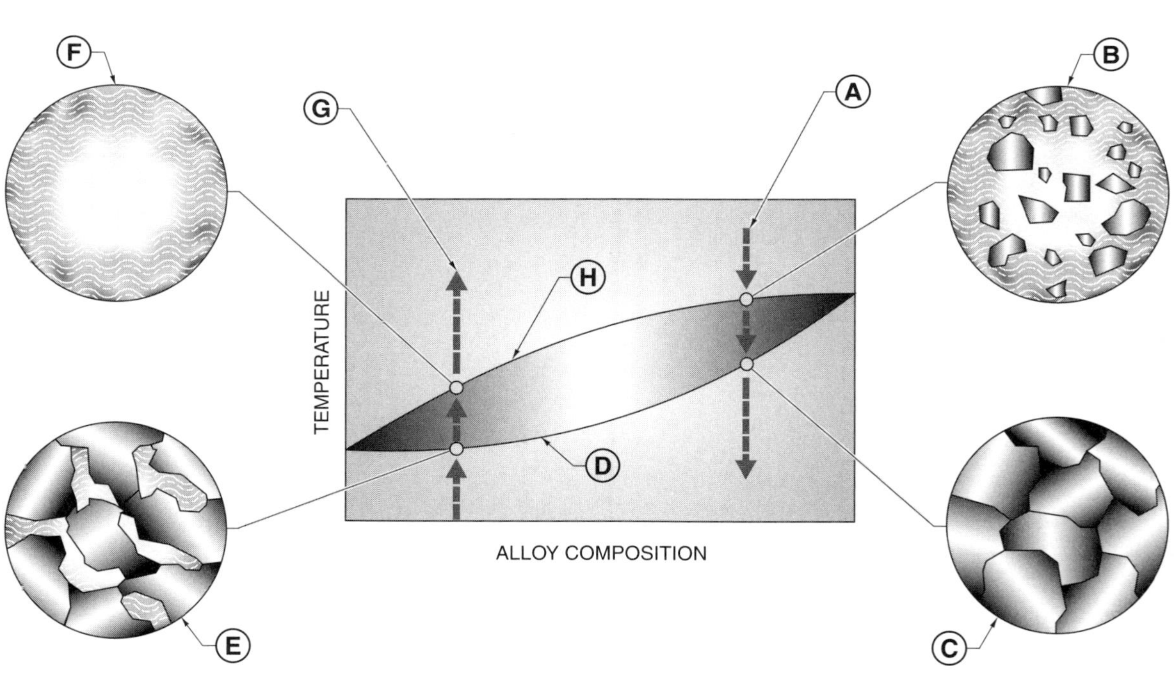

Refer to the Carbon Steel Alloy Families Table for the following questions.

______________ **58.** ___ is added to low-carbon steels and medium-carbon steels to increase strength.

______________ **59.** Adding nickel for low temperature toughness produces ___.

______________ **60.** Type 4140 is a(n) ___ alloy steel.

______________ **61.** Chrome-moly steels are produced by adding ___ and ___.

______________ **62.** Chromium and molybdenum increase the ___ of steels.

Carbon Steel Alloy Families

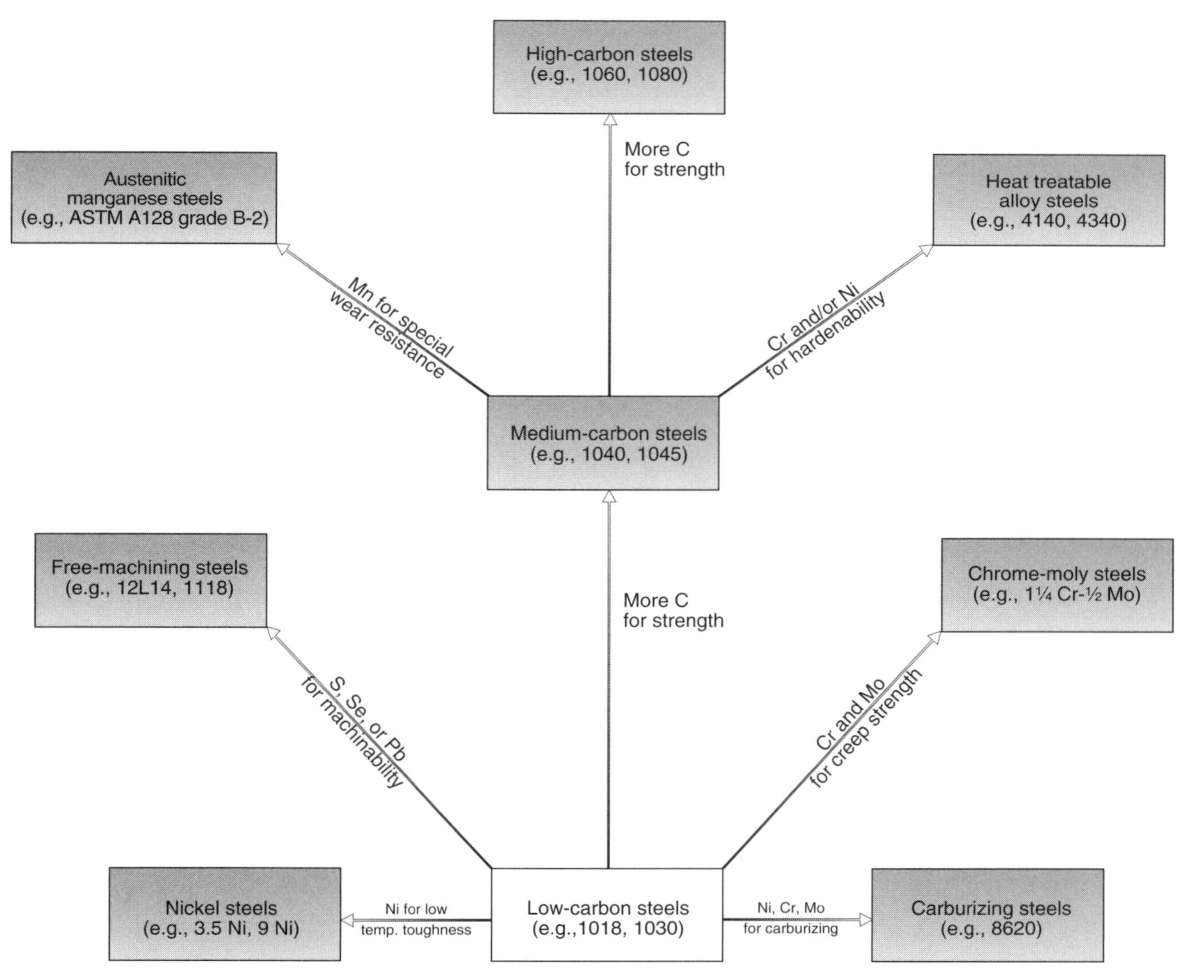

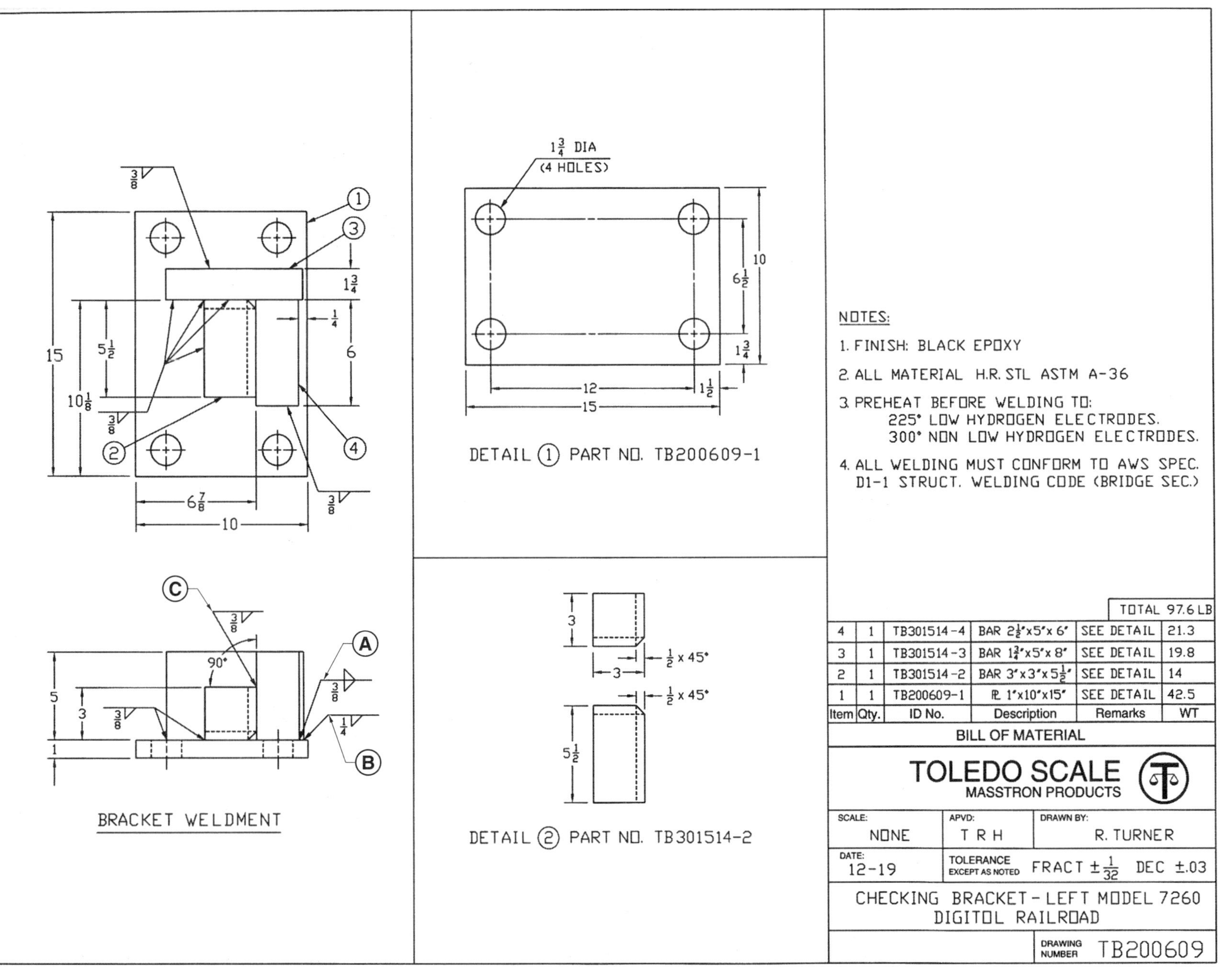

BRACKET WELDMENT
15
10 1/8
5 1/2
1 3/4
1/4
6
6 7/8
10
3/8
5
3
1
90°
1/4
A
B
C
DETAIL ① PART NO. TB200609-1
1 3/4 DIA
(4 HOLES)
12
15
1 1/2
6 1/2
10
DETAIL ② PART NO. TB301514-2
1/2 x 45°
NOTES:
1. FINISH: BLACK EPOXY
2. ALL MATERIAL H.R. STL ASTM A-36
3. PREHEAT BEFORE WELDING TO:
225° LOW HYDROGEN ELECTRODES.
300° NON LOW HYDROGEN ELECTRODES.
4. ALL WELDING MUST CONFORM TO AWS SPEC.
D1-1 STRUCT. WELDING CODE (BRIDGE SEC.)
TOTAL 97.6 LB
4 | 1 | TB301514-4 | BAR 2 1/2"x5"x 6" | SEE DETAIL | 21.3
3 | 1 | TB301514-3 | BAR 1 3/4"x5"x 8" | SEE DETAIL | 19.8
2 | 1 | TB301514-2 | BAR 3"x3"x5 1/2" | SEE DETAIL | 14
1 | 1 | TB200609-1 | ℞ 1"x10"x15" | SEE DETAIL | 42.5
Item | Qty. | ID No. | Description | Remarks | WT
BILL OF MATERIAL
TOLEDO SCALE
MASSTRON PRODUCTS
SCALE: NONE
APVD: T R H
DRAWN BY: R. TURNER
DATE: 12-19
TOLERANCE EXCEPT AS NOTED FRACT ± 1/32 DEC ±.03
CHECKING BRACKET - LEFT MODEL 7260
DIGITOL RAILROAD
DRAWING NUMBER TB200609

Refer to the Checking Bracket Bill of Material on page 368 for questions 63 through 70.

____________________ **63.** A preheat temperature of ___° is required when using low-hydrogen electrodes.

____________________ **64.** Welding must conform to the ___ code.

____________________ **65.** On the bracket weldment, a ___ weld is required at point A.

____________________ **66.** On the bracket weldment, a ___ weld is required at point B.

____________________ **67.** On the bracket weldment, a ___ weld is required at point C.

____________________ **68.** ___ piece(s) of 1¾″ × 5″ × 8″ bar are required.

____________________ **69.** All material is manufactured to ASTM ___.

____________________ **70.** The fractional tolerance of the checking bracket is ___.

Answer the following questions in the spaces provided. Show all work. See Appendix.

____________________ **71.** What is the tensile stress of a 25,000 lb force applied to a square steel rod with a cross-sectional area of 2.5 sq in.?

____________________ **72.** What is the compressive stress of a 44,250 lb force applied to a rectangular steel bar with a cross-sectional area of 6 sq in.?

____________________ **73.** What is the torque of a 2500 lb force applied over a distance of 4½″?

____________________ **74.** Find the compressive stress.

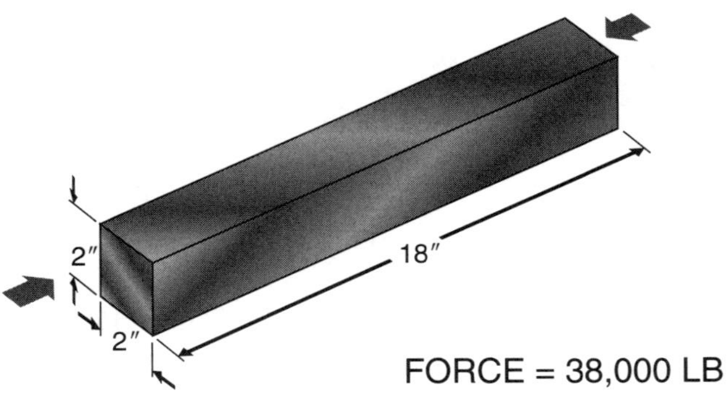

____________________ **75.** Find the tensile stress.

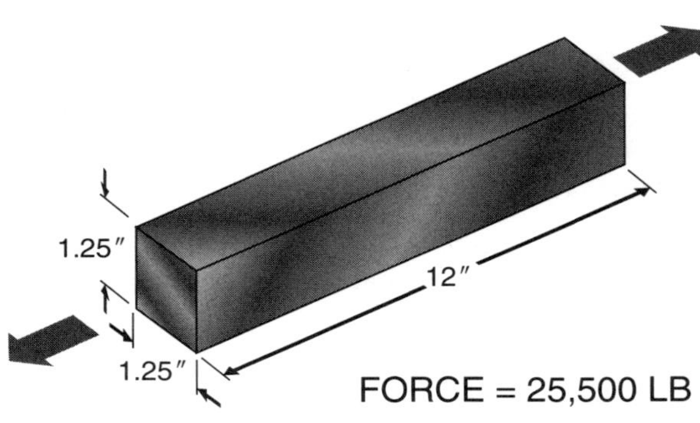

76. Using the procedure qualification test record in the appendix, perform the weld specified by the instructor.

Section 1 Exam

Introduction to Welding

Name Kyle Goodwin Date 9-15-16

True-False

(T)	F	**1.** Welding was first used in 2000 B.C.
(T)	F	**2.** One advantage of the oxyacetylene welding process is the mobility of the necessary equipment.
T	(F)	**3.** Shielded metal arc welding was developed after gas metal arc welding.
(T)	F	**4.** All accidents should be reported and documented.
(T)	F	**5.** A respirator should be used when welding metals that give off toxic fumes.
(T)	F	**6.** Hollow castings should not be welded without proper venting.
(T)	F	**7.** All injuries should be reported immediately to a supervisor.
(T)	F	**8.** Tack welds are used to hold members in proper alignment until welds are made.
(T)	F	**9.** A crater is a depression at the termination of a weld bead.
(T)	F	**10.** The weld face is the exposed surface of a weld, bounded by the weld toes on the side on which welding was done.
(T)	F	**11.** A material safety data sheet includes data about every hazardous component comprising 1% or more of a material's content and is used by a manufacturer, importer, or distributor to relay chemical hazard information to the employee.
T	(F)	**12.** A confined space can be used for continuous occupancy to complete weld jobs as needed.
T	(F)	**13.** Noninsulated electrode holders should be used for SMAW.
T	(F)	**14.** A flange joint is one of five basic weld joints.
(T)	F	**15.** Most weld joint designs are covered by AWS codes.

Multiple Choice

D 1. The ___ welding process involves fusing two pieces of metal together using heat and pressure.
   - A. shielded metal arc
   - B. oxyfuel gas
   - C. gas tungsten arc
   - D. resistance

C 2. The ___ welding process uses heat from the combustion of a mixture of oxygen and a fuel gas.
   - A. gas metal arc
   - B. gas tungsten arc
   - C. oxyfuel gas
   - D. resistance

C 3. The ___ welding process is used primarily for production welding.
   - A. shielded metal arc
   - B. gas metal arc
   - C. gas tungsten arc
   - D. resistance

B 4. ___ welding was the first type of welding used by man.
   - A. Oxyacetylene
   - B. Forge
   - C. Stick
   - D. Inert

D 5. ___ welding is the most commonly used oxyfuel gas welding process.
   - A. Propane
   - B. Methylacetylene-propadiene
   - C. Hydrogen
   - D. Oxyacetylene

B 6. Before welding on a container that has held an unknown substance, the ___ to prevent possible injury.
   - A. hot chemical solution cleaning method must be used
   - B. exact substance which was held by the container must be determined
   - C. chemical cleaning method with trisodium phosphate must be used
   - D. container should be partially filled with water

C 7. A Class ___ fire extinguisher is identified by a yellow star.
   - A. A
   - B. B
   - C. C
   - D. D

B **8.** A ___ weld is a type of weld.
A. lap
B. fillet
C. corner
D. square

D **9.** The portion of the groove face within the joint root is the ___.
A. weld toe
B. weld reinforcement
C. joint penetration
D. root face

10. The ___ welding process uses a nonconsumable electrode.
A. oxyfuel gas
B. submerged arc
C. shielded metal arc
D. gas tungsten arc

11. The shape within a deposited bead caused by the movement of the electrode is the ___.
A. toe
B. crown
C. ripple
D. throat

12. Stainless steels require special attention to ventilation because they produce ___ fumes when welded.
A. beryllium
B. zinc
C. manganese
D. hexavalent chromium

13. The exposed surface of a weld bounded by the weld toes is the ___.
A. weld face
B. root opening
C. actual throat
D. joint root

A **14.** Transferred arc ___ is used for welding high-strength, thin metal.
A. GTAW
B. PAW
C. FCAW
D. SMAW

__________ **15.** A fillet weld is used in a(n) ___.

A. butt joint
B. groove joint
C. T-joint
D. edge joint

__________ **16.** The maximum safe operating pressure for acetylene is ___ psi.

A. 2
B. 5
C. 10
D. 15

__________ **17.** A single progression of welding along a joint is a ___.

A. base metal
B. weld pass
C. crater
D. weld bead

__________ **18.** ___ welding is the most common method of welding metals.

A. Forge
B. Arc
C. Oxyfuel gas
D. Resistance

__________ **19.** The proper ___ is required to allow complete penetration to the root of the joint.

A. crater size
B. groove angle
C. single-V joint
D. root reinforcement

__________ **20.** A class ___ fire extinguisher is used for putting out fires caused by combustible metals.

A. A
B. B
C. C
D. D

Section 2 Exam

Oxyacetylene Welding (OAW)

Name ______________________________ Date ____________________

True-False

T F 1. Oxygen cylinders are charged to a pressure of 2200 psi.

T F 2. A flash arrestor is used to ignite a welding torch.

T F 3. The connecting nut used to mount an acetylene hose on a welding torch is notched to indicate left-hand threads.

T F 4. Acetylene becomes unstable at 250 psi.

T F 5. When overhead welding, heat should be increased to prevent the weld pool from falling out of the joint.

T F 6. Needle valves on a welding torch control the gas pressure from the cylinders.

T F 7. Purging welding hoses involves removal of residual gases.

T F 8. The thickness of the base metal being welded determines the size of the torch welding tip to use.

T F 9. A carburizing flame has excess oxygen.

T F 10. When opening oxygen and acetylene cylinder valves, the welder must stand to one side.

T F 11. Holes in joints are caused by holding a flame too long in one spot.

T F 12. A torch is ignited with the tip facing upward.

T F 13. A popping noise usually indicates an insufficient flow of gases to the welding tip.

T F 14. A cylinder valve regulates gas flow to a torch.

T F 15. Oxygen and acetylene cylinders must be chained in position to prevent tipping.

T F 16. When selecting a size of filler metal, the diameter should be equal to one-half the thickness of the base metal.

T F 17. A flashback flame causes metal to spatter around the weld pool.

T F 18. A straight torch movement should be used for horizontal and vertical welding.

T F 19. Soapy water can be used to test for leaks.

T F 20. GTAW skills can be developed by practicing oxyacetylene welding.

T F 21. If a weld pool becomes too large when welding in positions other than flat position, more heat should be directed to the weld area.

T F 22. The adjusting screw controls the flow of gas from a cylinder.

T F 23. When welding heavy steel thicker than ¼″, edge preparation is not required.

T F 24. Cast iron must be preheated to a dull red to minimize stresses from contraction and expansion.

T F 25. Cast iron should be cooled rapidly to prevent distortion caused from the heat of welding.

T F 26. When heated to its melting point, aluminum does not change color.

T F 27. Carpenter's chalk can be used to determine when the correct preheat temperature of aluminum has been reached.

T F 28. The edges of heavy aluminum must be notched before welding to ensure complete penetration.

T F 29. Aluminum must be cleaned thoroughly before it is welded to prevent impurities from weakening the weld.

T F 30. The forehand welding technique is typically used for metals up to ⅛″ in thickness.

Multiple Choice

______ D 1. It is not true that acetylene is ___.
- A. transported in a red hose
- B. less sensitive compared to MAPP gas
- C. ignited first when lighting the torch
- D. most commonly used for underwater welding

______ C 2. To move an oxygen or acetylene cylinder, ___.
- A. roll it on its side
- B. lift it by the protector cap
- C. use a cylinder cart
- D. roll it on its bottom edge

______ B 3. When opening an oxygen cylinder valve, it is necessary to ___.
- A. open the valve one-half turn
- B. open the valve all the way
- C. remove the handwheel
- D. open the valve with a wrench

A **4.** A(n) ___ flame is used for most welding operations.
- A. neutral
- B. oxidizing
- C. harsh
- D. carburizing

B **5.** An acetylene leak is usually easier to detect than an oxygen leak because ___.
- A. of a quicker drop in regulator pressure
- B. of its distinctive odor
- C. it has less pressure in the cylinder
- D. of its distinctive color

B **6.** When welding in flat position, the heat of the torch should be concentrated ___.
- A. on the filler metal
- B. on the base metal
- C. in front of the weld area
- D. toward the lower edge of the metal

C **7.** Overhead welding requires a ___ torch motion.
- A. back-and-forth
- B. circular
- C. semicircular
- D. straight

B **8.** When welding a butt joint in vertical position, more of the flame should be directed to the ___.
- A. weld pool
- B. area in front of the weld
- C. filler metal
- D. top of the plate

A **9.** A welding torch should be held at a(n) ___° angle to the weld joint.
- A. 45
- B. 60
- C. 85
- D. 90

B **10.** When welding aluminum with the oxyacetylene welding process, the welder does not ___.
- A. preheat the metal to approximately 300°F to 500°F
- B. use a tip larger than the tip used for welding mild steel
- C. clean the metal thoroughly before welding
- D. apply heat to the bottom workpiece when welding a lap joint

Matching

______________ **1.** A ___ prevents the possibility of a flashback reaching the manifold system.

______________ **2.** The ___ controls the flow of gases at the torch.

______________ **3.** A ___ is used to ignite the flame.

______________ **4.** The ___ converts cylinder gas pressure into working pressure.

______________ **5.** The ___ shuts off gas flow from the cylinder.

A. sparklighter
B. check valve
C. regulator
D. cylinder valve
E. needle valve

______________ **6.** The ___ cylinder is opened no more than one complete turn during welding.

______________ **7.** A balanced flame used for most welding applications is known as a ___ flame.

______________ **8.** The ___ flame has an excess flow of oxygen.

______________ **9.** After the flame is ignited, ___ is added to produce a neutral flame.

______________ **10.** The ___ flame has an extended feather beyond the inner core.

A. oxygen
B. acetylene
C. carburizing
D. oxidizing
E. neutral

Section 3 Exam

Shielded Metal Arc Welding (SMAW)

Name Kyle Goodwin Date ______________

True-False

T	F	**1.** In DCEP welding, current flows from the electrode into the workpiece.
T	F	**2.** Resistance occurs when material in a conductor opposes the passage of electrical current.
T	F	**3.** A transformer welding machine produces DC current only.
T	F	**4.** Safety glasses are not necessary if a helmet is worn when welding using SMAW.
T	F	**5.** High voltage and low current are used in the SMAW process.
T	F	**6.** A chipping hammer is used to remove slag from a weld.
T	F	**7.** The workpiece lead is positive when welding with DCEN.
T	F	**8.** A constant-current welding machine is used primarily for welding using GMAW.
T	F	**9.** Alternating current in North America is rated at 60 hertz.
T	F	**10.** A voltmeter is used to measure current.
T	F	**11.** The E6013 electrode is classified as a fill-freeze electrode from the F2 group.
T	F	**12.** The position of the weld is a factor in determining the type of electrode to use.
T	F	**13.** Fast-fill electrodes from the F1 group are used for welding in overhead position.
T	F	**14.** The coating on an electrode produces shielding gas that protects molten metal from atmospheric contamination.
T	F	**15.** An E6010 electrode from the F3 group can be used with alternating current.
T	F	**16.** The first two digits in the AWS electrode classification refer to the minimum tensile strength of the deposited weld metal.
T	F	**17.** Before striking an arc, the base metal must be free of any dirt, grease, or oil.
T	F	**18.** An electrode holder is used to hold the coated end of an electrode.

T F **19.** An arc length that is too long results in excess spatter and poor penetration.

T F **20.** A weld crater (weld pool) is formed as the arc comes in contact with the base metal.

T F **21.** Undercutting can be prevented by lowering the current and changing the work angle of the electrode.

T F **22.** A multiple-pass weld requires at least two or more passes to deposit the required amount of weld metal.

T F **23.** Welding is simplified if the work is positioned in flat position.

T F **24.** A lap joint requires edge preparation.

T F **25.** Welding speed can be increased if welding is done in flat position.

T F **26.** Downhill welding allows greater penetration than uphill welding.

T F **27.** A whipping motion should not be used with an E7018 electrode.

T F **28.** Welding in vertical position requires a longer arc length than welding in flat position.

T F **29.** An intermediate weld pass is deposited after a root pass and before a cover pass.

T F **30.** Positioners are used to secure workpieces when practicing welds.

Multiple Choice

A **1.** Electrical current that flows in one direction only is ___ current.
A. direct
B. alternating
C. static
D. constant

C **2.** ___ is the force that causes current to move.
A. Resistance
B. Current
C. Voltage
D. Polarity

D **3.** ___ is voltage present after an arc is struck.
A. Constant current
B. Resistance
C. Voltage drop
D. Arc voltage

C **4.** ___ voltage is produced when no welding is being done.
A. Variable
B. Arc
C. Open-circuit
D. Dropped

________ **5.** When welding using DCEN, the electrode is ___.

A. positive
B. negative
C. neutral
D. alternating

________ **6.** The groove angle of a single-V groove butt joint for SMAW should not exceed ___°.

A. 45
B. 60
C. 75
D. 90

________ **7.** A(n) ___ electrode is best suited for vertical and overhead welding.

A. fast-fill
B. fast-freeze
C. fill-freeze
D. iron powder

________ **8.** An ___ electrode is recommended for flat position and horizontal fillet welds only.

A. E6010
B. E6013
C. E7018
D. E7024

________ **9.** ___ provides an extra boost of current to help establish an arc when using electrodes that are hard to start.

A. Arc blow
B. Flashback
C. Hot start
D. A welding lead

________ **10.** ___ covers the weld metal as it cools to prevent atmospheric contamination.

A. Wire
B. Slag
C. The crater
D. Shielding gas

________ **11.** The ___ depends on the electrode diameter and welding position.

A. electrode angle
B. work angle
C. travel speed
D. arc length

______D______ **12.** The ___ is the rate at which an electrode is moved across the weld area.
A. travel direction
B. arc length
C. electrode angle
D. travel speed

______A______ **13.** The ___ is the distance between the electrode and the work.
A. arc length
B. work angle
C. electrode gap
D. weld bead

______B______ **14.** A ___ motion is used to control the heat of a weld by moving the arc quickly forward and backward along the weld joint.
A. forehand
B. whipping
C. backhand
D. sliding

______B______ **15.** ___ is the process of building up worn surfaces by depositing successive weld beads.
A. Undercutting
B. Surfacing
C. Overlapping
D. Remelting craters

______C______ **16.** ___ occurs when the current is set too high.
A. Overlapping
B. Weaving
C. Undercutting
D. Flashback

______B______ **17.** ___ is a technique used to increase the width of a weld bead.
A. Restarting
B. Weaving
C. Surfacing
D. Cover passing

______B______ **18.** Weld metal deposited with the current set too low results in ___.
A. undercutting
B. overlapping
C. excessive spatter
D. surfacing

D **19.** ___ are used to hold the workpieces in position before a root pass is deposited.
- A. Intermediate passes
- B. Intermittent passes
- C. Backstep welds
- D. Tack welds

C **20.** A ___° work angle is used when depositing the first pass of a multiple-pass fillet weld in a T-joint.
- A. 15
- B. 30
- C. 45
- D. 60

B **21.** When welding base metals of different thicknesses, more heat should be directed to ___.
- A. the thinner workpiece
- B. the thicker workpiece
- C. either workpiece
- D. both workpieces evenly

C **22.** A(n) ___ pass is deposited to achieve a smooth weld appearance.
- A. root
- B. intermediate
- C. cover
- D. final

A **23.** The ___ position is not one of the four main welding positions.
- A. forehand
- B. vertical
- C. overhead
- D. horizontal

D **24.** A rectifier changes ___ current into ___ current.
- A. AC; DCEP
- B. DCEP; DCEN
- C. DC; AC
- D. AC; DC

D **25.** Of the four welding positions used for SMAW, ___ is typically the easiest and ___ is typically the most difficult to master.
- A. horizontal; vertical
- B. flat; vertical
- C. overhead; flat
- D. flat; overhead

______C 26. The last digit in the AWS electrode classification refers to ___.
- A. tensile strength
- B. welding position
- C. type of coating and current
- D. improved impact strength

______D 27. ___ current is preferred for vertical and overhead welding.
- A. Constant
- B. Alternating/direct
- C. Direct
- D. Alternating

______B 28. Problems that occur during SMAW are typically the result of ___.
- A. grounding the workpiece
- B. improper settings on the welding machine
- C. using an electrode that has been stored in an oven
- D. improper filler metal selection

______C 29. ___ length should be kept as short as possible when using SMAW in overhead position.
- A. Joint
- B. Electrode
- C. Arc
- D. Travel

Section 4 Exam

Gas Tungsten Arc Welding (GTAW)

Name ______________________ **Date** ______________

True-False

T F **1.** Electrode stickout is primarily determined by the type of joint being welded.

T F **2.** GTAW uses a consumable steel electrode.

T F **3.** Electrode stickout for butt joints is ⅛″ to 3/16″.

T F **4.** During a welding operation, a shield of inert gas prevents contamination of the weld by protecting it from the atmosphere.

T F **5.** GTAW produces a slag-free weld.

T F **6.** DCEN is commonly used for welding stainless steel.

T F **7.** The DCEP portion of an AC cycle provides deep penetration.

T F **8.** GTAW torches used for welding over 200 A are typically water-cooled.

T F **9.** A ferrous metal is a metal that does not contain iron.

T F **10.** Tungsten electrodes are alloyed to increase their current capacity.

T F **11.** Argon is commonly used as a shielding gas for GTAW.

T F **12.** The gas nozzle size required is determined by the diameter of the tungsten electrode used.

T F **13.** When arc welding, a tungsten electrode should touch the base metal when striking an arc.

T F **14.** In the GTAW process, filler metal may be added to the weld pool with an in-and-out dipping motion.

T F **15.** A torch is held at a 45° angle when welding a T-joint in horizontal position.

T F **16.** A gas lens is used to control an arc at the torch.

T F **17.** Helium with oxygen is used as a shielding gas for welding stainless steel in the GTAW process.

T F **18.** Nonheat-treatable wrought aluminum alloys, such as 1000, 3000, and 5000 series, are readily weldable.

T F **19.** Using DCEN with helium as a shielding gas provides deep penetration but no surface cleaning.

T F **20.** A high-velocity ventilating system should be used when welding copper to remove the toxic fumes generated from welding.

Multiple Choice

__________ **1.** In the ___ welding process, the operator applies the weld by hand.
- A. manual
- B. mechanized
- C. semiautomatic
- D. automatic

__________ **2.** ___ is the shielding gas most commonly used in the GTAW process.
- A. Carbon dioxide
- B. Helium
- C. Oxygen
- D. Argon

__________ **3.** The ___ controls the rate of flow of shielding gas to a welding torch in cubic feet per hour.
- A. regulator
- B. flowmeter
- C. check valve
- D. gas nozzle

__________ **4.** ___ refers to the time in which gas continues to flow after an arc has been extinguished.
- A. Postflow
- B. Preshield
- C. Inert flow
- D. Stickout

__________ **5.** A ___ angle is commonly used in the GTAW process.
- A. drag
- B. pull
- C. push
- D. travel

__________ **6.** Arc starting is more difficult with ___ because of its high ionization potential.
- A. argon
- B. helium
- C. an argon-helium mixture
- D. oxygen

______________ **7.** ___ is recommended when welding aluminum with the GTAW process using a conventional square wave transformer-rectifier.

A. DCEP with HF set to continuous
B. DCEN with HF set to start
C. AC with HF set to start
D. AC with HF set to continuous

______________ **8.** ___ added to argon, when used as a shielding gas, increases the penetration achieved in the GTAW process.

A. Carbon dioxide
B. Hydrogen
C. Helium
D. Oxygen

______________ **9.** The operator can vary the welding current during a welding operation with a(n) ___.

A. jog button on the wire feeder
B. foot / hand operated control
C. change in electrode angles
D. turn of the check valve

______________ **10.** ___ is most commonly used when using GTAW on mild steel.

A. DCEP
B. DCEN
C. AC with HF set to start
D. AC with HF set to continuous

______________ **11.** The ___ should be reduced when overhead welding using the GTAW process.

A. electrode extension
B. filler metal length
C. current
D. work angle

______________ **12.** The ___ used for welding stainless steels is alloyed to prevent cracking problems.

A. base metal
B. filler metal
C. tungsten electrode
D. nonconsumable electrodes

______________ **13.** A ___ welding technique usually produces the best results when welding copper and copper alloys.

A. forehand
B. backhand
C. circular
D. whipping

_______________ **14.** Carbon steels with a metal thickness between ¼″ and ½″ should have a filler metal diameter of ___″ for GTAW.

A. 1⁄32
B. 1⁄16
C. ⅛
D. ¼

_______________ **15.** Grinding across the grain of a tungsten electrode creates ridges on the taper that may melt off and cause ___.

A. tungsten spitting
B. globular transfer
C. excessive slag
D. craters

_______________ **16.** ___ welding is used for welding very thin metals for critical control of metallurgical factors.

A. GTAW
B. Hot wire
C. Vertical
D. Pulsed GTAW

_______________ **17.** By preheating the filler metal in the ___ welding process, porosity can be eliminated.

A. GTAW
B. Hot wire
C. Vertical
D. Pulsed GTAW

_______________ **18.** Using ___ current results in a narrow, deep penetrating weld.

A. AC with HF
B. AC with HF
C. DCEP
D. DCEN

_______________ **19.** A(n) ___ torch is recommended when welding requires current levels over 200 A.

A. air-cooled
B. hot-wire
C. water-cooled
D. gas

_______________ **20.** The ___ directs shielding gas to the weld zone.

A. torch
B. collet
C. gas lens
D. gas nozzle

Section 5 Exam

Gas Metal Arc Welding (GMAW)

Name ______________________________ **Date** ____________________

True-False

T F **1.** Three main types of metal transfer can be used in the GMAW process.

T F **2.** DCEP is commonly used in the GMAW process.

T F **3.** Current level can be adjusted on a welding machine in the GMAW process.

T F **4.** In globular transfer, gravity causes a molten droplet to separate and transfer across the arc.

T F **5.** A push angle or drag angle can be used in the GMAW process.

T F **6.** Gas drift can cause weld defects by allowing nitrogen and oxygen to come in contact with the weld area.

T F **7.** Spray transfer is only used for metals less than ¼″ thick.

T F **8.** In pulsed spray transfer, peak and background current levels are used to obtain maximum penetration without heat buildup.

T F **9.** A common shielding gas mixture used for spray transfer on carbon steel is a mixture of 90% argon and 10% carbon dioxide.

T F **10.** When using GMAW on carbon steel, beveling is required for thicknesses over ¼″.

T F **11.** When welding aluminum using GMAW, it is recommended to use a narrower root opening and lower welding current than recommended for steel.

T F **12.** When welding aluminum using GMAW, an argon/nitrogen shielding gas mixture provides high heat input and good cleaning action.

T F **13.** When using GMAW on stainless steel, the welding gun is moved back and forth with a slight side-to-side movement.

T F **14.** Copper backing bars must be used when welding copper using GMAW.

T F **15.** Argon or an Ar/He gas mixture is commonly used as a shielding gas when welding copper using GMAW.

Multiple Choice

______________ **1.** The distance electrode wire extends beyond the gas nozzle of a GMAW gun is known as ___.
- A. arc length
- B. cold lap
- C. visible stickout
- D. whiskers

______________ **2.** The ___ conducts the welding wire, shielding gas, and welding current to the welding gun.
- A. welding gun cable
- B. welding machine
- C. manifold system
- D. wire feeder

______________ **3.** ___ transfer is most practical for welding with wire 0.045″ or less on thinner metals.
- A. Spray
- B. Globular
- C. Short circuiting
- D. Pulsed spray

______________ **4.** ___ transfer requires high current levels and argon-rich shielding gas, and is practical for thick metal.
- A. Spray
- B. Globular
- C. Short circuiting
- D. Pulsed spray

______________ **5.** ___ is most frequently used in the GMAW process.
- A. DCEP
- B. DCEN
- C. AC
- D. ACHF

______________ **6.** ___ transfer is prone to fusion defects and is limited to the flat position and horizontal fillet welds.
- A. Short circuiting
- B. Spray
- C. Pulsed spray
- D. Globular

______________ **7.** ___ occurs when the arc does not melt the base metal completely and weld metal flows beyond the weld toe without fusing.
- A. Porosity
- B. Overlap
- C. Whiskers
- D. Excessive penetration

_______________ **8.** Using the ___ technique for GMAW allows for greater penetration of the weld.
A. pushing
B. weaving
C. pulling
D. side-to-side

_______________ **9.** A(n) ___ welding machine is used in the GMAW process.
A. constant-current
B. constant-voltage
C. transformer
D. inverter

_______________ **10.** The ___ in a GMAW system is controlled by the wire feeder.
A. welding current
B. voltage
C. welding gun
D. gas flow

_______________ **11.** Starting an arc becomes more difficult with ___.
A. an increase in electrode extension
B. a decrease in work angle
C. a decrease in gas pressure
D. an increase in metal thickness

_______________ **12.** A ___ technique is used for welding stainless steel.
A. circular
B. forehand
C. backhand
D. whipping

_______________ **13.** ___ is often mixed with other shielding gases to improve their stability.
A. Hydrogen
B. Carbon dioxide
C. Argon
D. Helium

_______________ **14.** Steel from ___ to ___ thick can be butt welded with no edge preparation.
A. 1⁄64″; 0.035″
B. 0.035″; 1⁄8″
C. 3⁄16″; 1⁄4″
D. 1⁄4″, 1″

_______________ **15.** Misalignment of the welding wire in a wire feeder can cause ___.
A. slower feed speed
B. insufficient penetration
C. jamming
D. bird nesting

_______________ **16.** ___ transfer occurs when the welding current is low or below transition current.
A. Spray
B. Globular
C. Short circuiting
D. Pulsed spray

_______________ **17.** Using GMAW on copper is usually limited to ___ types of copper.
A. mixed
B. alloyed
C. pure
D. deoxidized

_______________ **18.** Inadequate gas shielding produces a popping sound and results in ___.
A. craters
B. porosity
C. whiskers
D. excessive penetration

_______________ **19.** ___ causes the most severe atmospheric contamination problems in a GMAW weld.
A. Oxygen
B. Carbon dioxide
C. Hydrogen
D. Nitrogen

_______________ **20.** When used for short circuiting transfer, a 75% argon/25% CO_2 gas mixture helps to minimize ___.
A. cold lap
B. spatter
C. porosity
D. whiskers

Section 6 Exam

Other Welding and Joining Processes

Name ____ **Date** ______________________

True-False

T F **1.** Carbon dioxide is used as a shielding gas in the FCAW-S process.

T F **2.** Flux cored arc welding commonly uses DCEP current.

T F **3.** Brazing can be used to join some dissimilar metals that cannot be joined by welding.

T F **4.** Surfaces to be brazed must be cleaned thoroughly to obtain the necessary flow of filler metal.

T F **5.** A brazed joint is stronger than a welded joint but weaker than a soldered joint.

T F **6.** An oxidizing flame is used for most brazing applications.

T F **7.** The amounts of tin and lead in solder determine its melting point.

T F **8.** In braze welding, the filler metal is deposited in a manner similar to that in fusion welding.

T F **9.** When brazing, the base metal must be melted to obtain the necessary penetration.

T F **10.** Flux must be used when brazing and soldering.

T F **11.** Metal is deposited more quickly in weld overlay than in thermal spraying.

T F **12.** Preheat and postheating are necessary when surfacing.

T F **13.** The SMAW process is used extensively for weld overlay because of its high deposition rate.

T F **14.** Thermal spraying deposits fine semimolten metal particles or metal powder onto the surface of a metal.

T F **15.** Pipe thicker than 5⁄16″ is considered thick-wall pipe.

T F **16.** Uphill welding permits deep weld penetration and is used on thick-wall pipe.

T F **17.** A root pass should have a 1⁄16″ crown on the inside of a pipe.

T F **18.** Downhill welding is a slower method compared to uphill welding.

T F **19.** Fast-freeze electrodes are commonly used for welding pipe.

T F **20.** The cover pass reinforces a pipe weld and gives it a neat appearance.

T F **21.** Six tack welds should be deposited on a pipe joint to ensure proper root opening and alignment before welding.

T F **22.** MAPP gas, natural gas, propane, or acetylene gas with oxygen can be used to provide the necessary flame for flame cutting.

T F **23.** For best results, the preheat and cutting flame must be neutral when flame cutting.

T F **24.** An air carbon arc cutting torch uses carbon electrodes and compressed air.

T F **25.** Air jet orifices must be positioned above an electrode for maximum metal removal.

T F **26.** Preheat holes are located around the oxygen cutting hole of a flame cutting torch tip.

T F **27.** Adhesive bonding requires close contact between the surfaces to be joined.

T F **28.** Failure modes and effects analysis determines how a part failed by looking only at the failed part.

T F **29.** Friction welding is a form of resistance welding used in production settings.

T F **30.** Carbon dioxide is used as plasma gas in plasma arc welding.

T F **31.** Roller-type electrodes are used in projection welding.

T F **32.** Gas tungsten arc spot welding produces a weld with deeper penetration than conventional spot welding.

T F **33.** Ultrasonic welding uses vibratory energy to disperse moisture, oxides, and irregularities between weld pieces.

T F **34.** The beam-in-air EBW process requires a vacuum chamber to prevent atmospheric contamination of the weld metal.

T F **35.** The submerged arc welding process is used primarily for welding thin metals.

T F **36.** Two types of robot manipulators are the rectilinear and the articulating.

T F **37.** A workpiece positioner positions a workpiece in a predetermined location.

T F **38.** A robot controller is used to program specific weld functions.

T F **39.** The type of filler material used for plastic welding is determined by the properties of the base material and the type of joint being welded.

T F **40.** A welding gun with a heating element and compressed gas are used in hot gas welding to soften base metal and filler metal.

Multiple Choice

_______________ **1.** The ___ welding process uses a tubular electrode.
A. GMAW
B. FCAW
C. SMAW
D. GTAW

_______________ **2.** The ___ welding process does not require a shielding gas.
A. FCAW-S
B. FCAW-G
C. GMAW
D. GTAW

_______________ **3.** The heat used for ___ brazing is generated in the same manner as for spot welding.
A. resistance
B. dip
C. manual
D. furnace

_______________ **4.** The filler metal is drawn into the brazed joint by ___ action.
A. flow
B. capillary
C. flux
D. free-flowing

_______________ **5.** ___ temperature is the highest temperature a base metal can reach and still remain in a solid state.
A. Liquidus
B. Molten
C. Solidus
D. Peak

_______________ **6.** Of the types of solder listed, ___ has the highest melting point.
A. 50/50
B. 70/30
C. 5/95
D. 10/90

_______________ **7.** ___ soldering joins two metal pieces without any solder being visible.
A. Seam
B. Torch
C. Furnace
D. Sweat

____________________ **8.** In brazing, ___ is used to prevent oxides from forming when heat is applied to the base metal.
A. filler metal
B. flux
C. sal ammoniac
D. shielding gas

____________________ **9.** A metal to be brazed should be preheated until a ___.
A. dull red color shows
B. bright red color shows
C. melting point is reached
D. bright white color shows

____________________ **10.** Filler metals suitable for brazing are those that begin to melt, or change to a liquid state, above ___°F.
A. 675
B. 780
C. 840
D. 950

____________________ **11.** Pipe weld positions are identified as ___ positions.
A. work
B. certifying
C. trade
D. test

____________________ **12.** Braze welding should be done in ___ position.
A. flat
B. horizontal
C. vertical
D. overhead

____________________ **13.** ___ is a type of wear is caused by oscillatory movement between the surfaces of two materials, usually metal.
A. Erosion
B. Gouging
C. Adhesive wear
D. Fretting

____________________ **14.** When applying weld overlay using SMAW, the welder should ___.
A. use a high current setting
B. remove dust, scale, or dirt with the heat of the electrode
C. arrange the work in flat position
D. maintain a long arc length

______________ **15.** Surfacing and hardfacing overlays are usually applied in the ___ position.
A. uphill
B. downhill
C. overhead
D. horizontal

______________ **16.** The air carbon arc cutting process uses a ___ electrode.
A. stainless steel
B. tungsten
C. carbon
D. copper

______________ **17.** ___ is a mechanical repair method in which a thinned, pitted, or cracked part is smoothed to create a gentle transition with the unaffected surface.
A. Cold mechanical repair
B. Sleeving
C. Blend grinding
D. Wallpapering

______________ **18.** ___ is a repair weld process that consists of methods that join failed parts or restore their surfaces.
A. Thermal spray coating
B. Weld repair
C. Mechanical repair
D. Adhesive bonding

______________ **19.** ___ atmospheres in confined spaces can be life-threatening.
A. Oxygen-deficient
B. Oxygenated
C. Ventilated
D. Flowing

______________ **20.** ___ welding refers to welding performed in the flat, vertical, and overhead positions on pipe that is fixed in place and not rotated.
A. Roll
B. Tack
C. Downhill
D. Position

______________ **21.** Of the welding, soldering, and brazing methods of joining metals, ___ requires the most heat and ___ requires the least heat.
A. soldering; brazing
B. brazing; welding
C. soldering; welding
D. welding; soldering

Section 7 Exam

Weld Evaluation and Testing

Name ______________________________ **Date** ____________________

True-False

T	F	**1.** Visual examination is used to detect grain growth.
T	F	**2.** Fill lighting is a lighting method that uses a small region of a brighter light to increase detail on a dark area of a subject.
T	F	**3.** A welding procedure specification provides formal documentation of all welding variables.
T	F	**4.** In tensile testing, the weld piece is twisted until it breaks.
T	F	**5.** Lighting has the greatest overall effect on the appearance of a surface.
T	F	**6.** The guided bend and wraparound guided bend tests are types of impact tests used to determine soundness.
T	F	**7.** When measuring ductility, round specimens are more accurate for calculating percent reduction of area.
T	F	**8.** Nondestructive testing does not require a part to be permanently removed from service.
T	F	**9.** The liquid penetrant examination method is used to detect defects in a weld by outlining surface defects.
T	F	**10.** Microscopic examination is conducted at high magnification.
T	F	**11.** Visual examination is a form of nondestructive testing.
T	F	**12.** Discontinuities that amplify stress are usually more detrimental than discontinuities that concentrate stresses.
T	F	**13.** The magnetic particle examination method requires a weld piece to be magnetized.
T	F	**14.** A penetrant is a liquid substance used between the search unit and the test surface in ultrasonic examination to permit or improve the transmission of the ultrasonic energy.
T	F	**15.** Specimen mounting is usually temporary.
T	F	**16.** Different etchants are used to reveal specific types of microstructural details.
T	F	**17.** A tensile test is a dynamic test in which the weld is broken by a single blow.

T F **18.** A weld discontinuity is an interruption in the typical structure of a weld.

T F **19.** Qualified welding procedures consist of welding procedure specifications and procedure qualification records.

T F **20.** Rough polishing is a polishing process that is performed on a series of rotating wheels covered with a low-nap cloth.

T F **21.** The percent elongation of a weld specimen is found by fitting the broken ends of the tested piece and measuring the new gauge length.

T F **22.** Mock-up tests can be used to simulate difficult or restricted welding conditions.

T F **23.** Qualification under one fabrication code or standard does not necessarily qualify a welder to weld under another, even though the tests appear to be identical.

T F **24.** The purpose of microscopic examination is to look for clues as to how a metal was made.

T F **25.** A WPQR is usually developed for the easiest position expected during welding or brazing.

Multiple Choice

______________ **1.** ___ magnetization is a concentric magnetic field produced by a straight conductor, such as a piece of wire, carrying an electrical current.
- A. Longitudinal
- B. Conducted
- C. Circular
- D. Transverse

______________ **2.** When preparing a specimen for microscopic examination, ___ is the last stage performed before polishing the mount.
- A. sanding
- B. rough grinding
- C. fine grinding
- D. cleaning

______________ **3.** ___ is the last stage of metallographic preparation before microstructural examination.
- A. Polishing
- B. Mounting
- C. Cleaning
- D. Etching

______________ **4.** ___ is the maximum stress at which stress is directly proportional to strain.
- A. Proportional balance
- B. Median tensile strength
- C. Stress limit
- D. Proportional limit

_______________ **5.** To prevent overheating while preparing a specimen for microscopic examination, flame or plasma cutting must be performed at a minimum distance of ___″ away from the area to be examined.
A. ⅛
B. ¼
C. ½
D. 1

_______________ **6.** The ___ test is a nondestructive test type.
A. guided bend
B. peel
C. ultrasonic
D. Charpy V-notch

_______________ **7.** Typically, one face specimen and one root specimen are used for a ___ test on plate up to ⅜″ thick.
A. Brinell hardness
B. tensile
C. standard guided bend
D. fillet weld break

_______________ **8.** A(n) ___ is a person who is qualified and certified to conduct certain types of NDE processes.
A. welder
B. technician
C. inspector
D. contractor

_______________ **9.** A ___ serves as a "recipe" for making a weld.
A. welding procedure qualification record
B. welding procedure specification
C. welder performance qualification test record
D. welding procedure variable sheet

_______________ **10.** ___ is the most serious type of discontinuity in weldments.
A. Slag inclusion
B. Porosity
C. Melt-through
D. Cracking

_______________ **11.** ___ is cooling that occurs when gas expands, as with the sudden release of gas from a pipe or piece of equipment.
A. Auto-refrigeration
B. Air conditioning
C. Quick freezing
D. Nick-break freezing

_______________ **12.** When using the ___ method, surface discontinuity indications can be easily seen if colored iron particles are applied sparingly to the area of a test surface between the poles.
A. prod
B. yoke
C. dry magnetization
D. wet magnetization

_______________ **13.** ___ examination cannot be used to detect slag inclusions.
A. Visual
B. Radiographic
C. Ultrasonic
D. Microscopic

_______________ **14.** A Bourdon tube is used to measure the ___.
A. hydraulic load applied to a test specimen
B. tensile strength of a test specimen
C. hardness of a base metal
D. density of a weld

_______________ **15.** ___ joints may be especially susceptible to lamellar tearing.
A. Butt
B. Lap
C. T-
D. Corner

_______________ **16.** A ___ wave is a transverse wave that represents wave motion in which the particle oscillation is perpendicular to wave propagation direction.
A. shear
B. longitudinal
C. perpendicular
D. horizontal

_______________ **17.** A ___ is the chemical composition or industry specification of a base metal.
A. base metal material specification
B. base metal weldability classification
C. base metal thickness range
D. filler metal specification

_______________ **18.** A ___ is a document in which the actual values for the welding variables used to produce an acceptable test weldment as well as the results of tests conducted on the weldment are recorded.
A. welding procedure qualification record
B. welder performance qualification test record
C. welding procedure specification
D. welding procedure report

_______________ **19.** Welder ___ is the demonstration of a welder's or welding operator's ability to produce welds meeting prescribed standards.
A. performance qualification
B. certification
C. performance registration
D. accreditation

_______________ **20.** Groove weld qualifications usually qualify the welder to weld both ___ welds and groove welds in the positions qualified.
A. seam
B. fillet
C. plug
D. slot

Matching

Choose the best answer for each question.

_______________ **1.** A(n) ___ requires a new welding procedure specification for metals where impact testing is required.

_______________ **2.** A(n) ___ may be changed in a WPS without requalification of the WPS.

_______________ **3.** A(n) ___ affects the mechanical properties of the weld.

A. nonessential variable
B. essential variable
C. supplementary essential variable

Identify the following destructive tests.

_______________ **4.** The ___ test indents a metal with a 136° square-base cone.

_______________ **5.** The ___ test forces a steel ball into the surface of a metal.

_______________ **6.** The ___ test bends a piece of welded metal around a U-shaped die.

_______________ **7.** The ___ test uses hammer blows, stretching, or bending to break a small, notched specimen.

_______________ **8.** The ___ test uses a dynamic load to measure the energy needed to break a small machine-notched specimen.

A. Vickers hardness
B. Charpy V-notch
C. Brinell hardness
D. nick-break
E. standard guided bend

__________ **9.** The ___ test is a test in which a tensile load is placed on a fillet weld specimen so that the load shears the weld.

__________ **10.** The ___ test is a test in which a specimen is gripped in a vise and then bent and peeled apart with pincers to reveal the weld.

__________ **11.** The ___ test is a test in which a specimen is bent around a stationary mandrel.

__________ **12.** The ___ test is a test that uses two loads to form an indentation on a metal test specimen.

A. wraparound guided bend
B. fillet weld shear
C. peel
D. Rockwell hardness

Identify the examination methods used for nondestructive examination.

__________ **13.** ___ examination uses electromagnetic energy having frequencies less than visible light to yield information on the quality of the part being tested.

__________ **14.** ___ examination uses a strong magnetizing current and very fine colored (dyed) iron particles.

__________ **15.** ___ examination introduces ultrasonic waves through the surface of a part and determines various attributes of the material from its effects on the waves.

__________ **16.** ___ examination uses dyes suspended in high-fluidity liquids to penetrate solid materials and indicate the presence of discontinuities.

__________ **17.** ___ examination uses X rays or gamma rays to detect various types of internal and external discontinuities in a material.

A. Electromagnetic
B. Ultrasonic
C. Liquid penetrant
D. Radiographic
E. Magnetic particle

Section 8 Exam

Welding Technology

Name ______________________________ Date ____________________

True-False

T F **1.** Austenitic stainless steels are not susceptible to hot cracking.

T F **2.** Molybdenum is added to austenitic stainless steels to improve corrosion resistance.

T F **3.** Hardness is the ability of a metal to resist penetration or indentation.

T F **4.** Duplex stainless steels exhibit the highest strengths of all stainless steels.

T F **5.** Alpha titanium alloys are generally the lowest strength titanium alloys.

T F **6.** Qualitative identification is metal identification that applies a physical stimulus to an unknown metal to produce a signal that is interpreted against a set of standards.

T F **7.** The ability of a structure to support a load may be reduced during heat shaping, possibly leading to failure.

T F **8.** Mechanical properties determine the behavior of metals under applied loads.

T F **9.** A strongback is a mechanical restraint device that is attached to one side of a weld joint to hold workpieces in alignment during welding.

T F **10.** Qualitative identification is metal identification by a qualified person to confirm the identity of an unknown metal.

T F **11.** Carbon steels are alloys of iron, carbon, and manganese.

T F **12.** Alteration is any repair that does not restore a mechanical component to its original design.

T F **13.** Medium-carbon steels are typically weaker than low-carbon steels.

T F **14.** A user enquiry is a formal procedure developed by standards committees and code-creating organizations to help users interpret issues and offer suggestions.

T F **15.** Tool steels are generally the hardest and strongest steels available.

T F **16.** Tool steels are always preheated for welding.

T F **17.** The ductility of a given metal is its ability to yield plastically under load instead of fracturing.

T F **18.** Stainless steels are the least versatile family of metals.

T F **19.** Beta titanium alloys have exceptional work hardening characteristics.

T F **20.** Duplex stainless steels have higher strength than austenitic stainless steels.

T F **21.** A nameplate identifies the design, pressure and temperature rating, test pressure, and materials of construction of fabricated equipment.

T F **22.** Nickel is incorporated as a major or minor constituent in approximately 3000 alloys.

T F **23.** Joint cleanliness is the single most important requirement for welding nickel alloys.

T F **24.** Magnesium is one of the heaviest commercial metals.

T F **25.** Distortion is the undesirable dimensional change of a fabrication.

T F **26.** Quantitative identification methods analyze for every chemical element that may be present in a metal.

T F **27.** Distortion control is necessary to overcome poor fit-up and undesirable stresses.

T F **28.** Austenitic stainless steels are less prone to distortion during welding than carbon steels.

T F **29.** The arrow side is the surface that is obstructed from the vision of the welder.

T F **30.** When a definite number of spot welds are needed in a joint, this number is indicated in parentheses either above or below the reference line.

T F **31.** When brazing titanium alloys, the brazing temperature must be below 750°F to prevent reduction in mechanical properties.

T F **32.** The method of examination required can be specified on a separate reference line of the welding symbol or as a separate NDE symbol.

T F **33.** Low-carbon steels are significantly easier to weld than other carbon steels.

T F **34.** Materials standards are classified according to the kind of information they contain.

T F **35.** Strain is the change in dimension of a metal that results from stress.

Multiple Choice

____________________ **1.** A ___ is a document that serves as a model for the measurement of a property or the establishment of a procedure.

A. standard
B. specification
C. recommended practice
D. code

____________________ **2.** Commercially pure coppers are ___.

A. hard
B. strong
C. very ductile
D. heat crack resistant

__________ **3.** ___ is the ability of a metal to absorb energy, such as impact loads, by deforming instead of cracking.
A. Toughness
B. Hardness
C. Strength
D. Ductility

__________ **4.** ___ groups consist of water hardening, cold work, shock resisting, hot work, high-speed, mold, and special purpose.
A. Aluminum alloy
B. Stainless steel
C. Tool steel
D. Nickel alloy

__________ **5.** The ___ used in a fillet weld determines the weld joint strength.
A. welding technique
B. depth of penetration
C. effective throat
D. filler metal density

__________ **6.** A(n) ___ is an incandescent (glowing) streak that traces the trajectory (path) of a particle (spark) in a spark stream.
A. carrier line
B. fork
C. arrowhead
D. burst

__________ **7.** ___ causes the loss of corrosion resistance in stainless steels by making less chromium available for the protective chromium oxide film.
A. Sensitization
B. Heat tint
C. Buttering
D. Creep

__________ **8.** ___ uses measurement of the electric potential generated when two metals are heated.
A. Electrical resistivity
B. Thermoelectric potential sorting
C. Chemical spot testing
D. Electrographic testing

__________ **9.** A solid semicircle on the reference line opposite a weld symbol indicates ___.
A. spot welds
B. projection
C. surface welds
D. melt-through

_______________ **10.** Cast stainless steels are ___ prone to hot cracking than cast steel, so the bevel angle should be ___ than that used for cast steels.
A. more; narrower
B. more; wider
C. less; narrower
D. less; wider

_______________ **11.** Some factors affecting the ___ of base metals and the selection of filler metals include solubility, mechanical properties, and thermal expansion.
A. buttering
B. resistance
C. ductility
D. weldability

_______________ **12.** ___ steels are the easiest to weld since no special welding preparations are necessary.
A. Low-carbon
B. Medium-carbon
C. High-carbon
D. Free-machining

_______________ **13.** ___ tool steels are typically the least costly tool steels.
A. Cold work (Groups O, A, and D)
B. Hot work (Group H)
C. Water hardening (Group W)
D. Mold and special purpose (Groups L, P, and F)

_______________ **14.** For chrome-moly steels, ___ is recommended for stress relief.
A. a low temperature
B. postheating
C. preheating
D. a rapid welding speed

_______________ **15.** ___ is the most significant alloying element that is added to iron in steels.
A. Magnesium
B. Carbon
C. Nickel
D. Chromium

_______________ **16.** High-speed tool steels are resistant to softening at temperatures up to ___°F.
A. 500
B. 750
C. 1000
D. 1500

_______________ **17.** ___ is a surfacing technique wherein weld metal is applied to one or more joint surfaces to provide compatible base metal for welding.
- A. Dilution
- B. Buttering
- C. Residual welding
- D. Intermittent surfacing

_______________ **18.** ___ irons are a ductile form of iron produced by heat treating white iron.
- A. Gray
- B. Ductile
- C. Malleable
- D. Compacted graphite

_______________ **19.** ___ is the precipitation of chromium carbides in stainless steels from exposure to high temperatures.
- A. Chroming
- B. Carburization
- C. Sensitization
- D. Spalling

_______________ **20.** The color change caused by welding stainless steel is known as ___.
- A. pigmentation
- B. heat tint
- C. colorization
- D. carburization

_______________ **21.** Aluminum alloys have ___.
- A. high density
- B. poor corrosion resistance
- C. poor weldability
- D. required high welding currents

_______________ **22.** ___ is resistant to oxidation, moisture, and organic chemicals, making it safe to use in electrical conductors and chemical equipment.
- A. Phosphorus
- B. Copper
- C. Titanium
- D. Lead

_______________ **23.** ___ welds are made before a groove weld is deposited to prevent excessive penetration of the weld metal.
- A. Back
- B. Backing
- C. Melt-through
- D. Surfacing

______________ **24.** Using ___ eliminates increased joint contraction and the need for excessive filler metal.

A. square butt joints
B. proper fit-up and edge preparation
C. large-diameter filler metal
D. presetting

______________ **25.** ___ is slow, plastic elongation that occurs in some metals during extended service under elevated temperatures.

A. Malleability
B. Fatigue
C. Creep
D. Ductility

______________ **26.** A ___ symbol has instructions attached as to the type of weld required, the location of the weld, whether it is a field weld or a shop weld, and other reference data.

A. reference
B. welding
C. seam
D. field weld

______________ **27.** A ___ is a type of standard that provides instructions for performing one or more repetitive technical functions.

A. standard
B. specification
C. recommended practice
D. code

______________ **28.** A ___ is a type of standard that indicates the technical and commercial requirements for a product.

A. standard
B. specification
C. recommended practice
D. code

______________ **29.** ___ standards cover automatic, semiautomatic, and manual welding processes.

A. American Welding Society (AWS)
B. ASTM International (ASTM)
C. American National Standards Institute (ANSI)
D. ASME International (ASME)

______________ **30.** A ___ is certification issued by a primary manufacturer verifying the chemical analysis and mechanical test properties of stock obtained from a starting ingot or billet of metal.

A. certification
B. certificate of compliance
C. mill test report
D. filler metal approval

Appendix

AMERICAN WELDING SOCIETY
Welding Symbol Chart

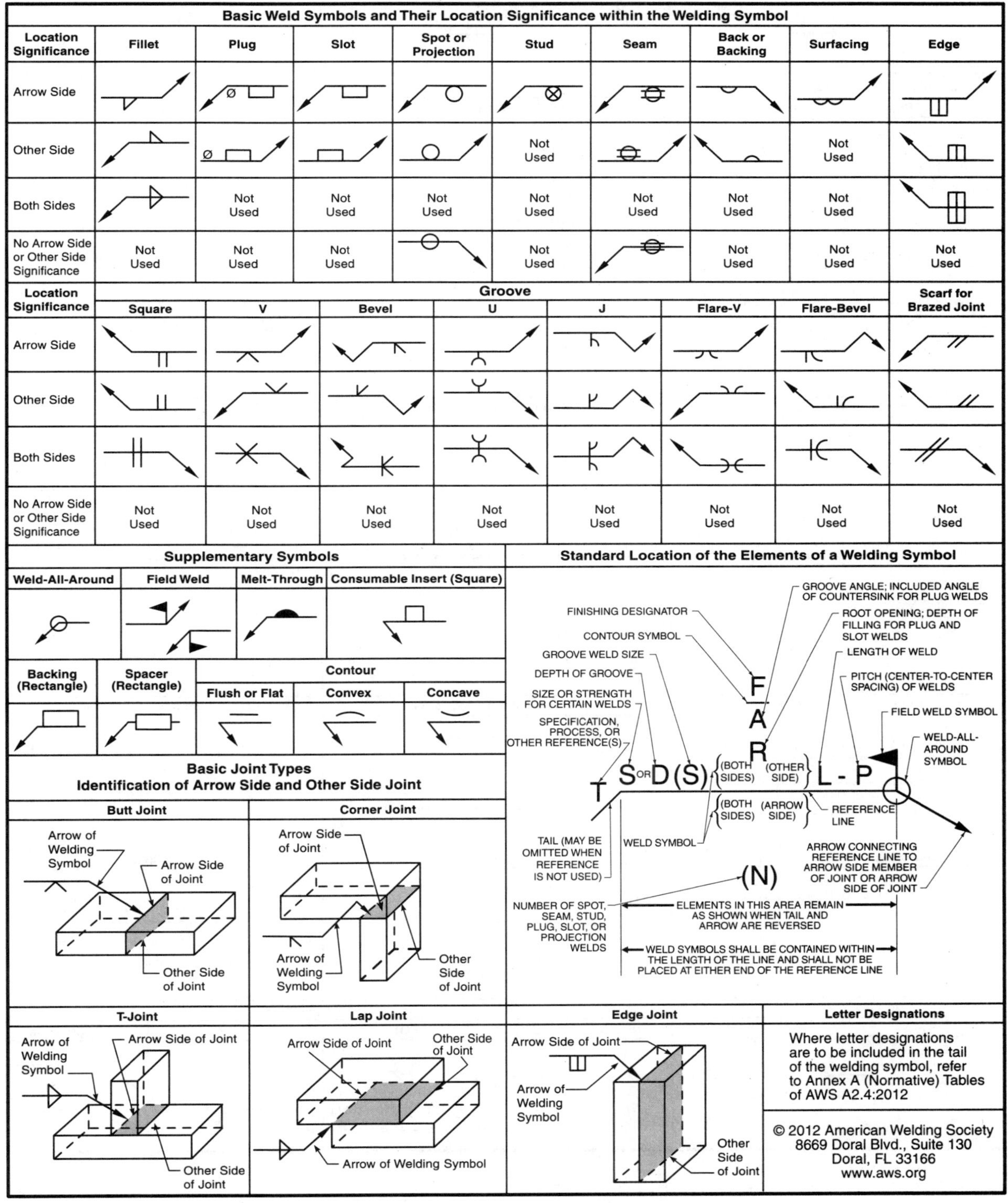

AMERICAN WELDING SOCIETY
Welding Symbol Chart

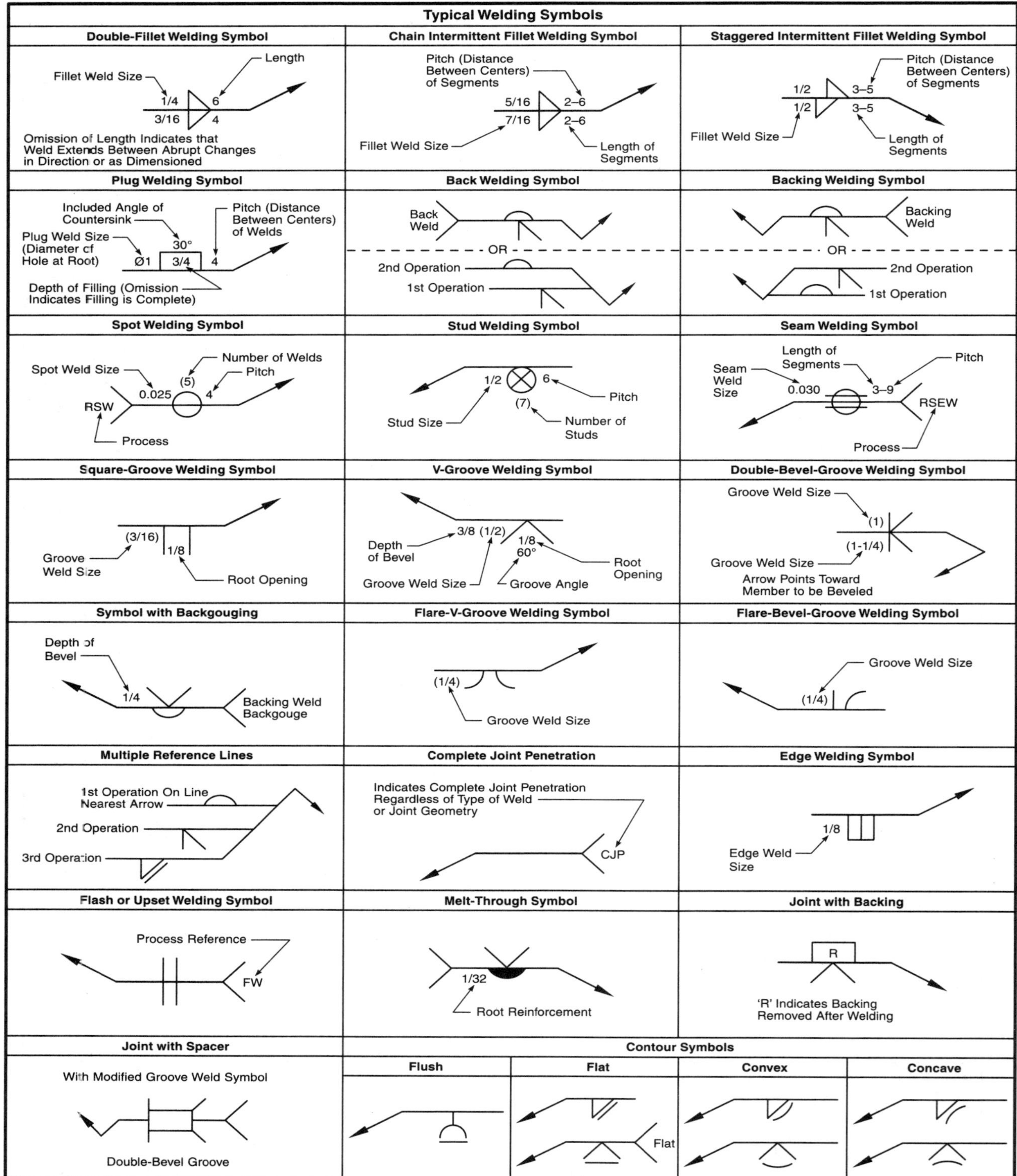